Proceedings of the 27th Solvay Conference on Physics

The Physics of Living Matter: Space, Time and Information

Proceedings of the 27th Solvay Conference on Physics

The Physics of Living Matter: Space, Time and Information

Brussels, Belgium, 19 – 21 October 2017

Editors

David Gross
University of California at Santa Barbara

Alexander Sevrin
Vrije Universiteit Brussel & International Solvay Institutes, Belgium

Boris Shraiman
University of California at Santa Barbara

World Scientific

NEW JERSEY · LONDON · SINGAPORE · BEIJING · SHANGHAI · HONG KONG · TAIPEI · CHENNAI · TOKYO

Published by

World Scientific Publishing Co. Pte. Ltd.
5 Toh Tuck Link, Singapore 596224
USA office: 27 Warren Street, Suite 401-402, Hackensack, NJ 07601
UK office: 57 Shelton Street, Covent Garden, London WC2H 9HE

Library of Congress Control Number: 2020008289

British Library Cataloguing-in-Publication Data
A catalogue record for this book is available from the British Library.

THE PHYSICS OF LIVING MATTER: SPACE, TIME AND INFORMATION
Proceedings of the 27th Solvay Conference on Physics

ISBN 978-981-3239-24-1 (hardcover)
ISBN 978-981-3239-25-8 (ebook for institutions)
ISBN 978-981-3239-26-5 (ebook for individuals)

For any available supplementary material, please visit
https://www.worldscientific.com/worldscibooks/10.1142/10964#t=suppl

In memory of

Dr. Suzanne Eaton (1959–2019)

whose visionary and inspiring participation in the

27th Solvay Conference is captured in these Proceedings.

Suzanne was a scientific leader and a wonderful human being.

We mourn her loss deeply.

The International Solvay Institutes

Honorary Members

Prof. Franz Bingen
Emeritus Professor VUB, Former Vice president and Treasurer
of the Solvay Institutes

Prof. Jean-Louis Vanherweghem
Former President of the Administrative Board of the ULB

Prof. Irina Veretennicoff
Emeritus Professor at the VUB

Guest Members

Prof. Marc Henneaux
Director and Professor ULB

Prof. Alexander Sevrin
Deputy Director for Physics, Professor VUB
Scientific Secretary of the Committee for Physics

Prof. Lode Wyns
Deputy Director for Chemistry
Former Vice-rector for Research VUB

Prof. Anne De Wit
Professor ULB and Scientific Secretary of the Committee for Chemistry

Prof. Franklin Lambert
Emeritus Professor VUB

Ms Marina Solvay

Prof. Hervé Hasquin
Permanent Secretary of the Royal Academy of Sciences, Letters and
Fine Arts of Belgium

Prof. Freddy Dumortier
Permanent Secretary of the Royal Flemish Academy of Belgium for
Sciences and the Arts

Director

Prof. Marc Henneaux
Professor at the ULB

Solvay Scientific Committee for Physics

Prof. David Gross (Chair)
Kavli Institute for Theoretical Physics (Santa Barbara, USA)

Prof. Roger Blandford
Stanford University (USA)

Prof. Steven Chu
Stanford University (USA)

Prof. Robbert Dijkgraaf
Director of the Institute for Advanced Study (Princeton, USA) and
Universiteit van Amsterdam (The Netherlands)

Prof. Bert Halperin
Harvard University (Cambridge, USA)

Prof. Giorgio Parisi
Università la Sapienza (Roma, Italy)

Prof. Pierre Ramond
University of Florida (Gainesville, USA)

Prof. Gerard 't Hooft
Spinoza Instituut (Utrecht, The Netherlands)

Prof. Klaus von Klitzing
Max-Planck-Institut (Stuttgart, Germany)

Prof. Peter Zoller
Universität Innsbruck (Austria)

Prof. Alexander Sevrin (Scientific Secretary)
Vrije Universiteit Brussel (Belgium)

Hotel Métropole (Brussels), 19–21 October 2017

The Physics of Living Matter:
Space, Time and Information in Biology

Chair: Professor Boris Shraiman

The 27th Solvay Conference on Physics took place in Brussels from October 19 through October 21, 2017. Its theme was "The Physics of Living Matter: Space, Time and Information in Biology" and the conference was chaired by Boris Shraiman. The conference was followed by a public event entitled *Frontiers of Science — From Physics to Biology*. David Gross and Eric Wieschaus each delivered a lecture and a panel of scientists — led by David Gross and consisting of Daniel Fisher, Holly Goodson, Ottoline Leyser, Boris Shraiman, Aleksandra Walczak and Eric Wieschaus — answered questions from the audience.

The organization of the 27th Solvay Conference has been made possible thanks to the generous support of the *Solvay Family*, the *Solvay Group*, the *Université Libre de Bruxelles*, the *Vrije Universiteit Brussel*, the *Belgian National Lottery*, the *Foundation David and Alice Van Buuren*, the *Brussels-Capital Region*, the *Communauté française de Belgique*, de *Actieplan Wetenschapscommunicatie* of the *Vlaamse Regering*, and the *Hôtel Métropole*.

Participants

Uri	**Alon**	Weizmann Institute of Science, Rehovot, Israel
Alexander	**Aulehla**	EMBL, Heidelberg, Germany
William	**Bialek**	Princeton University, Princeton, USA
Roger	**Blandford**	Stanford University, Palo Alto, USA
Clifford	**Brangwynne**	Princeton University, Princeton, USA
Lars	**Brink**	Chalmers U. Gothenburg, Sweden
Arup	**Chakraborty**	MIT, Cambridge, USA
Steven	**Chu**	Stanford University, Palo Alto, USA
Michael	**Desai**	Harvard University, Cambridge, USA
Robbert	**Dijkgraaf**	Institute for Advanced Study, Princeton, USA
Suzanne	**Eaton**	Max Planck Institute of Molecular Cell Biology and Genetics, Dresden, Germany
Michael	**Elowitz**	California Institute of Technology, Pasadena, USA
Daniel	**Fisher**	Stanford University, Palo Alto, USA
Irene	**Giardina**	Sapienza U. di Roma, Italy
Albert	**Goldbeter**	The Solvay Institutes & ULB, Brussels, Belgium
Holly	**Goodson**	U. Notre Dame, Notre Dame, USA
Isabel	**Gordo**	Instituto Gulbenkian de Ciência, Oeiras, Portugal
Thomas	**Gregor**	Princeton University, NJ, USA
Stephan	**Grill**	Technische Universität Dresden, Germany
David	**Gross**	Kavli Institute for Theoretical Physics, Santa Barbara, USA
Bertrand	**Halperin**	Harvard University, Cambridge, USA
Marc	**Henneaux**	The Solvay Institutes & ULB, Brussels, Belgium
Jonathon	**Howard**	Yale University, New Haven, USA
James	**Hudspeth**	The Rockefeller University, New York, USA
Terence	**Hwa**	University of California at San Diego, La Jolla, USA
Anthony	**Hyman**	Max Planck Institute of Molecular Cell Biology and Genetic, Dresden, Germany
Frank	**Jülicher**	Max-Planck-Institute for the Physics of Complex Systems, Dresden, Germany
Nicole	**King**	University of California, Berkeley, USA

Eugene	**Koonin**	NCBI, Bethesda, USA
Thomas	**Lecuit**	Collège de France, France
Herbert	**Levine**	Rice University, Texas, USA
Ottoline	**Leyser**	Sainsbury Laboratory, University of Cambridge, UK
Jennifer	**Lippincott-Schwartz**	HHMI, Janelia Research Campus, VA, USA
L.	**Mahadevan**	Harvard University, Cambridge, USA
Cristina	**Marchetti**	Syracuse University, NY, USA
Satyajit	**Mayor**	NCBS, Bangalore, India
Edwin	**Munro**	University of Chicago, USA
Andrew	**Murray**	Harvard University, Cambridge, USA
Daniel	**Needleman**	Harvard University, Cambridge, USA
Richard	**Neher**	Biozentrum, University of Basel, Switzerland
Nipam	**Patel**	University of California, Berkeley, USA
Alan	**Perelson**	LANL, Los Alamos, NM, USA
Rob	**Phillips**	California Institute of Technology, Pasadena, USA
Stephen	**Quake**	Stanford University, Palo Alto, USA
Paul	**Rainey**	ESPCI Paris, France
Sharad	**Ramanathan**	Harvard University, Cambridge, USA
Pierre	**Ramond**	University of Florida, Gainesville, USA
Alexander	**Sevrin**	The Solvay Institutes & VUB, Brussels, Belgium
Boris	**Shraiman**	Kavli Institute for Theoretical Physics, Santa Barbara, USA
Eric	**Siggia**	The Rockefeller University, New York, USA
Ben	**Simons**	Cambridge University, UK
Jan	**Skotheim**	Stanford University, California, USA
Gürol	**Süel**	UC San Diego, La Jolla, USA
Massimo	**Vergassola**	UC San Diego, La Jolla, USA
Klaus	**Von Klitzing**	Max-Planck-Institut, Stuttgart, Germany
Aleksandra	**Walczak**	Ecole Normale Supérieure, Paris, France
Eric	**Wieschaus**	Princeton University, NJ, USA
Ned	**Wingreen**	Princeton University, NJ, USA
Kurt	**Wüthrich**	The Solvay Institutes & Scripps Institute & ETH Zurich
Peter	**Zoller**	Innsbruck University, Austria

Auditors

Enrico	**Carlon**	Katholieke Universiteit Leuven
Ben	**Craps**	Vrije Universiteit Brussel
Sophie	**De Buyl**	Vrije Universiteit Brussel
Geneviève	**Dupont**	Université Libre de Bruxelles
Rouslan	**Efremov**	Vrije Universiteit Brussel
Abel	**Garcia-Pino**	Université Libre de Bruxelles
Pierre	**Gaspard**	Université Libre de Bruxelles
Lendert	**Gelens**	Katholieke Universiteit Leuven
Didier	**Gonze**	Université Libre de Bruxelles
Thomas	**Hertog**	Katholieke Universiteit Leuven
Tom	**Lenaerts**	Université Libre de Bruxelles
Remy	**Loris**	Vrije Universiteit Brussel
Bortolo	**Mognetti**	Université Libre de Bruxelles
Han	**Remaut**	Vrije Universiteit Brussel
Jacques	**Tempere**	Universiteit Antwerpen
Christian	**Van Den Broeck**	Universiteit Hasselt
Sabine	**Van Doorslaer**	Universiteit Antwerpen
Lode	**Wyns**	Vrije Universiteit Brussel

Contents

Opening Session

The Physics of Living Matter: Space, Time and Information in Biology

Opening Address by Marc Henneaux,
Director of the International Solvay Institutes

Dear Mrs. Solvay,
Dear Members of the Solvay Family,
Dear Members of the International Scientific Committee for Physics,
Dear Colleagues,
Dear Friends,

It is my great honour and pleasure to open the 27th Solvay Conference on Physics, entitled "The Physics of Living Matter: Space, Time and Information in Biology".

This Conference belongs to a long and prestigious series that began more than one century ago.

- Solvay 1 (1911): La théorie du rayonnement et les quanta
- Solvay 2 (1913): La structure de la matière
- Solvay 3 (1921): Atomes et électrons
- Solvay 4 (1924): Conductibilité électrique des métaux
- Solvay 5 (1927): Electrons et photons
- Solvay 6 (1930): Le magnétisme
- Solvay 7 (1933): Structure et propriétés des noyaux atomiques
- Solvay 8 (1948): Les particules élémentaires
- Solvay 9 (1951): L'état solide

- Solvay 10 (1954): Les électrons dans les métaux
- Solvay 11 (1958): La structure et l'évolution de l'univers
- Solvay 12 (1961): La théorie quantique des champs
- Solvay 13 (1964): The Structure and Evolution of Galaxies
- Solvay 14 (1967): Fundamental Problems in Elementary Particle Physics
- Solvay 15 (1970): Symmetry Properties of Nuclei
- Solvay 16 (1973): Astrophysics and Gravitation
- Solvay 17 (1978): Order and Fluctuations in Equilibrium and Nonequilibrium Statistical Mechanics
- Solvay 18 (1982): Higher Energy Physics: what are the possibilities for extending our understanding of elementary particles and their interactions to much greater energies?
- Solvay 19 (1987): Surface Science
- Solvay 20 (1991): Quantum Optics
- Solvay 21 (1998): Dynamical Systems and Irreversibility
- Solvay 22 (2001): The Physics of Communication
- Solvay 23 (2005): The Quantum Structure of Space and Time
- Solvay 24 (2008): Quantum Theory of Condensed Matter
- Solvay 25 (2011): The Theory of the Quantum World
- Solvay 26 (2014): Astrophysics and Cosmology

While focused at the beginning on quantum mechanics and the constituents of matter, the subjects of the conferences gradually expanded. The 27th conference is dedicated to biophysics. This is a real first in the history of the Solvay Conferences since, as you can see from the list, it is the first time that this subject is addressed.

What is biophysics? Since I am not an expert, I went to Wikipedia, according to which — and I quote — "it is an interdisciplinary science that applies the approaches and methods of physics to study biological systems." We shall hear more later in this session about biophysics and its history from Professors Boris Shraiman and James Hudspeth, who are leading authorities in the field. Therefore I will only give one additional piece of information, taken again from Wikipedia.

It is that the International Biophysical Society was founded in 1958 and will thus celebrate next year its sixtieth birthday. 60 years, this is very young for a scientific society — chemical societies or physical societies are much older. But still, this indicates that the field is not in its infancy anymore and that it has matured.

It was thus high time to organize a Solvay conference on the challenging questions raised already by Schrödinger in 1944 in its little book "What is life?": "How can the events in space and time which take place within the spatial boundary of a living organism be accounted for by physics and chemistry?" This will be one of the central topics of this year's Solvay Conference.

The International Solvay Institutes are grateful to its International Scientific Committee for Physics, chaired by Professor David Gross, that this theme was chosen for the 27th Conference on Physics.

The role of the Solvay International Scientific Committees is central in the scientific organization of the Solvay Conferences, because it is the Committees that choose their subjects. The Committees have complete freedom in doing so. They have "carte blanche". This requires a perfect and broad knowledge of the most promising directions pursued in each field and at their frontiers. This is where the help and expertise of the International Committees are crucial. Without the International Committees, the Solvay Conferences would not be the Solvay Conferences.

We are very fortunate and grateful that Professor David Gross has been actively helping us by chairing the Solvay Committee for Physics for so many years. He is playing today the same role as Lorentz played in the first years of the Institutes. In fact, this is the fifth conference for which David is the Chair of the Committee. This number equals Lorentz' number (1911, 1913, 1921, 1924 and 1927). Only Lawrence Bragg did as well (1948, 1951, 1954, 1958 and 1961).

We are also extremely grateful to Professor Boris Shraiman, who accepted the very demanding task of chairing and organizing this Conference, in coordination with the Chairs of the various scientific sessions. The Solvay Conferences have a very special format. These are conferences by invitation-only, with a limited number of participants. There are few presentations but a lot of discussions. Scientific interactions are privileged. For the discussions to be fruitful, an extremely careful preparation is needed. This requires an enormous amount of work and is very time-consuming.

As I just recalled, discussions are a key element of the Solvay Conferences. For that reason, they are included in the proceedings — another distinctive feature. We have a scientific secretariat in charge of carrying this essential work. Our deep thanks go to all our colleagues involved in this mission, and in particular, to Professor Alexander Sevrin, scientific secretary of the Committee, who is coordinating the entire work leading to proceedings of the highest quality.

Thank you very much for your attention.

**Opening Address by David Gross,
Chair of the Solvay Scientific Committee for Physics**

Welcome everybody to this, first-of-its-kind, Solvay Conference on the Physics of Living Matter. As you all know, the Solvay Conferences are a very special kind of scientific meetings. They began in 1911 as a result of a remarkable collaboration of Ernest Solvay and Hendrik Lorentz and had an extremely innovative structure. First, the conference was international and open to the best physicists from anywhere in the world, which was not at all common in those days. They had a very interesting format which we have tried very hard to preserve. The conference organizers invited a small, highly selective group of scientists. The first conference had only 34 participants. Today we have a little more than double that number, even though science has grown by a factor of 100. The nature of the conference was also very special: it consisted of short rapporteur talks and much discussion. The discussion was carefully recorded by hand and then published. Consequently, the proceedings of the Solvay Conferences provide a wonderful history of 20th century modern physics.

The Solvay Conferences began in 1911, at a very special time in physics, a time when classical physics was under attack and quantum physics was being born. To give you a sense of the atmosphere at that time, I will quote from the opening address of Lorentz at the first Solvay Conference in 1911 which reverberates with the anguish that this master of classical physics felt at the first glimpse of the quantum world: "Modern research has encountered more and more serious difficulties when attempting to represent the movement of smaller particles of matter and the connections between these particles and phenomena that occur in the aether. At the moment we are far from being completely satisfied that with the kinetic theory of gases, gradually extended to fluids and electron systems, physicists could give an answer in ten or twenty years. Instead we feel that we have reached an impasse. The old theories have shown to be powerless. It appears that darkness is surrounding us at all sides." That was the state of physics in 1911 and the Solvay Conference offered a perfect setting for intense discussions about the most fundamental questions that physicists were faced with: the emergence of quantum mechanics. The Solvay Conferences played an enormously important role in the development of quantum mechanics, culminating 16 years later in the famous 1927 5th Solvay Conference where some of the basic conceptual foundations of quantum mechanics were discussed at length. Since then the Solvay Conferences have continued to play a very important role in the development of elementary particle physics, the Standard Model, astrophysics, quasars and many other areas.

Since 2004, I have been closely involved in organizing these meetings. Every three years a conference takes place and their preparation takes almost three years. The primary feature of these conferences is the spontaneous discussion, in which you all will engage in over the next few days, but as I learned: spontaneity requires much preparation and organization. When we began reviving these conferences 13 years ago, one of our goals was to reproduce the atmosphere of the original conferences. We wanted to hold the conferences in Brussels, in the Metropole Hotel, indeed in the same room that the original 1911 conference was held. It might not look so from the photograph but this is the same room as the first conference. The other thing I wanted was to have everyone seated around a big table, as they were in 1911. Well that proved to be difficult, it is hard to find a big enough table. It is a great advantage to have everyone sitting around a round table. With a round table everyone is equal, as contrasted to having a speaker talking to an audience and questions going back and forth between the audience and the speaker. Thus, we came up with the idea of having a topologically round table with the added bonus of a hole in the middle where we can project the inevitable slides and power-point presentations that we are all accustomed to nowadays. You should think of this conference as being held around a round table where everyone is equal and there is no center. After this opening session we are going to squeeze a bit and hopefully that will stimulate the kind of discussions as those which took place back then.

This is the fifth conference in the new series that started in 2005 with the "Quantum Structure of Space and Time". In 2008, the conference was devoted to "The Quantum Theory of Condensed Matter", chaired by Bert Halperin. In 2011 the 25th Solvay Conference in Physics celebrated 100 years of Solvay Conferences in Physics and was devoted to the "Quantum Theory of the World", covering all of quantum physics. Three years ago, Roger Blandford chaired the Solvay Conference on "Astrophysics and Cosmology".

When we decided to strike out in a new direction and have a conference devoted to the physics of living matter, the Physics Committee was very supportive and unanimously agreed that this was a good direction and an exciting topic to address. In fact, almost the whole physics committee is present at this conference, which is rather unusual. This is a strong indication of how much and how broad the interest among physicists is in the structure and dynamics of living matter. I think it is fair to say that in this field, unlike in 1911, there is no crisis. There is opportunity! That is the reason it is so exciting.

I reread Schrödinger's very influential book, *What Is Life? The Physical Aspect of the Living Cell*, written in the middle of World War II in Ireland, a few months ago. It is really marvelous reading. His description at the beginning of what the book is dealing with, is really quite appropriate: How can the events in space and time which take place within the spatial boundaries of a living organism be accounted for by the laws of physics and chemistry? He definitely believed that the answer could be provided by physics and chemistry and he offered some fascinating ideas. I believe the book was so influential for many great physicists and biologists because the typical arrogant, ambitious point of view of a physicist, that motivated many to follow his lead. Especially his argument that the information carrying locus of living organisms had to be embodied in something which he called an aperiodic crystal, which can easily be thought of as a crude model of what turned out to be the Double Helix. He came to that conclusion by understanding that the kind of order that exists in biological systems cannot be the kind of order that emerges from the statistical behavior of macroscopic systems in ordinary physical systems; but rather had to be embodied in a very stable but yet aperiodic structure and that quantum mechanics could explain how.

The second topic of discussion in Schrödinger's book was the attempt to understand the information growing, or entropy reducing, behavior of self-organizing systems. He put forward some ideas, which were subsequently heavily criticized, as to how that could take place. Schrödinger's first hypothesis, the aperiodic crystal, was absolutely brilliant. It supposedly stimulated Francis Crick and others. It is imaginable that in some variation of history he could have hypothesized that that information carrying aperiodic crystal was a polymer and even a double polymer. The second issue that he discussed about how open, out of equilibrium, systems can exhibit self-organized complexity, and pattern formation remains one of the basic unresolved issues behind the understanding of the physics of living matter.

When we decided that the physics of life was worthy of a Solvay Conference, we thought that maybe in 100 years it would be seen as memorable as the Solvay Conferences of a hundred years ago. The issue was then: how to organize such a conference? I had a secret weapon in my pocket (who just gave me a nasty look) because I had brought Boris Shraiman to Santa Barbara 15 years ago (or more) to try to incorporate in the Institute for Theoretical Physics a component devoted to the physics of living matter. Boris has done that extraordinarily well in creating and fostering a new community of people trained in physics and neighboring fields who are fascinated by the problems of living matter, biology. These scientists are bringing the tools that are perhaps needed to take advantage of the enormous advancements in observational and experimental biology and turn it into a truly quantitative and theoretical science. This was an experiment at Santa Barbara. The experiment has succeeded wonderfully, better than any of my expectations. Much of this is due to Boris and his untiring creativity. So, I turned to Boris and it only took about half a year to persuade him to undertake what is an extremely difficult job of putting together this great collection of people from different areas and trying to organize the self-organized spontaneity that we hope will emerge in the next three days.

So, with no further ado I, turn over the floor to Boris.

Opening Address by Boris Shraiman,
Chair of the 27th Solvay Conference of Physics

Let me begin by thanking our hosts, the illustrious Solvay family and the Scientific Leadership of the Solvay Institutes for continuing the venerable tradition of Solvay Conferences in Physics: the tradition carried on from the legendary times when scientific giants roamed the earth to this very day. The frontier of scientific knowledge has advanced spectacularly since then, yet there is still a frontier, as challenging and exciting as ever, and the future will surely bring more fundamental breakthroughs.

For all its association with tradition, Solvay Institutes' view of Physics is quite untraditional. Just look around this room, nearly half full with biologists. We are gathered here to brainstorm the fundamental challenges of understanding Living Matter — a frontier of knowledge at the interface of Physics and Biology. The problem of life always had intellectual attraction for physicists. Niels Bohr, for example, thought that living matter could not be explained by physics and chemistry alone without a drastically new insight, just like the quantum atom could not be explained by classical physics. In contrast Erwin Schrödinger came down strongly on the side of life being in the domain of physics and chemistry — a view, by the way, also held by Ernest Solvay. Schrödinger's little book, "What is Life", in 1944, anticipated many challenges in understanding Living Matter as a physical system and advanced an idea of an "aperiodic solid" as a repository of genetic information — much like DNA turned out to be, when it's structure was solved a decade later. In contrast to fundamental physics — the realm of symmetry (think periodic crystals!) — Schrödinger saw Living Matter as a system ruled by information imbedded in a complex (aperiodic) but orderly structure.

Living matter involves complex but orderly spatial structures on all scales, from submicroscopic DNA and protein structure, to cells and tissues, forming visible and familiar shapes of plants and animals. These structures come about through programmed self-assembly processes guided by genetic information, the heritable memory that controls self-replication and encodes functional capabilities, including the capacity for information processing and adaptive response to the environment. Evolution, which is a uniquely biological phenomenon, is the trial and error learning process that leads over time to the accumulation of that heritable information. The way in which biological information organizes the structure and the dynamics of Living Matter — Space, Time and Information — will be the main theme of our discussions here.

At this point one could discuss at length whether Living Matter thus defined belongs to Physics and what Physics can contribute to Biology, or conversely how Biology can extend the intellectual domain of Physics. It is not necessary here, as we agree that understanding Living Matter is an intellectual frontier and Biologists and Physicists should both study it, preferably together, as we are gathered here. In this spirit, our interpretation of Physics of Living Matter will go far beyond direct physical aspects of biology. For better or worse we will be addressing all — perhaps we will not have time for all — but in any case many of the fundamental problems in biology with no holds barred.

Let me give you a quick birds eye tour of our sessions, to the extent that we can foresee. We will start on the subcellular scale and over five sessions proceed all the way to evolutionary dynamics.

Session 1: Subcellular Structure and Dynamics

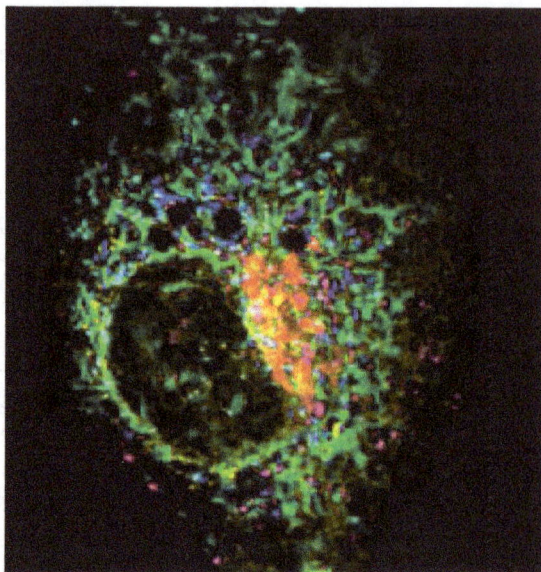

Fig. 1.   Live-cell, 6-colour 4D Lattice Light Sheet microscopy to characterize organelle distribution in space and time from. Valm AM *et al.*, *Nature* **546**(7656): 1620167 (2017).

Every improvement in microscopy uncovers more complexity inside the cells. Subcellular organelles like mitochondria, for example, are not well-defined isolated entities, as we learned in high school, but rather a connected network forming a structure that would be a nightmare to a plumber, all the more so because it is highly dynamic, continuously reshuffled by fission and fusion of membranes. Quite generally, what in a snapshot looks like a structure, live video microscopy reveals to be a dynamical process. So we have here the first glimpse of a common theme, of how space and time are intertwined in biology. The structure that we see is a result of self-assembly dynamics. How is this dynamics and the resulting structure encoded genetically? A common theme that we shall see in many different manifestations is that biology harnesses physical interactions and instabilities (phase segregation, for example) to generate its structures.

Session 2: Cell Behavior and Control

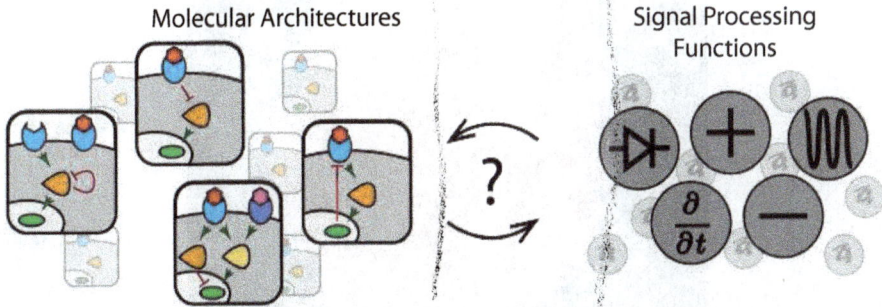

Fig. 2.   Signaling pathways and signal processing. Antebi *et al.*, *Curr. Opin. Syst. Biol.* **1**: 16–24 (2017).

In the second session we will turn to cell behavior. Cells do not just blindly execute genetic orders. They are programmed to sense the environment and control their behavior, changing their state accordingly. We now know a great deal about molecular circuits involved in controlling cell behavior. Yet knowing molecular players is not always enough to uncover the system level function. There is plenty of room, a great need really — for a complimentary "top down" description of the emergent, system level behavior. Niels Bohr thought that new and complementary ideas would be needed to describe living matter. I wonder, perhaps he was thinking of the need for phenomenological models that focus on the collective behavior which emerges at a certain level of complexity and which cannot be seen through the dense forest of molecular details. This sort of phenomenological approach is of course well within the purview of physics! The top-down approach naturally complements the engineering ideas that are being used in the bottom-up dissection of specific molecular circuits of cellular control and information processing. Can it provide deeper insight into the role these circuits play in coordinating cellular processes and optimizing utilization of resources, I am pretty sure this will come up in our discussions.

Session 3: Intercellular Interactions and Collective Behavior

In session three, we move further up in scale and focus on intercellular interactions and collective behavior. Even bacteria, the ultimate single cell individualists, form biofilms that have fascinating collective behavior. Intercellular signals and interactions are still more important for multicellular organisms that fully rely on the separation of labor between cells. So we will discuss key mechanisms that control and coordinate cell fate determination in embryonic stem cells starting on the path of development.

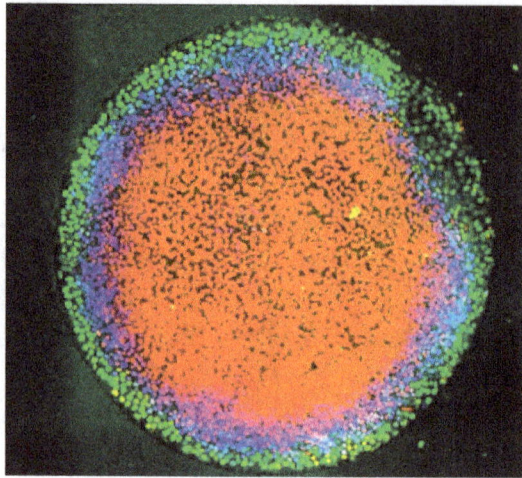

Fig. 3.   Self-organized patterning in 2D micropatterning colonies. Siggia ED, Warmflash A, *Curr. Top. Dev. Biol.* **129**: 1–23 (2018).

Characteristically, even as we try to pigeon-hole our subject into well defined scales, the idea of collective behavior transcends scale, and common ideas describe collective behavior of crawling cells and of flying birds. Talking about signals and cell differentiation will bring us to the problem of animal and plant development and we will take on that problem explicitly in the following session on Morphogenesis.

Session 4: Morphogenesis

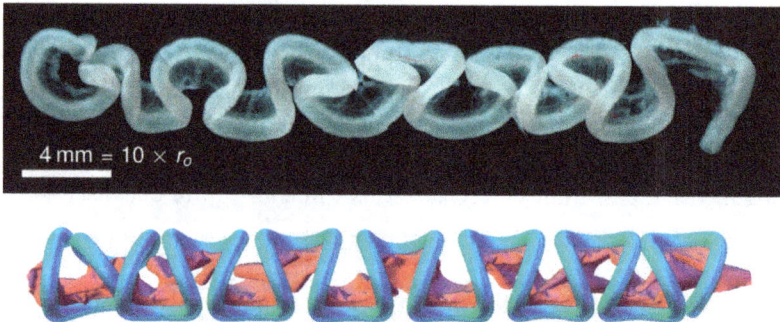

Fig. 4.   A model of gut development in a chick. Savin T. *et al.*, *Nature* **476**(7358): 57–62 (2011).

How does the process of development define the shape of an organ or an organism? Thanks to fantastic progress of developmental biology we may already know which gene would switch a fly antenna into a leg. But we understand very little about the process that actually defines a leg-shaped appendage as opposed to an antenna.

How do we go from genes to geometry? Perhaps physics can help understand how shape is encoded in the dynamics of tissue growth: think of a dynamical equation — a genetically encoded executable program — supplied with initial data courtesy of maternal factors in the egg. Guided by genetic information, developmental dynamics defines spatial structure: an example of space/time and information — the subtitle of our meeting — playing out in biology.

Session 5: Evolutionary Dynamics

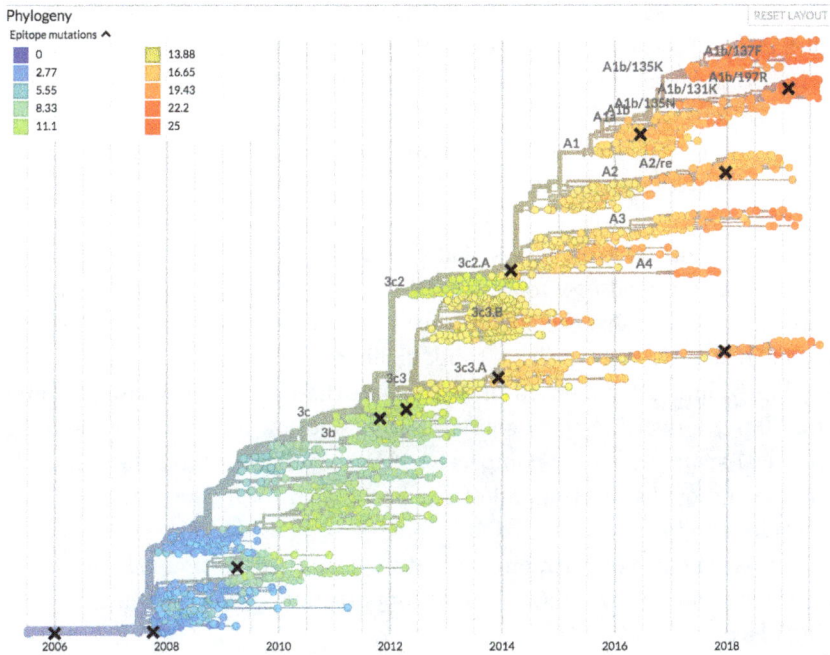

Fig. 5. Nextstrain.org: Real time tracking of influenza evolution. Hadfield *et al.*, *Bioinformatics* **34**: 4121–4123 (2018).

Finally, in our pursuit of big questions, we will move to the subject of Evolutionary Dynamics. Evolution is a uniquely biological dynamical process. Yet physics ideas have been remarkably useful in thinking about it. There has been a recent paradigm shift in the subject: whereas people used to think that adaptive events were quite rare, we now think that they are ubiquitous and natural selection is ever present, continuously at work sorting through multitudes of small competing mutations. The first challenge is to characterize and explain observed evolutionary dynamics. Going beyond that, can we predict evolution? Prospects are pretty good for short term prediction, much like weather forecasting. Predicting long term is much more difficult. What is the space of possibilities available to evolution?

Organisms do not live in isolation as they shape their environment and engage in ecological interactions. A big question is to explain the remarkable diversity of life, understand what drives the diversification and what limits it.

Conclusion of Introduction

As you see we aim to cover a vast field. Yet we have sacrificed so many interesting topics that it would take me even longer probably to list what we could not include in this meeting. I will not go that way, except for mentioning the interface of physics and neuroscience, which has been very fertile, intellectually, in the last couple of decades, and is most promising going forward. Perhaps something that Solvay Scientific Committee may want to consider in the future?

We have organized sessions by increasing scale, from small to large. Yet the "big picture" questions of how Living Matter uses information to organize itself in space and time are not easily confined to individual sessions. They will spill out, cutting across scales and connecting different sessions with each other. And that is not a bad thing!

Most importantly, let me stress that Solvay Conference is not a usual meeting and we have encouraged everyone to focus on the future more than on reporting their latest results. Ideally, the outcome of this meeting would be a roadmap of challenging but still achievable goals — think "Hilbert problems"! — that could inspire young people joining the field to propel it into the future. For example, jumping the gun, I'll put forward my own entry: figuring out how animals and plants encode their physical shape. Or more compactly: "How do Genes encode Geometry?" We hope by the end of the meeting, all of our participants will contribute to the list of challenges for the future.

Finally, as we move on with our scientific agenda, let me again thank the Solvay Institutes for giving us the extraordinary opportunity to meet in this historic and inspiring place.

Prologue: The Evolving Interface of
Physics and Biology

A. J. Hudspeth

For many decades, the observational richness of biology and the mathematical rigor of physics seemed incompatible. In recent years, however, the improved precision and accuracy of biological measurements have resulted in data that both provoke and test rigorous physical theories. As a consequence, we stand at the outset of a period of unparalleled interactions between the two disciplines.

1. Introduction

This meeting represents the first Solvay Conference dedicated to the interface between physics and biology. In introducing the meeting, I shall take the perspective of an experimental biologist who recognizes the value of physical insight and techniques in dealing with the puzzles posed by matter that is not only soft and condensed, but also quite alive. My hope is to transcend the subject called "biophysics" in a narrow sense — the use of techniques developed by physicists for the investigation of living systems — and to focus on the application of a physical point of view to biological problems.

This overview cannot be comprehensive, but has two, more modest goals. First, I shall endeavor to introduce several subjects meant to illustrate the growing links between physics and biology. The sessions throughout the meeting will amplify on some of these topics and introduce many others. Second, and more importantly, I shall try to point out areas in which further connections can be made: subjects in which biological phenomena have passed the stage of description and are ready to yield to the powers of novel experimental techniques, mathematical analysis, statistics, and modeling. These will be among the subjects in which physicists newly entering into the pursuit of biological objectives will make their mark.

2. Biophysics and the Physics of Living Matter

The term "biophysics" now bears two significantly distinct meanings. To many biologists, biophysics represents a suite of technical approaches: NMR spectroscopy, X-ray crystallography, FRET, single-channel recording, and so on. This terminology makes sense: these techniques represent progressive innovations in the breadth and precision of data acquisition in biology, and have been occasioned by the application of principles first developed in physics — sometimes decades before their biological application.

The second sense of the word "biophysics" involves the physics of living matter. Here we mean contemporary approaches to biological problems by physicists, or by others with an interest in and some expertise in physics. Although speaking in generalities is fraught with misinterpretation, the approach of physicists, in contrast with that of biologists, is generally mathematical, rather than anecdotal; reductionist, rather than integrative; and predictive, rather than descriptive.

Although the present gathering includes biophysicists of both varieties, those of the latter type will be of the greater interest to the distinguished physicists in our audience. We have reached a point at which more and more physicists, including many of the most innovative, recognize biological subjects as the greatest challenges for their research. This meeting will show why.

3. Historical Perspective on Scientific Cultures

Before growing specialization separated "natural history" into disciplines such as physics and biology, many outstanding researchers engaged in both spheres and gained insights from their intersection. Robert Hooke, for example, conducted many of the experiments that underlay the physical laws for which Robert Boyle is known. As an expert in optics, he developed an effective microscope, made observations of the moon's craters, and antagonized Isaac Newton by correcting flaws in the latter's optical theory. In the biological realm Hooke wrote the great book *Micrographia*, in which he depicted the structures of numerous miniscule animals and discovered and named the cell.[1] Hermann von Helmholtz provided the first clear statement of the conservation of energy, clarified the concept of free energy, and taught many other great physicists. His curiosity about the nature of perfect pitch and frequency discrimination in the human ear led him to the physical theory of cochlear action that underlies our current understanding of human hearing.[2] Ernst Mach gained fundamental insights into the nature of inertial mass — insights that contributed to Einstein's theory of general relativity. Owing to his interest in the perception of acceleration, he also undertook experiments that defined the properties of our sense of equilibrium.[3] Max Delbrück, to whom the author is endebted for communicating his first paper as an independent investigator, worked in astrophysics and statistical physics. His greatest contribution, though, was to conceive a statistical means of demonstrating the random nature of biological mutation (Figure 1), an experiment that helped to consolidate the modern synthesis of evolutionary theory.[4] And perhaps most famously, Erwin Schrödinger in his book *What is Life?* offered concrete physical definitions of a variety of biological problems that have proven to be fruitful subjects of investigation.[5]

One source of the subsequent separation between physicists and biologists is cultural, akin to the division between hunters and farmers. Theoretical physicists in particular are — or wish to be — hunters who track elusive prey and dispatch them. This metaphor is supported by the language of the discipline: it is a high compliment to say that an investigator has killed a problem! A corollary of this approach is that a theoretician can make valuable contributions to many distinct subjects over a career, or that an Einstein can in a single year offer fundamental

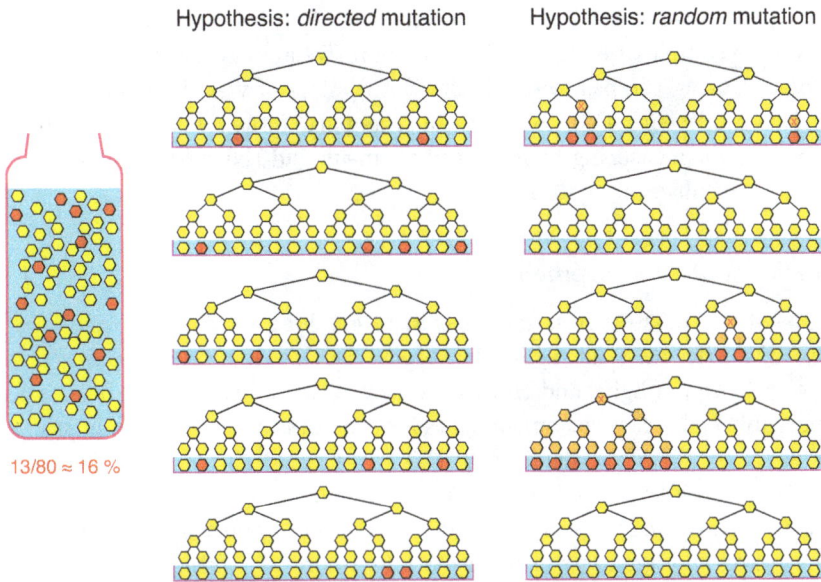

Fig. 1.　The experiment of Luria and Delbrück. [4] When placed in a medium containing pathogenic virus (*left*), a modest fraction of bacteria (red) survives owing to an advantageous mutation lacking in control cells (yellow). This phenomenon might reflect directed mutation (*center*), in which the bacteria — after growing for generations in virus-free conditions — somehow become resistant only upon exposure to the virus-containing medium. By this hypothesis, the generations of bacteria prior to exposure lack any protective mutations. An alternative possibility is that a random mutation (red X) can occur with a low probability in any generation (*right*), giving rise to latently resistant bacteria (orange) whose capability becomes apparent only upon exposure to the virus-containing medium. The two hypotheses yield distinct statistical predictions for the distribution of resistant bacteria: directed mutation would follow Poisson statistics, whereas random mutation early in the geneology would occasionally produce much larger clusters of resistant organisms. The experimental evidence for the latter hypothesis supports the Darwinian model of natural selection.

insights into the photoelectric effect, molecular diffusion, mass-energy equivalence, and special relativity.

Biologists, in contrast, tend to be farmers: they devote years, sometimes their entire careers, to specific experimental subjects. This agrarian orientation stems from the complexity of organisms and of the methods used to study them. It takes years to master the genetics and molecular biology of roundworms, fruit flies, zebrafish, mice, or humans; to develop electrophysiological recording techniques for a novel experimental system; or to comprehend the social behavior of creatures ranging from ants and bees to porpoises and primates. In most instances, a deep understanding of biological principles emerges slowly and incrementally. Again to cite a prominent example, Darwin required more than twenty years to develop and publish his theory of natural selection.

These two distinct approaches have led to ample misunderstanding. On the one hand, physicists have sometimes regarded biologists as "stamp collectors" obsessed with enumeration and anecdote, but lacking in the skills of quantification and prediction. And on the other hand, biologists have often been suspicious of physicists who elide the complexities of life with such putative approximations as the massless point cow. During the past few decades both points of view have shifted radically — and in a positive direction — owing in large part to the remarkable improvement in techniques for measuring biological phenomena and the consequent availabity of data worthy of physical analysis.

4. Precision of Measurement

The capacity to make precise measurements has long underpinned the success of physics. Eratosthenes's determination of the Earth's diameter, Rømer's measurement of the speed of light, and Hubble's confirmation of cosmic expansion all stood as remarkable technical accomplishments for their eras and changed mankind's view of the universe. In a continuation of this effort, modern determinations of some physical constants have reached precisions as great as 14 significant figures. Measurements of this quality provide a solid foundation upon which physical theories can be erected and against which they can be rigorously tested.

Measurements in biology are frequently harder, or anyway cruder. Living things are always heterogenous, often flexible, usually wet, and forever changing — properties not conducive to precise quantification. Moreover, inasmuch as the cell is the fundamental unit of life, researchers require techniques that operate at the cellular or even molecular level: data from a whole, homogenized creature are usually unsatisfactory. As a result, biological experimentation is limited primarily by technical issues.

Several methodological innovations spurred by physics have contributed over the past few decades to the increasing precision of biological research. To take but one example, fluorescence techniques have been of immense value. Molecular fluorescence can be modulated by dozens of distinct processes, and light microscopy provides the resolution necessary to measure fluorescence signals from individual cells and even parts of cells. Reagents have been developed to monitor the intracellular concentrations of a variety of ions, especially Ca^{2+}, and of small organic molecules. Some of these fluorescent reporters are genetically encoded proteins, which experimenters can target exclusively to those cells whose characteristics are to be measured.[6]

Fluorescence recovery after photobleaching — FRAP — provides an effective means of probing the physical environment within a cell. After introducing a suitable fluorophore and allowing it to spread throughout the cell, an experimenter uses bright light to bleach the fluorophore in a specified pattern. The reappearance of fluorescence in the bleached region then permits calculation of the effective diffusion coefficient of a mobile fluorophore through the crowded cytoplasm. The same approach operates on a slower timescale to quantitate the turnover of immobilized proteins.

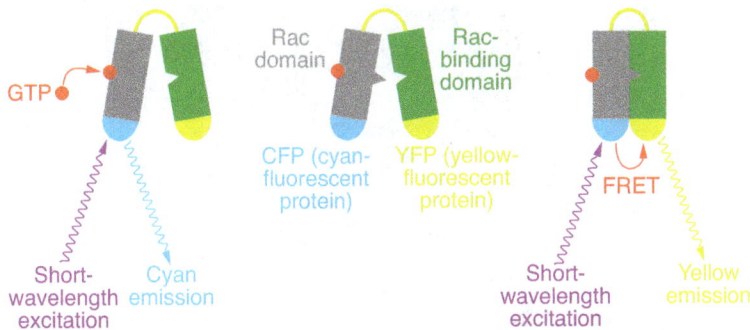

Fig. 2. Fluorescence resonance energy transfer (FRET). A probe for the intracellular second messenger GTP comprises linked domains from the actin-activating protein Rac and from a protein with which it normally interacts. The former domain is attached to cyan-fluorescent protein, the latter to yellow-fluorescent protein. Until GTP binds, the two assemblies are somewhat separated by diffusion; under this condition, short-wavelength illumination evokes principally cyan fluorescence (*left*). The binding of GTP causes a molecular rearrangement in the Rac domain that leads to interaction between the two assemblies (*center*). The consequence of closer interaction is the radiationless transfer of excitation between the nearby fluorophores, culminating in the emission of yellow light (*right*).[7]

Fluorescence (or Förster) resonance energy transfer — FRET — offers an invaluable yardstick for short-range interactions between biomolecules (Figure 2).[7] In this technique, one of a pair of distinct fluorophores is excited by light; nonradiative energy transfer then occurs to the other fluorophore, which subsequently emits light of its characteristic wavelength. Inasmuch as the efficiency of transfer scales inversely to the sixth power of the distance between the donor and acceptor fluorophores, the approach is highly sensitive on the distance scale of 1–10 nm associated with individual protein molecules, small molecular assemblies, and organelles.

Measurement and Manipulation of Single Molecules

The advent of techniques for the manipulation of individual molecules has greatly enhanced research in numerous areas of biology. The study of elastic proteins and molecular motors, for example, now makes routine use of optical traps — laser tweezers — that can measure nanometer displacements and piconewton forces with millisecond resolution (Figure 3). It is possible to evaluate the behaviors of individual molecules of myosin, kinesin, and dynein, as well as their average behaviors and the statistical characteristics of ensembles.[8] A related tool, magnetic tweezers, has proven comparably useful in the study of the enzymes that process DNA and RNA, cutting and pasting these long molecules to effect changes in their topology. These technologies have had a revolutionary effect, for earlier investigations of molecules free in solution neglected the most important aspect of motors, their capacity for sensing and exerting force. Our understanding of the ensuing data is still in its infancy: long time series and more sophisticated analysis should yield insights into the details of motor activity.

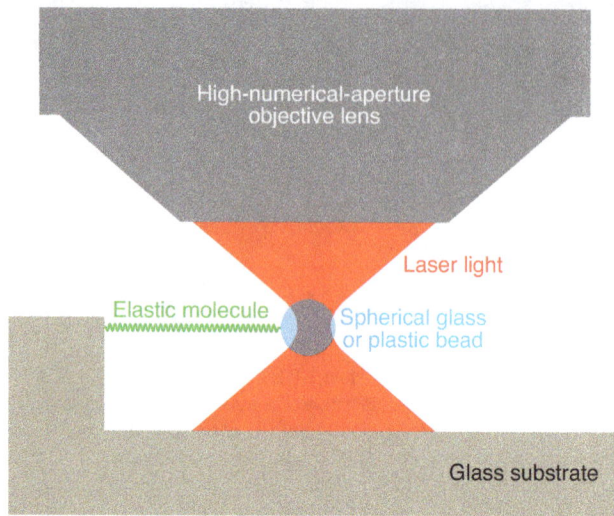

Fig. 3.   An optical trap or laser tweezers.[8] Photonic force pulls a bead of high dielectric constant toward the focus of an intense laser beam. If a single protein or nucleic-acid molecule links the bead to a fixed pedestal, the tension in that molecular tether can displace the bead from the center of the optical trap. Calibration of the trap's strength allows an experimenter to measure the stiffness or unfolding of an elastic protein or the magnitude of steps by a molecular motor.

A challenge for the future involves the deployment of these tools, not just on purified molecules, but on proteins within cells as well. Can we find a means of measuring force production by molecular motors and DNA-processing machinery *in situ*? Are we able to ascertain how ensembles of motors operate, for example whether they exert cooperative or anticooperative effects? Can we determine how the molecules and organelles engaged in intracellular trafficking are addressed so that they reach their appropriate destinations?

Electrophysiology or membrane biophysics has been revolutionized simultaneously by the advent of single-channel recording.[9] Tight-seal microelectrodes have largely overcome the problem of thermal (Johnson) noise in electrical recordings; field-effect transistors have provided sensitivity to miniscule voltages. These tools allow measurements of currents well below a picoampere in magnitude with sub-millisecond resolution, a capability matching the specifications of individual ion channels. With access to the statistics of channel activity and thus a means of inferring possible schemes of channel gating, there is much more to be learned. Can we detect the fleeting intermediates of the gating process? By correlating the results with structures obtained by X-ray crystallography and cryo-electron microscopy, can we infer precisely how gating occurs?

5. Nanotechnology

Our ability to conduct experiments on a nanometer scale is no longer limited to naturally occuring systems. Microfabrication and chemical synthesis have reached a point at which well-defined structures can be created at a subcellular scale. We are now encountering a first generation of nanobots, machines of sub-micrometer dimensions with numerous working parts. It is not difficult to foresee that such units will be packaged in lipid membranes to allow their incorporation into cells.

An important issue, and for now a fundamental limitation of this approach, is that of interfacing and communications. Even if a device positioned within a cell were able to record the value of some important parameter — the concentration of an ion, signaling molecule, or metabolite; the force exerted by a molecular motor; or the state of a genetic switch — how might that information be communicated to an experimentalist? Would it be possible, for example, to orchestrate fluorescent signals that encode a few bits of information? To create an assembly that could be interrogated like an RFID tag? To write a sequence of values onto a DNA "recording tape" that could later be recovered, replicated, and read?

The transmission of information in the opposite direction poses a similar challenge. Although a passive nanobot might be a useful reporter, more valuable still would be a device that could intervene in cellular activities, for example by working with or against molecular motors in mitosis, meiosis, cellular division, or the transport of organelles; by generating molecular signals under spatial and temporal control; or by repositioning organelles. Can we develop a means of communicating with such a device — as already seems plausible through the use of photonic techniques — and at the least switching it between two states, or preferably orchestrating a range of activities?

6. Discretization and Statistics

The classical approach of biochemistry is to construe molecular events in terms of continua. Cellular metabolism, for example, involves enough molecules that their discrete nature is immaterial. One therefore deals in terms of concentrations and macroscopic rate constants — a good approximation when a 20 μm cell contains over a billion glucose molecules and a million copies apiece of the enzymes necessary for their metabolism.

The continuum approximation must fail for many important processes in biology. Consider the signals transmitted between cells during embryonic development. These morphogens occur at low concentrations and must diffuse for hundreds of micrometers, yet they exert effects across many cellular diameters. The effectiveness of signaling by this means evidently requires the temporal integration of signals, as is known to occur in bacterial chemotaxis. Over an adequate period of time, even sparse encounters of receptors with morphogens can provide useful information. Eukaryotic cells may have developed multiple solutions to this problem, but none has been fully analyzed. How might an experimenter enumerate the actual number of morphogen molecules that a given cell encounters? How is temporal integration accomplished: at the level of receptors, of intracellular second messengers, or of genetic switching? What is the statistical effect of small numbers of signaling molecules on the timecourse and accuracy of cellular responses? [10]

7. Experimental and Theoretical Ecology

The behavior of modest numbers of discrete objects is also the province of ecology, the study of relationships within a species and between a given species and its environment, including other, coexisting species. Although ecology is classically a descriptive subject dealing with the characteristics of macroscopic creatures such as zebras and orangutans, similar principles apply on a smaller, even a microscopic scale. It has recently become possible to conduct ecological experiments in well-defined cultures created with several species or strains of bacteria and a fixed amount of nutrients. The organisms not only grow, mutate, and divide, but additionally attack one another by producing antibiotics and defend themselves — and other bacteria — by inactivating those substances. Because numerous experiments can be run in parallel, it is possible to replicate the results and determine statistics. A rich variety of behaviors emerges over tens of thousands of generations, including oscillatory solutions and seeming chaos. [11] The data have spurred theoretical models that attempt to explain the circumstances under which particular organisms survive and others vanish.

A similar approach can be extended to ensembles that include eukaryotic organisms. As an example, one may coculture a single species of bacteria, a line of photosynthetic algae, and a predatory ciliate. [12] Confined in a sealed ampoule, such a community persists for years, or hundreds of generations, during which time an experimenter can continually monitor the number of organisms of each type. Again replication is straightforward and the parallel systems are found to evolve in similar ways that can be characterized by eigenmodes. Can we understand this behavior in terms of the characteristics of the participating cell types? Are we able to predict the effects of changes in the environment or of the introduction of additional species? Can ecological theories be rigorously tested and extended through the use of such systems?

8. Equilibrium — or the Lack Thereof

Life operates out of thermodynamic equilibrium. So universal is this characteristic that one might define a living organism as a membrane-bounded aggregation of molecules — most importantly information-rich macromolecules such as proteins and nucleic acids — that functions far from equilibrium and is capable of accurate reproduction. The extent of dysequilibrium is staggering: while dissipating as much as 10 MJ of energy a day, each of us consumes an amount of ATP equal to his or her total body mass!

Neither physicists nor biologists have fully come to terms with this issue. The mindset of biochemistry, for example, is penny-wise and pound-foolish. We learn from textbooks that, through the complex apparatus of oxidative phosphorylation by mitochondria, a cell achieves an efficiency of about 40% in creating 36 molecules of ATP from the free energy in a single molecule of glucose. And having accomplished this miracle of parsimony, the cell then......burns through its wealth like a drunken sailor, dissipating energy on the futile treadmilling of microtubules and actin filaments, the idle shuffling of molecular motors, and the Sisyphean pumping of ions against concentration and voltage gradients. Evolution has presumably reached an optimal tradeoff between economy, flexibility, and rapidity of responsiveness, and the optimum evidently lies well in the direction of dissipation. How might we quantify this optimum, and how might we demonstrate it experimentally? Inasmuch as analyses of biological phenomena by physicists are usually couched in the terms of equilibrium thermodynamics, are these approaches deficient in capturing essential features of life? Are there better ways to proceed?

9. Symmetry — or its Absence

Every biological edifice is erected from asymmetrical bricks. With the exception of some bacteria, organisms construct their proteins from Lamino acids and consume Dhexoses as energy sources. Despite the chirality of these molecules, though, most organelles, most cells, and most animals appear bilaterally, pentamerously, or radially symmetrical. It is unclear just how this comes about: must the evolution of a protein both achieve a molecular function and also find a structure that can be assembled with the appropriate symmetry? What rules govern this process, and what restrictions do they impose on molecular structure?

Some organelles, though superficially symmetrical, make use of their intrinsic asymmetry. Cilia, including eukaryotic flagella, are the organelles responsible for motility by protozoa, spermatozoa, and many epithelia. Although each cilium displays an apparently ninefold rotatory symmetry of its components, the microtubule doublets, in many instances its operation breaks that symmetry to achieve a nearly planar beat. Strokes in the alternate directions are produced by a few dynein motors on one side of the organelle, then by a few on the opposite side; the motors in the perpendicular plane contribute nothing. Despite its seemingly simple appearance, a cilium contains some 200 proteins; how do they orchestrate this peculiar violation of the intrinsic symmetry?

At a still coarser level, organisms make use of the way in which ciliary beating deviates from a planar waveform. A protozoan such as *Paramecium*, for example, is covered with thousands of neatly aligned cilia. Each beats with a complex waveform: the active stroke of an extended cilium is nearly planar and propels the organism; the recovery stroke features a highly flexed cilium moving near the organism's surface (Figure 4). Such an asymmetrical pattern of beating is necessary at a low Reynolds number: symmetrical forward and backward strokes would simply cancel one another. What is the mechanical and hydrodynamic basis of this beating pattern?

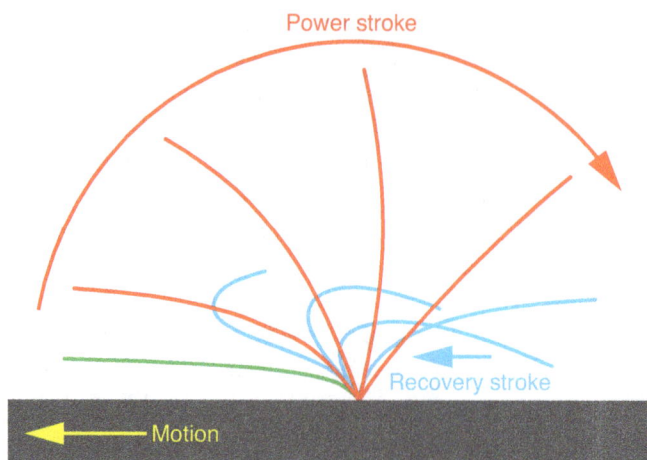

Fig. 4.   The ciliary beat. In many eukaryotes, for example, ciliates such as *Paramecium*, each cilium has distinct power and recovery strokes despite its apparent ninefold symmetry. Even in an environment of very low Reynolds number, the resultant asymmetry propels the organism.

Asymmetrical ciliary motion plays a critical role in the morphogenesis of vertebrates, including ourselves. An early embryo is bilaterally symmetrical, and indeed can develop stochastically into an organism of the ordinary handedness — heart on the left, liver on the right, and so forth — or into its mirror image. The normal handedness is dictated by the beating of cilia of the primitive (Hensen's) node, a small cluster of epithelial cells near the base of the nascent spine. As a result of their asymmetrical ciliary beating, these cells produce a directional flow of the surrounding liquid, which in turn conveys an as-yet-uncertain morphogenetic signal to the left side of the embryo and thus triggers a cascade of biochemical signals that consolidate the asymmetry and control the subsequent development of chiral organs.[13] As a consequence, diseases in which ciliary beating is compromised, such as primary ciliary dyskinesia (Kartagener Syndrome), are associated with situs inversus — mirror reflection of the ordinary positions of internal organs. Are there other instances of broken symmetry that serve the needs of cells and organisms?

10. Information

One of Schrödinger's greatest insights was the importance of information in biology.[5] This point had largely escaped most biologists, and until recently had not figured explicitly in biological education. Biologists tend to be awed by the adaptations of living things, the solutions that evolution has found for the myriad challenges of development, growth, sustenance, behavior, and reproduction. These issues, however, often do not lend themselves to quantitation of the information involved.

The most obvious manifestation of biological information is the genome, the compressed code for what we are. Recent advances in our understanding of this subject have stemmed from the ever-growing processing power and memory capacity of computers. Because specifying a human genome requires 2 GB of information, modern computers are absolutely essential for examining the statistics of genetic coding, for comparing genomes between individuals and across species, and for identifying the roots of genetic diseases.

Still more exciting applications likely lie ahead. We know, for instance, that only 3% of our genome directly encodes proteins; what does the other 97% do? Puffer fishes have dispensed with most of this seemingly extraneous DNA, yet function perfectly well — at least in their capacities as sushi and as barroom decorations. Formerly denegrated as "junk DNA," some of the noncoding material is now known to be essential for programming gene expression. But can we say how much? Can we progress beyond accepting genomes as givens and adduce the principles behind their organization? As Jacques Monod asked in a related context, what parts of our genomic organization represent chance, and what parts necessity?[14]

Another great challenge lies in relating the organization of the genome to an organism's development. Our bodies contain some ten trillion cells; our nervous systems encompass a hundred billion neurons with a hundred trillion connections. How is all this complexity specified by only twenty thousand genes? Moreover, those genes are so highly conserved that, owing to the shared presence of a Y chromosome, the genome of a man more closely resembles that of a male chimpanzee than that of a woman! Although the choreography of gene expression is clearly of absolute primacy, we know almost nothing about the subject. Once again, we cannot say whether fundamental principles underlie our genetic organization, or whether it simply represents the accumulation of some three billion years of accidents. This issue also bears on hopes for regenerative therapies and the potential requirement for prosthetic organs. Is it possible, for example, to rewire a mature brain after a stroke or other injury, or can proper connections develop only through the elaborate ballet of neural development?

11. Controlled Instability and Bifurcation

For the most part, living organisms — ourselves among them — crave stability. Nature is relentlessly competitive, and evolution constantly eliminates lifeforms that are less successful than their competitors. This selective pressure implies that there is ordinarily little tolerance of instability: a species must adopt the most effective strategy to deal with any challenge that it faces and then operate consistently in accordance with that principle. In keeping with this requirement, a fundamental principle of physiology at the level of the individual organism is homeostasis, the maintenance of an appropriate steady state. Insulin and glucagon, for example, mediate negative-feedback control of our blood-sugar levels; thyroid hormone regulates energy consumption; aldosterone and vasopressin ensure proper concentrations of Na^+, K^+, and H^+ in bodily fluids; and so on.

There are nonetheless exceptions, circumstances in which organisms exploit controlled instability. For example, on a cool morning one may encounter insects in search of a bifurcation. Unlike human muscle fibers, each of which contracts in a one-to-one relation with the firing of its associated nerve fiber, the flight muscles of many insects can oscillate autonomously. As a consequence, an active muscle fiber can operate at a frequency up to 4 kHz, far above the limit of our muscles, and thus can sustain the rapid wingbeats of tiny flies and midges. Two parameters control the bifurcation from quiescence to self-sustained activity: the intracellular Ca^{2+} concentration, which is set by neural activity, and the temperature. A cold insect preparing for flight consequently beats its wings vigorously, yet futilely insofar as levitation is concerned, until its body temperature reaches a level at which the muscles traverse the bifurcation and can sustain autonomous activity. What other physiological processes make use of bifurcations to switch between discrete functional states?

The ear's sensory receptors, or hair cells, also illustrate the benefits of controlled instability. As a result of the cooperative gating of mechanically sensitive ion channels, the mechansensitive hair bundle of each hair cell is dynamically unstable, a situation that fosters amplification or oscillation. A bundle's transition from a quiescent to an active state constitutes a Hopf bifurcation that confers three valuable properties on the hair cell (Figure 5).[15] First, near the bifurcation the system's gain

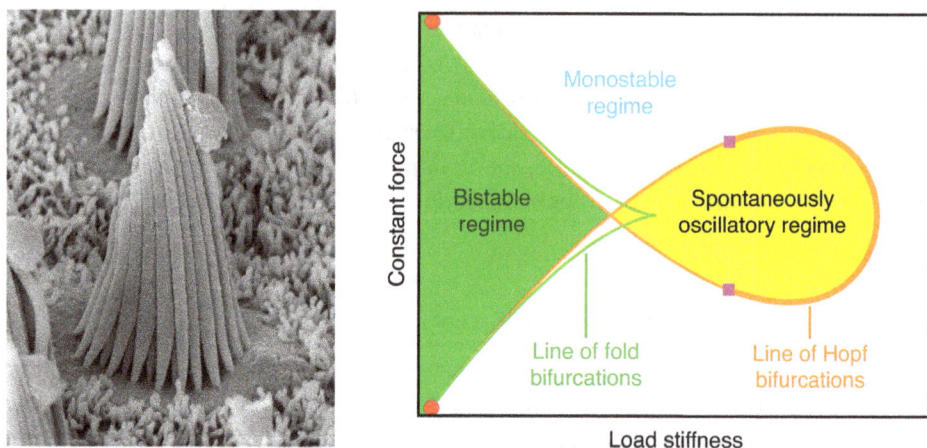

Fig. 5.   Bifurcation in the response of hair cells. Sounds and accelerations are detected in the human internal ear by sensory hair cells, each of which is marked by a mechanically sensitive hair bundle protruding from its apical surface. A hair bundle from the frog's ear (*left*) comprises approximately 60 stiff, actin-filled stereocilia and a single, tubulin-based kinocilium with a bulbous swelling at its tip. The bundle's hexagonal symmetry is broken by its beveled top surface. Deflection of the bundle toward its tall edge depolarizes the cell; movement in the opposite direction causes hyperpolarization. A state diagram (*right*) portrays the bundle's behavior as a function of the elastic load and offset force to which it is subjected.[15] In the monostable regime the bundle is sensitive to mechanical stimulation with a gain that rises as the system approaches the line of Hopf bifurcations at the right edge of the diagram. Beyond the bifurcations the hair bundle exhibits limit-cycle oscillations. Optimal responsiveness to weak signals occurs close to the line of bifurcations.

formally diverges, and our ears in fact display mechanical amplification that can exceed 1000X. Next, amplification is accompanied by enhanced frequency selectivity that allows each hair cell to reject noise at extraneous frequencies. This feature endows our hearing with a frequency resolution of 0.2% or even less, one thirtieth of the interval between successive piano keys. And finally, the relation of an ear's mechanical output to its input follows a power law with an exponent of one-third. As a consequence, our hearing compresses six orders of the magnitude in input sound pressure — twelve orders of magnitude in sound power — into the two orders of magnitude that can be encoded by the rate of neural firing. The capacity of hair cells to traverse a bifurcation to instability also underlies a striking epiphenomenon: the ears of 70% of normally hearing individuals emit pure tones in an ultraquiet environment! How can this transition be controlled? How do our ears adjust to the ambient level of sound to ensure comfortable and effective performance?

These examples hint that biology embraces more instances of controlled instability than we yet appreciate and point to the importance of dynamical-systems analysis and bifurcation theory in the field. Do buckling instabilities help to shape growing organs and organisms? Do our brains operate close — sometimes too close — to an instability of excitability? Many important examples of bifurcation must occur during cellular differentiation. Although intermediates are known to exist, most cells fall into specific categories defined by their morphologies and molecular properties. We construe differentiation as a series of steps — normally irreversible steps — between these configurations. How accurate is this picture? Are the transitions between distinct cell types truly step changes? If so, what biochemical machinery ensures that the transitions are decisive and militates against indeterminate states that might lead to cell death — or worse, to cancer?

12. Conclusion

Although this Prologue is meant to provide a sense of the prospects for the application of physical approaches to living matter, the effort must inevitably fall short. The greatest contributions will doubtlessly be those that cannot be predicted at present, the insights and techniques that will lead to shifts in biological paradigms. These advances will surely come: as organisms are better understood and as mathematical and experimental tools improve, the field of physics will devour biology, just as it previously consumed chemistry. The present is an extraordinarily attractive time for those interested in attending the feast.

Acknowledgments

The author is an Investigator of Howard Hughes Medical Institute.

References

1. R. Hooke, *Micrographia: or Some Physiological Descriptions of Minute Bodies Made by Magnifying Glasses, With Observations and Inquiries Thereupon* (The Royal Society, London, 1665).
2. H. Helmholtz, *Die Lehre von den Tonempfindungen als physiologische Grundlage für die Theorie der Musik* (Friedrich Vieweg und Sohn, Braunschweig, 1863).
3. E. Mach, *Grundlinien der Lehre von den Bewegungsempfindungen* (Wilhelm Engelmann, Leipzig, 1875).
4. S. E. Luria and M. Delbrück, Mutations of bacteria from virus sensitivity to virus resistance, *Genetics* **28**, 491–511 (1943).
5. E. Schrödinger, *What is Life?* (Cambridge University Press, Cambridge, 1944).
6. E. A. Rodriguez, R. E. Campbell, J. Y. Lin, M. Z. Lin, A. Miyawaki, A. E. Palmer, X. Shu, J. Zhang, and R. Y. Tsien, The growing and glowing toolbox of fluorescent and photoactive proteins, *Trends Biochem. Sci.* **42**, 111–129 (2017).
7. E. Kardash, J. Bandemer, and E. Raz, Imaging protein activity in live embryos using fluorescence resonance energy transfer biosensors, *Nat. Protoc.* **6**, 1835–1846 (2011).
8. A. Ashkin, K. Schütze, J. M. Dziedzic, U. Euteneuer, and M. Schliwa, Force generation of organelle transport measured *in vivo* by an infrared laser trap, *Nature* **348**, 346–348 (1990).
9. E. Neher and B. Sakmann, Single-channel currents recorded from membrane of denervated frog muscle fibers, *Nature* **260**, 799–802 (1976).
10. T. Gregor, D. W. Tank, E. F. Wieschaus, and W. Bialek, Probing the limits to positional information, *Cell* **130**, 153–164 (2007).
11. B. H. Good, M J. McDonald, J. E. Barrick, R. E. Lenski, and M. M. Desai, The dynamics of molecular evolution over 60,000 generations, *Nature* **551**, 45–50 (2017).
12. D. R. Hekstra and S. Leibler, Contingency and statistical laws in replicate microbial closed ecosystems, *Cell* **149**, 1164–1173 (2012).
13. H. Song, J. Hu, W. Chen, G. Elliott, P. Andre, B. Gao, and Y. Yang, Planar cell polarity breaks bilateral symmetry by controlling ciliary positioning, *Nature* **466**, 378–382 (2010).
14. J. Monod, *Chance and Necessity: Essay on the Natural Philosophy of Modern Biology* (Alfred A. Knopf, New York, 1971).
15. J. D. Salvi, D. Ó Maoiléidigh, B. A. Fabella, M. Tobin, and A. J. Hudspeth, Control of a hair bundle's mechanosensory function by its mechanical load, *Proc. Natl. Acad. Sci. USA* **112**, E1000–E1009 (2015).

Session 1

Intra-cellular Structure and Dynamics

Chair: *Anthony Hyman*, MPI-CBG, Dresden, Germany
Rapporteurs: *Jennifer Lippincott-Schwartz*, HHMI, Janelia, Virginia, USA and *Clifford Brangwynne*, Princeton University, USA
Scientific secretaries: *Enrico Carlon*, KULeuven, Belgium and *Sabine Van Doorslaer*, UAntwerpen, Belgium

A. Hyman Welcome to the first session of the Conference. It is my job to conduct us through the first session of a type of meeting that, I think, none of us has ever been to before. We have rapporteurs and also some prepared remarks, but we also want to leave enough time for discussions. The important point here is that our focus is discussion. After people give talks we are not asking them questions, but we want to discuss as a group to try to develop ideas related to presentations. We will have two rapporteurs Clifford Brangwynne and Jennifer Lippincott-Schwartz and then we will have some prepared remarks. I will give just a quick introduction on my side. The purpose of this meeting is to articulate fundamental questions which will inspire future generations. This session will focus on fundamental open questions on how cells organize their biochemistry. This question is an old one. The biochemists of the 19th and the early 20th centuries understood that one could invoke principles of collective behavior to understand biochemistry. As an example Hopkins, who won the Nobel Prize for discovering vitamins, had a famous address to the Royal Academy in 1913 — for biochemists Hopkins is as famous as Schrödinger is for physicists. He said on that occasion that the cell is simply a particular dynamic equilibrium of a polyphasic system and invoked principles of physical chemistry. But in that address he also said that there is too much physical chemistry here and what we need is more organic chemists who can get in and understand the molecules that allow to test somebody's ideas. And really what happened then was that organic chemists flooded into biochemistry and described the molecules. This individual description was focused on individual molecules and did not require the theoretical approaches of physics and chemistry and so few biologists learned physics or physical chemistry. But rather as Jim Hudspeth explained, there was the technical approach of experimental physics, which was so important for biology. As an example the thermodynamic description of Flory–Huggins had actually very little effect on thinking about cell biology. At the end of the cataloging era, for example, myself as a biochemist, and others actually did begin to look around ourselves in a similar way as Bragg looked around physics, and we said "The old way we learned to describe individual molecules and all have been trained on, is powerless to explain the phenomena of how the cell organizes its collective properties of biochemistry". And now biologists are returning to think about collective properties, which brought us back to approach physics and more specifically also physical chemistry. So now, we are going to try to articulate some of the fundamental problems of how the cell organizes its biochemistry using physical approaches. We are going to start with Clifford Brangwynne.

Rapporteur Talk by Clifford P. Brangwynne: Self-Assembly of Intracellular Matter

1. Introduction

Thanks so much to the organizers for putting together this great meeting. I must say it is simultaneously a dream come true and a great honor to be here, but also given the assembled scientific dignitaries and brainpower of this audience it is something of a nightmare to give a talk in front of you all! But I am looking forward to the discussion. I am going to tell you about some of the work that we are doing, and some of the things in this field in general, getting at the question of how intercellular matter is organized in space and time.

Let us imagine ourselves within a living cell, say a human cell. We can feel the constant buffeting of random microscopic motion, as thousands of different types of agitated molecules slam around at extremely high and highly fluctuating speed. Somehow in this sea of different molecules, crowded within the intracellular environment, order must somehow emerge, with information propagating up to larger length scales of biological organization. This truly awe-inspiring organizational process enables the cardiomyocyte to contract, the neutrophil to chase and ultimately capture the bacterium, and embryonic blastomeres to coordinate their pushing and pulling to drive gastrulation. But how the molecular-scale information propagates up to larger-length scales, and the principles by which this facilitates the self-assembly of living matter, are very much still mysterious.

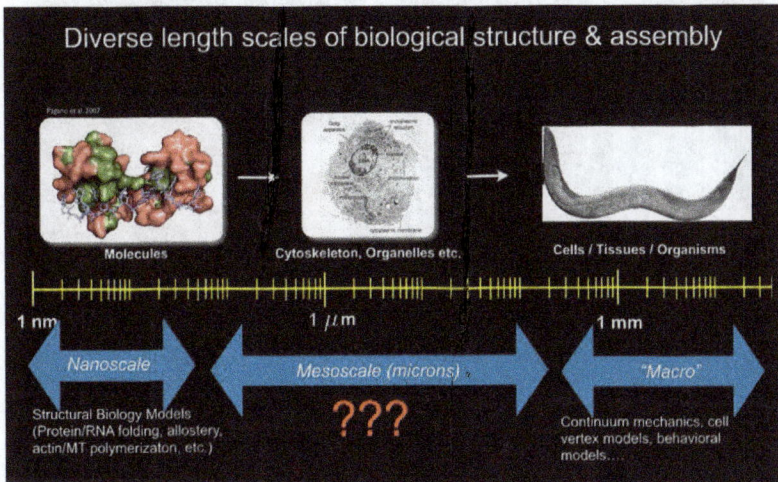

Fig. 1.   Diverse length scales in biological self-assembly.

One can view this organization as occurring on different length scales. We have nanometer-sized buildings blocks of the cell that encode information in these inter- actions on how to build larger-length scale assemblies within the cell — I'll refer to this as mesoscale organization. One can think about things like the cytoskeleton or the organelles and sub-compartments of the cell. The properties of these mesoscale assemblies are, in turn, dictating information on larger-length scales, whether a cell is going to grow, divide and differentiate, and so forth. On the nanometer length scale, with the molecular building blocks, we have a suite of models (and certainly there is much still to be done) — one can think about structural biology models, protein folding as occurring along energy landscapes, which dictate the conforma- tional states of proteins, and their interactions with other biomolecules, for example, via surfaces of complementary charge and so forth.

These examples underscore the point that we have a number of models for describing and understanding the organization at the nanoscale level. Similarly, at the other end — at the level of cells, tissues and organisms, which I'll refer to as the macroscale — we can think about a variety of models, continuum mechani- cal models, cell vertex models, perhaps one could even include physically-inspired behavioral models. And so we have some approaches for addressing organization at this macroscale level. I should emphasize that later in this meeting we are going to hear about developmental morphogenesis, and some of the things that are ongoing in that area, and clearly there is still much work to be done at this scale. But I think it is pretty clear that at the mesoscale, which is the focus of my talk today, concerning the micron-level organization within living cells (here we'll be focused

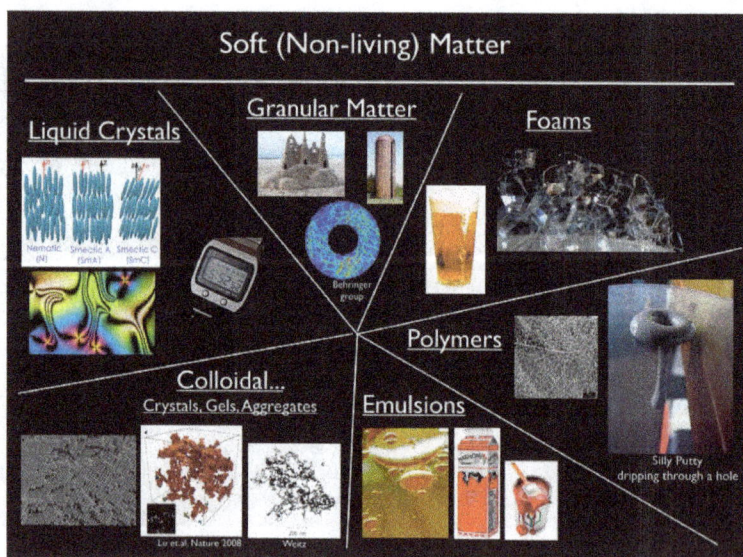

Fig. 2.   Soft matter as inspiration for understanding biological organization.

mostly on eukaryotic cells, although the same issues are certainly relevant within prokaryotes), it is much less clear how to think about it, particularly from a fundamental, biophysical perspective. What are the kinds of models, approaches and ways of thinking about organization at this level that one should be using?

Much of what we have been doing in my group, which is also true for a number of other people in this field, has been inspired by soft (non-living) matter physics. Soft matter deals with materials that are "squishy", i.e. relatively easy to deform under external perturbation. A prevailing theme in soft matter is the concept of emergent behavior. For example, one can think about foams, which are roughly 95 percent gas and 5 percent liquid, and yet come together to form a solid scaffold. Polymeric materials are another form of soft matter, in which the rich viscoelastic dynamics, which can be readily seen in materials like silly putty, arise from the fact that there are these huge interacting macromolecular chains which can slither past one another. Of course, the building blocks of a living cell — DNA, RNA, and protein — are all polymers, and so it is easy to see the connection with biology. But it is also the case for things like emulsions or colloidal assemblies or liquid crystals, or the concept of metastable liquids or fragility in granular matter — these are ways of thinking about molecular organization that are just as relevant for biological systems. And so we take a lot of inspiration from what is happening in this field, and try to bring it to bear in our approaches to understanding living matter. This cross-hybridization is turning out to be particularly interesting and helpful for tackling the problem of mesoscale organization within living cells.

2. Key Questions

I am going to open up with what I think are key questions in this field focused on living matter at the mesoscale. I will give you some of the ways that we and others are thinking about this problem, and provide a summary at the end:

1. What kinds of physical models can describe mesoscale structural organization?
2. Can such models elucidate distinct states of living matter — liquids, crystalline solids, glassy solids/gels?
3. How can we think about interplay between equilibrium & non-equilibrium driving forces?
4. Can answers to the above inform our understanding of cell function and dysfunction?

To begin, what types of models should we be thinking about to understand organization at this sort of intermediate micron length scale inside the cell, where it is much less clear than at the other extremes, what sorts of models should we be thinking about. I'd also like to touch on this idea of the distinct states of living matter, and how to elucidate what we now know are versions of liquids, crystal and solids, glassy solids and gels that occur within living cells. And so, can the sort of models that we would like to build give us insight into the biophysical principles underlying these different states of matter inside a living cell? The other aspect of this, which Jim Hudspeth has touched on, and I'm sure we are going to hear much more throughout this meeting, is how can we think about the interplay between equilibrium and non-equilibrium driving forces. This is something that I'll expand on and share some of my thoughts. And then finally, what we would like to know is can answers to the above inform our understanding of cell function and dysfunction? Can physical models of these interchanging states of biomolecular matter give us insights into the way in which molecular-level organization propagates up larger length scales, ultimately building the soft squishy robots that you and I represent at the organism level. And, of course, if we can understand the principles by which things "go right" to perform this magic act, we should simultaneously shed light on why things "go wrong", as they do in many cases, e.g. from cancer to Alzheimer's disease.

The first thing I'd like to discuss is this idea of equilibrium versus non-equilibrium. From equilibrium statistical mechanics, we consider the thermal energy scale kT, and the question is whether this is relevant in cells, or should be thinking about some effective temperature, or perhaps that is not even a relevant concept here. Because as many physicists have pointed out, equilibrium is death, or to paraphrase Rob Phillips, "only Napoleon is in Equilibrium".[1] Indeed, we know that there are many non-equilibrium signatures in living cells, and so this is just one example of a type of breakdown of the fluctuation-dissipation relation. Presented in the simplest way, we can write $D = k_B T/f$, where D is the diffusion coefficient, and f a dissipative drag coefficient. And so we think about this as a fundamental relation, the Einstein relation as it is often called. But what one can see in living cells is that if you look at fluctuating motion (i.e. of the diffusivity) of particles in the cell, some native bodies or introduced probe particles, what one sees is that the amplitude of those fluctuations depends sensitively on biological activity. So if we turn off motor activity or totally deplete ATP, the amplitude of these fluctuations goes down, with many documented examples of this in the cytoplasm.[2] So they seem to have a non-equilibrium origin in the chaotic molecular activity inside of the cell. This also seems to be the case in the nucleus, where people have looked at the apparent diffusion coefficient for various genomic loci, as a function of temperature.[3] And what you see is that there is this nonlinearity in diffusion coefficient in the native cells, and that seems to go away or comes down significantly when you deplete ATP, again consistent with non-equilibrium driving forces inside the cell that manifest in the fluctuation dynamics of the cytoplasm. Other recent examples include direct assessments of detailed balance violation, where the concept of detailed balance — the equality of individual transition probabilities, forward and back reactions — can break down, and one can have net circulation in loops in the state space that are detectable in livings cells, again underscoring the non-equilibrium driving forces in cells.[4]

These examples make it clear that our equilibrium concepts can certainly break down within living cells. Given these non-equilibrium features, much has been done in the active matter area to understand structures like the mitotic spindle, the apparatus that separates the chromosomes in dividing cells. Active matter approaches to understanding structures like this explicitly describe the non-equilibrium features which are put into the models for the spindle, or for an in vitro system like the 2-D active Nematic crystal formed from microtubules and kinesin motors, or for describing contractile actin networks. These emergent behaviors are described by non-equilibrium models — we are going to hear more about the experimental and theoretical approaches for these kinds of systems from Stephan Grill and others throughout the day, so I won't dwell on this point. Instead, I'd like to discuss the extent to which we can make progress on understanding aspects of intracellular organization using equilibrium ideas.

The studies I highlighted above make the point about deviations from equilibrium, but then again it may be the case that some intracellular processes operate close enough to equilibrium that we need not worry too much about our equilibrium conceptual frameworks failing us. I want to think about this possibility in the context of organelles, which we can view as mesoscale structuring compartments. We all have some sense for organelles; they are to the cell what organs are to our body. I know Jennifer Lippincott-Schwartz is going to discuss membrane-bound vesicle-like organelles, things like the Golgi apparatus, endoplasmic reticulum, secretory vesicles and so forth. What I'd like to discuss are these membrane-less assemblies, of which there are dozens of different types. These are a class of organelles, which are dynamic assemblies, typically comprised of both RNA and protein, that form in the cytoplasm, and in many cases in the nucleoplasm as well. In the cytoplasm, these include processing bodies which are involved in mRNA turnover and decay. In the nucleus, there are nuclear bodies like these nucleoli, which I'll tell you more about later, that assemble around nascent ribosomal RNA (rRNA) transcriptional sites. Much like conventional membrane-bound organelles, we view these sorts of structures as concentrating reactants and speeding up reaction rates within the cell, although in other cases they may serve to stop reactivity. But little has been known about how to think about what these things are, biophysically, i.e. what are rules that govern their assembly.

We became interested in this problem when studying these structures using the worm *C. elegans*. The *C. elegans* embryo is about 50 microns across, and contains the membrane-less organelles called P granules. As the embryo polarizes over the course of five or ten minutes, these structures end up in the posterior end of the cell. That is important because when the embryo divides in two, these structures are found in the posterior cell but not in this anterior cell. And so they are implicated in specifying the lineage of the cell, i.e. in specifying that this is the first progenitor germ cell. And so this is a really interesting example because this localization is tightly coupled to a very intricate reaction-diffusion system; in fact, the P granule localization is one of the consequences of some supremely interesting biophysics involving a reaction-diffusion-advection system, in which a stable reaction-diffusion pattern appears to be triggered by advective flows. These PAR polarity proteins assemble on the posterior cortex and ultimately signal into the cytoplasm to set up a gradient within the cell, specifically forming a gradient of a protein called Mex-5, among others. This leads to gradients in the concentration of unbound RNA, which ultimately impacts P granule assembly and disassembly.

3. Membraneless Organelles as Liquid-Liquid Phase Separation

In 2009, we showed that these P granules are liquid-like assemblies of RNA and protein, and that P granule localization to the posterior appears to be a spatially-regulated liquid-liquid phase separation process.[5] Here I'm showing an experiment where we applied shear stresses across a tissue that contains these P granules, and you see that they flow and drip and wet, much like conventional organelles. This and other data I'm not showing here led to a model wherein the anterior/posterior axis of this embryo spans a phase boundary, so the posterior end of the embryo dips into a two-phase liquid-liquid coexistence region. And thus this gradient in P granules stability, or to be more precise in the molecular interaction parameter, is mediated by the upsteam PAR/Mex-5/RNA reaction-diffusion-advection system. So this is one example of a clearly non-equilibrium system that sets up gradients in the embryo that we think ultimately dictates an equilibrium-like process, this liquid-liquid phase separation of P granule components. There is now a lot of evidence supporting this picture from our lab,[6] and work which Tony Hyman's lab

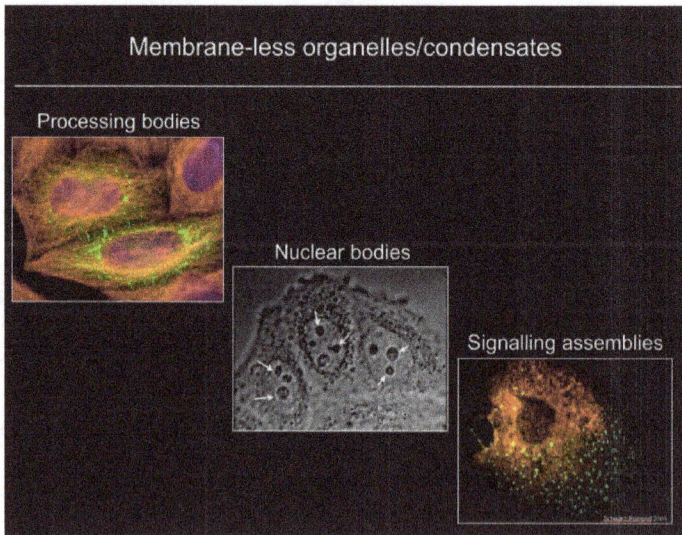

Fig. 3.   Examples of membrane-less organelles/condensates within living cells.

has continued,[7] and also from the lab of Geraldine Seydoux[8] and others. And so we have some handle on how this system works, although no doubt the system still contains many discoveries to be unlocked. In any case, it is a wonderful example of the intimate connection between equilibrium and non-equilibrium, and power of using concepts from both of these domains.

There are now a large number of examples of similar liquid-liquid phase separation, or more generally examples of phase transitions in living cells, and again in many cases these processes are coupled to the non-equilibrium biological activity of the cell. Nucleoli, for example, these prominent nuclear bodies, which my lab has done quite a bit of work on over the last few years, to examine the applicability of this pseudo-equilibrium liquid phase concept. It turns out that much of what happens in the assembly dynamics of these structures can be modeled using classical phase transition theoretical approaches, provided that one takes into account the non-equilibrium modulations of the phase behavior.[9,10] My Princeton colleague Eric Wieschaus has a nice study on nucleoli in *Drosophila*, which emphasizes some of this interplay using temperature control.[11] There are many other nuclear bodies, and we can start to ask what does it mean if there is condensation of liquid phase droplets in the nucleus of living cells. And in many cases these seem to be again examples of some near-equilibrium phase transition process inside the nucleus. In the cytoplasm, there are these structures called stress granules, which are RNA protein bodies that assemble in cytoplasm in response to stress, and these puncta exhibit have signatures of a liquid-liquid phase separation process. Another example comes from Mike Rosen's lab at UT Southwestern, work which has underscored the importance of these modular, repetitive multi-valent protein domains, which are found, for example, in many signaling assemblies and can drive phase separation into these liquid-like assemblies.[12] Multi-valent proteins can nucleate the polymerization of actin filaments, which is important for receptor clustering in the immune response of T-cells.[13] So this is a place where one can start to think about links between active matter theories for organization of the cytoskeleton, for example, coupled into liquid-liquid phase separation, in this case in a quasi-two-dimensional membrane system.

We now understand a fair amount about the molecular driving forces that promote condensation/phase separation inside the living cells. Proteins are of course linear polymeric chains of amino acids, but we usually think of all the proteins in a cell as being well-folded, with their amino acid sequence dictating their folding into a three-dimensional structure that in turn determines protein function. But many proteins in cells are not well-folded, and instead are conformationally heterogeneous. One of the proteins we've worked with is the P granule protein Laf-1, which has a disordered N-terminal domain, which is commonly referred to as an intrinsically disordered protein or region (IDR/IDP)[6,14]; IDRs are closely related to Prion-like domains, or low-complexity sequences. Work from our group and many groups in the field has shown that these disordered proteins are in many cases necessary and sufficient for driving condensation into these liquid states when they are purified in vitro. And there is quite some evidence that these same proteins and protein domains are responsible for driving phase separation into these liquid condensates in living cells.

One question that one can start to ask here is what sorts of theoretical approaches should we be taking to try and understand phase separation within the context of a living cell? Perhaps the simplest thing one can write down is a version of the Flory–Huggins/Regular Solution free energy:

$$\frac{F}{k_B T} = \phi \ln \phi + (1 - \phi) \ln (1 - \phi) + \chi \phi (1 - \phi)$$

This is the free energy as a function of concentration or mole fraction (ϕ) in a binary system. The first two terms represent the entropic contributions, which want the system to be well-mixed. But the third term is the contribution that arises from protein-protein interaction, which are encoded in this Flory χ parameter, which reflects the relative heterotypic versus the average homotypic interaction energy. If χ is large enough, i.e. if heterotypic interactions are energetically costly compared to homotypic interactions, then the system will want to phase separate into a two-phase equilibrium state.[15] The dynamics of the phase-separation process are often described by the Cahn–Hilliard equation, which one can view as a kind of diffusion equation which takes into account not only the entropy (as in usual Fickian diffusion) but also the molecular interactions. So this is probably the simplest formalism that one can think about, and one of the questions that people in this field are asking is how can we build on these models to incorporate the high degree of complexity that's intrinsic to these systems, and seem to be a feature of essential importance to what is happening? And then how can we start to add in, for example reaction terms to tie into the biological reactivity that modulates all of these interactions in the phase separation process?

Can one start to build models that take into account the complexity within living cells, and still remain tractable? Building up incrementally, one can certainly think about ternary extensions of this sort of Flory–Huggins foundation. Recently, there is some nice work by Daan Frenkel's group showing that one can start to make sense of highly multi-component systems, which seem essential to begin to connect with what is going on inside of living cell, with their thousands of different types of molecules.[16] And so the Frenkel group, building on earlier work by Sear and Cuesta,[17] have been using random matrix approaches to ask the question of what is the phase behavior of highly multicomponent systems within an equilibrium framework; what would the structure of the phase diagram look like? The key concept is that what you see depends on the variance of the interactions encoded in the matrix: for high variance one tends to get full demixing where you have many compositionally distinct assemblies, while for low variance one would just have a single condensate, where the relative amounts of the components are the same but just in a highly concentrated state. This work is just the beginning of what I view as an essential need to push these theoretical approaches. It will yield not only insights about the biology but I believe also new physics with wide ranging implications.

This is something that is clearly important to think about in the living cell, because it is a highly multi-component system. A simple example of the rich behaviors that we can start to understand using these ideas comes from some work that we have done with multi-component biomolecular systems, where we have proteins that form condensed liquid phases that co-exist with another condensed liquid phase.[18] This coexistence seems to be at play in the nucleolus. We have one set of biomolecules, particularly enriched in the protein fibrillarin shown here in green, and another set of biomolecules, particularly enriched in a protein called nucleophosmin (Npm1), which represent two condensed phases that are immiscible with one another. This is actually then a system exhibiting an apparent three-phase equilibrium, including the dilute, low-concentration phase, the red phase and the green phase. As with other multiphase liquid system, a key parameter here is the surface tension, which is encoded by the molecular interactions. The differences in surface tension dictate this core-shell architecture, which we think is biologically important for sequential processing of newly transcribed ribosomal RNA from the core outwards. It is likely that this is a general concept, in other words, surface tension likely plays a broadly important role in structuring these multiphase condensates. There are hints not only for the nucleolus, but also for other kinds of core-shell membrane-less condensates like P bodies or stress granules, for which we expect to have a similar surface tension inequality that drives the engulfment of one phase within the other. But there also appear to be many cases where the surface tensions are such that these droplets do not want to interact at all. And then

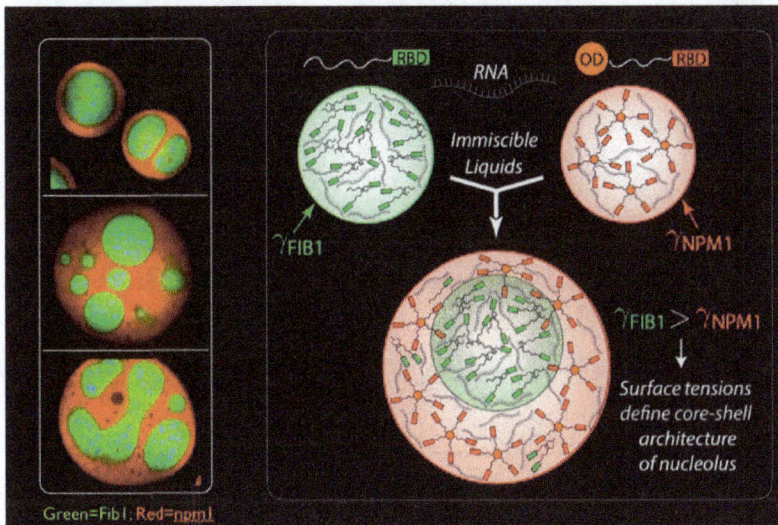

Fig. 4.   Coexisting RNA/protein liquids in living cells. Adapted from Ref. 18.

we think about cases where we have partial engulfment where the surface tensions are relatively equal, for example, in the beautiful electron micrographs of Cajal bodies from Joe Gall and colleagues. Again, these are open, non-membrane bound structures, co-existing in direct contact with one another. Here too we think this surface-tension mediated structuring is important for the facilitation of biological function in sequential or independent ways.

Another aspect of this, that I think is really important and something that is relevant for the question of biological function and dysfunction is the link between different molecular states. I've been discussing transitions from soluble molecular states — if you like a sort of gas state — which can condense into these liquid states. And then think that there are transitions from the liquid state into solid states which are associated with pathology. These may be amyloid states, or in some cases appear to be disordered, glassy-like states. There is evidence from a number of different groups, for example in some of the work we have done with Amy Gladfelter with the WHI3 protein she has studied, where you have droplets that phase separate in vitro, but over time they seem to nucleate fibers within the droplets.[19] And so there is a meta-stability of this liquid state in transitioning to the solid state, which likely relates the particular amino acid repeats — in this case polyglutamine, that are found in protein aggregation diseases. Another set of interesting examples is found in proteins associated with ALS — Stephen Hawking's disease. Here too, proteins like hnRNPA1 and FUS phase separate into liquid states that then transition into a solid fibrous state.[20,21]

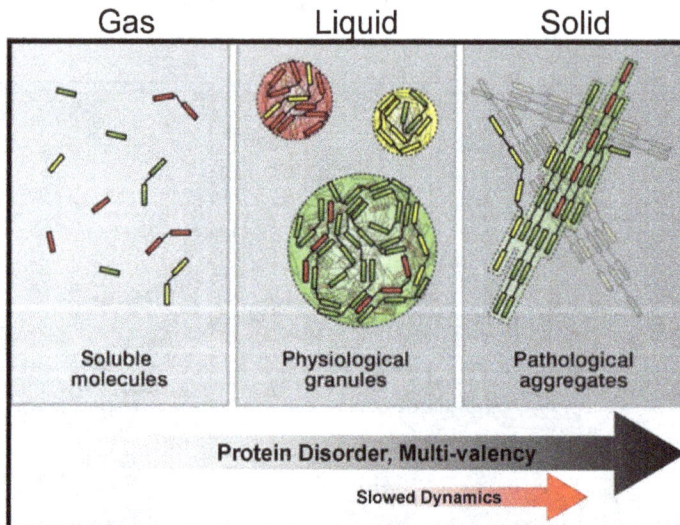

Fig. 5.   Transitions between soluble biomolecules, condensed liquid-like states, and solid-like pathological states are increasingly recognized within living cells. These are associated with an increased degree of protein disorder and multivalency, and a corresponding slowing of the molecular dynamics. Adapted from Ref. 22.

4. Equilibrium Concepts for Biological Activity?

So we think that these ideas of using equilibrium concepts and trying to see how they are modulated and how they occur in living cells are valid for a number of different reasons, including those I've already mentioned. My group is working to develop approaches to control phase transitions within living cells, using light-activatable proteins. We've been using these systems to start to map out phase diagrams within living cells.[23,24] It turns out that you can map out the location of the binodal and spinodal phase boundaries. We see exactly the kinds of phase diagrams that we can map in vitro where we know the system is in equilibrium. The big question is how and why can we do this in a living cell, where we know that there is so much distinctly non-equilibrium biological activity? Again, I think this is a key area to look into in more detail, and these tools are going to be very powerful for addressing the interplay — in other words, how does biological activity modulate the shape of this phase diagram.

Jim Hudspeth introduced fluorescence recovery after photobleaching (FRAP) already, so I just want to mention that our work with these "optogenetic" system suggests that where you are in the phase diagram dictates whether you are in a liquid state or in a less-dynamic, solid-like state. I just want to mention that one thing that is very interesting in this field is the question of these nuclear bodies, these membrane-less assemblies within the nucleus of living cells, of which there are dozens of different types. There is a lot of activity in the area of genomic architecture and organization. I believe Arup Chakroborty is going to speak to this point, so I won't go into it in a lot of detail, but the three-dimensional organization of the genome is really critical for the expression of the information encoded in the genome. There is a lot of activity just in the last year, for example, these three papers that just came out, one from Arup and colleagues at MIT on a phase separation model for transcriptional control where the organization of the genome is hypothesized to be impacted by local phase separation which brings together genomic loci.[25] Something very similar seems to be happening with heterochromatin. There is a series of papers that have come out suggesting phase separation and sort of local condensed states of this protein HP1A control condensation of chromatin.[26,27] And so we think this is maybe a useful framework for understanding organization inside the cell, although there is much that needs to be sorted out with respect to these questions I have been highlighting on the multi-component non-equilibrium nature of living cells.

And so, just to wrap up, I tried to introduce some of the questions and some of the ways that we and others in this field are thinking about this, and what kinds of physical models can describe meso-scale intercellular organization. I told you about some of the approaches that are being taken, some of the ways in which this is starting to elucidate links between these distinct states of living matter. I've discussed this interplay between equilibrium and non-equilibrium driving forces, and then how this is starting to give us some insights into function and dysfunction in a living cell. In the interests of time, I will not go through the list of everyone to thank, but I've had the opportunity to interact with many brilliant folks as collaborators, students and post docs. Thank you for your attention.

References

1. R. Phillips, Napoleon is in equilibrium. *Ann. Rev. Condens. Matt. Phys.*, **6**: 85–111.
2. M. Guo, *et al.*, Probing the stochastic, motor-driven properties of the cytoplasm using force spectrum microscopy. *Cell*, **158**(4): 822–832 (2014).
3. S.C. Weber, A.J. Spakowitz, and J.A. Theriot, Nonthermal ATP-dependent fluctuations contribute to the in vivo motion of chromosomal loci. *Proc. Natl. Acad. Sci. USA*, **109**(19): 7338–7343 (2012).
4. C. Battle, *et al.*, Broken detailed balance at mesoscopic scales in active biological systems. *Science*, **352**(6285): 604–607 (2016).
5. C.P. Brangwynne, *et al.*, Germline P granules are liquid droplets that localize by controlled dissolution/condensation. *Science*, **324**(5935): 1729–1732 (2009).
6. S. Elbaum-Garfinkle, *et al.*, The disordered P granule protein LAF-1 drives phase separation into droplets with tunable viscosity and dynamics. *Proc. Natl. Acad. Sci. USA*, **112**(23): 7189–7194 (2015).
7. S. Saha, *et al.*, Polar positioning of phase-separated liquid compartments in cells regulated by an mRNA competition mechanism. *Cell*, **166**(6): 1572–1584 e16 (2016).
8. J. Smith, *et al.*, Spatial patterning of P granules by RNA-induced phase separation of the intrinsically-disordered protein MEG-3. *eLife*, **5** (2016).
9. C.P. Brangwynne, T.J. Mitchison, and A.A. Hyman, Active liquid-like behavior of nucleoli determines their size and shape in Xenopus laevis oocytes. *Proc. Natl. Acad. Sci. USA*, **108**(11): 4334–4339 (2011).
10. J. Berry, *et al.*, RNA Transcription Modulates Phase Transition-Driven Nuclear Body Assembly. *Proc. Natl. Acad. Sci. USA*, **112**(38): E5237–E5245 (2015).
11. H. Falahati and E. Wieschaus, Independent active and thermodynamic processes govern the nucleolus assembly in vivo. *Proc. Natl. Acad. Sci. USA*, **114**(6): 1335–1340 (2017).
12. P. Li, *et al.*, Phase transitions in the assembly of multivalent signalling proteins. *Nature*, **483**(7389): 336–340 (2012).
13. X. Su, *et al.*, Phase separation of signaling molecules promotes T cell receptor signal transduction. *Science*, **352**(6285): 595–599 (2016).
14. M.-T. Wei, *et al.*, Phase behavior of disordered proteins underlying low density and high permeability of liquid organelles. *Nature Chemistry*, **9**(11): 1118–1125 (2017).
15. C.P. Brangwynne, P. Tompa, and R.V. Pappu, Polymer physics of intracellular phase transitions. *Nature Physics*, **11** (2015).
16. W.M. Jacobs and D. Frenkel, Phase transitions in biological systems with many components. *Biophys. J.*, **112**(4): 683–691 (2017).
17. R.P. Sear and J.A. Cuesta, Instabilities in complex mixtures with a large number of components. *Phys. Rev. Lett.*, **91**(24): 245701 (2003).
18. M. Feric, *et al.*, Coexisting liquid phases underlie nucleolar subcompartments. *Cell*, **165**(7): 1686–1697 (2016).
19. H. Zhang, *et al.*, RNA controls PolyQ protein phase transitions. *Molecular Cell*, **60**: 220–230 (2015).
20. A. Molliex, *et al.*, Phase separation by low complexity domains promotes stress granule assembly and drives pathological fibrillization. *Cell*, **163**(1): 123–133 (2015).

21. A. Patel, *et al.*, A Liquid-to-solid phase transition of the ALS protein FUS accelerated by disease mutation. *Cell*, **162**(5): 1066–1077 (2015).

22. S.C. Weber and C.P. Brangwynne, Getting RNA and protein in phase. *Cell*, **149**(6): 1188–1191 (2012).

23. Y. Shin, *et al.*, Spatiotemporal control of intracellular phase transitions using light-activated optoDroplets. *Cell*, **168**(1-2): 159–171 e14 (2017).

24. D. Bracha, *et al.*, Mapping local and global liquid-liquid phase behavior in living cells using light activated multivalent seeds. *bioRxiv*, 2018, under review.

25. D. Hnisz, *et al.*, A phase separation model for transcriptional control. *Cell*, **169**(1): 13–23 (2017).

26. A.R. Strom, *et al.*, Phase separation drives heterochromatin domain formation. *Nature*, **547**: 241–245 (2017).

27. A.G. Larson, *et al.*, Liquid droplet formation by HP1alpha suggests a role for phase separation in heterochromatin. *Nature*, **547**: 236–240 (2017).

Discussion

S. Chu I would like to make a comment. It is not so surprising, I would think, that you have exponential differences in rates versus temperatures because biology goes from local minimum to local minimum and there are thermal fluctuations around those bumps. So small change in temperature would mean that hopping from one to another would be in the exponent, so is that a big mystery?

C. Brangwynne I think you are exactly right. Biological systems are fundamentally out of equilibrium. I would like to argue though that there is reasonable evidence that in many cases one can use near to equilibrium approaches to understand self-assembly. I agree that the fact we see non-equilibrium signatures in living cells is not surprising.

D. Fisher Everybody always says that biology is out of equilibrium, but a lot of the processes run incredibly close to equilibrium: ATP synthesis, a lot of motors run close to equilibrium, and so on. It seems that there is a difference between some organization on many different scales and whether it is close to equilibrium.

A. Hyman I think we will pick up on that afterwards. Stefan Grill will discuss this. Anyone else would like to say anything else on out-of-equilibrium?

A. Hyman Thank you Cliff, I would like to call on Arup Chakraborty to make some prepared remarks.

Prepared comment

A. Chakraborty: Phase separation may regulate key genes that control healthy and diseased cell states

Enhancers are ubiquitous regulatory elements that control gene transcription in higher organisms. Recently, it was discovered (Whyte *et al.*, 2013) that the transcription of genes that play a key role in maintaining cell identity is regulated by large clusters of enhancers, called super-enhancer (SEs). Genes associated with many diseased cell states, such as cancer, are also regulated by SEs. High densities of co-activators, nucleic acids, transcription factors, etc are localized at SEs. SE elements are also known to interact with each other. SEs exhibit several unusual properties; for example, they are extraordinarily sensitive to drugs that disrupt the binding of certain co-activators, entire SEs can collapse upon deletion of a small number of elements or SEs can form by addition of a few elements.

Using simple theoretical and computational approaches, it was proposed that SEs form by phase separating into liquid-like droplets due to weak cooperative interactions between the participating moieties (Hnisz *et al.*, 2017). Thus, as is typical for phase transitions they form when upstream cues exceed a sharp threshold. The liquid-like droplet can concentrate various key molecules important for transcription in a robust fashion. This model provides an explanation for many observed features of SEs, including their unusual sensitivity to drugs which disrupt key co-activator interactions (Figure 1). This theoretical model seems to have been validated by recent high-resolution microscopy experiments (Sabari *et al.*, 2018) that show the formation of liquid-like droplets at SEs in live cells (Figure 2). These droplets are co-localized with transcribed RNA showing their importance for transcription, and they are dissolved by drugs that interfere with co-activator interactions.

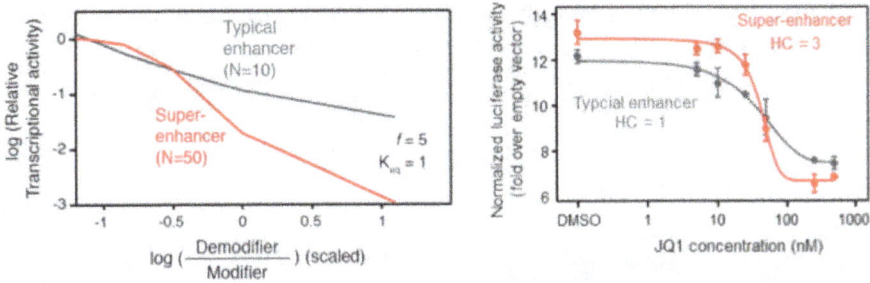

Fig. 1. The left panel shows model predictions for the enhanced sensitivity of SEs (red curve) to drugs compared to that of a typical enhancer (gray curve). The abscissa is a proxy in the model for drug concentration. The right panel shows experimental measurements showing the same phenomenon. The enhanced sensitivity of SEs is characterized by a Hill coefficient (HC).

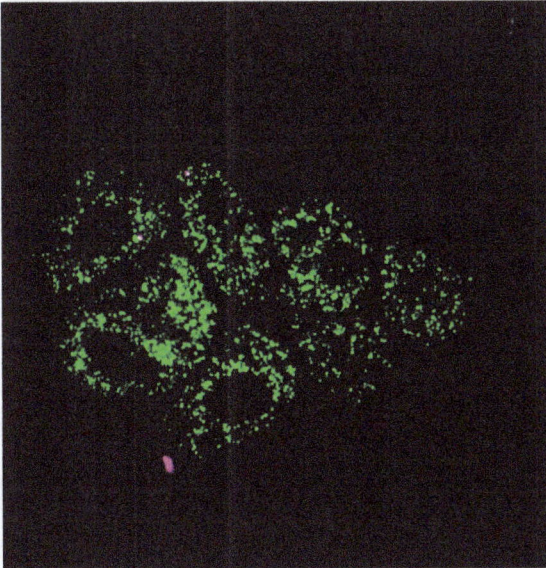

Fig. 2. High-resolution microscopy image showing that SE components (green) form phase separated droplets in the nucleus of the cell, and co-localize with RNA (purple) at transcriptionally active SEs.

The conceptual issues that need to be addressed to establish the principles underlying how gene regulation in mammals is regulated by phase separation include: (i) The dynamics of non-equilibrium phase transitions that lead to the formation and dissolution of finite size droplets at SEs; (ii) The molecular code that underlies the ability of the participating molecules to form droplets to regulate transcription in response to specific cues; (iii) Transport of proteins and transcriptional machinery into the phase separated droplets to regulate function; (iv) How diseased cell states co-opt this mechanism? (v) The selection forces that led to the evolution of weak cooperative interactions as a means to mediate specific biological functions. The search for the principles underlying these issues will require physics approaches grounded in biology.

Discussion

A. Hyman Thank you very much. In the talk by Cliff and the remarks by Arup we have heard that phase separation is a driving force for organizing a cell. This is a classic example where a theory in physics does really help us in understanding a fundamental problem. Does anyone have remarks?

S. Eaton Of course people have thought about phase transition as an important organizer of spatial dynamics in membranes for a long time. One problem is that phase transitions are temperature sensitive, so cold-blooded animals developed these mechanisms to control the lipid composition so that they can avoid unwanted phase transitions. I wondered if any of you have thought about whether, in cold-blooded animals, there might somehow be similar, analogous mechanisms that would maintain homeostasis of these phase transitions at different temperatures.

A. Chakraborty All I can say is that we have looked at some of the compositional biases in a key component of these superenhancers as you go through the evolutionary lineage. You find that in vertebrates, when they first started to evolve, the compositional biases are quite a bit different compared to what we see in other animals that are not vertebrates. We do not understand yet what this compositional bias means. But I have not looked at cold-blooded versus warm-blooded.

A. Hyman So that is a key point. We know how temperature sensitive these phase transitions are. And the question is whether warm-bloodedness gave the cells some evolutionary advantage by no longer having the phase transition systems having to cope with different temperatures.

C. Marchetti I have a comment regarding the liquid-liquid phase separation that Cliff was discussing. In equilibrium we understand that phase separation is essentially driven by attractive interactions. In active systems where the particles/entities are driven by internally generated forces, the dynamics is very different from a brownian one. We call it persistent in the sense that the particles go in a straight line before having the direction randomized by noise or interactions. In these systems, we know that this type of dynamics itself can generate phase separation without any attractive interactions. I was wondering whether you think that this kind of traffic-jam type of effect plays any role in the phenomena you are seeing.

C. Brangwynne I can just say that analogies with motility-induced phase separation are very interesting in the context of cytoplasm. If you think about the non-equilibrium fluctuations (I showed you some images of data on fluctuation/dissipation breakdown), I think that there are likely relevant points of contact between those theoretical approaches and active matter. At this stage we do not think that it is quite the same thing as, for example, collective dynamics of bacterial swarms. I think those are useful things to look at.

J. Howard One general question I have about these liquid-like domains is: How much structure is there inside these domains? I mean, there is quite some controversy about looking for structural elements within those domains. To what extent are the domains made up from polymers, etc? We can draw the analogy to the mitotic spindle which has liquid-like properties, but as we know there is a lot of structure inside.

A. Hyman Does anybody want to say something else?

F. Julicher I would like to comment on these points. Of course in simple phase separation one usually generates homogeneous phase. In the context of biology we are far from equilibrium, as mentioned before. So we couple these phases to non-equilibrium pattern-forming processes and therefore we get structures that are dynamic and at the same time based on phase coexistence and self-assembly. A nice example of this is the centrosome, which can be thought of as a phase, but also with the help of the centriole, as an organizer of active processes one gets structure, much more structure than a simple droplet-like object.

D. Fisher I just want to make a general comment. In some sense, some of the general features we are talking about may also occur in geology and geophysics. The crucial thing about biology is the control...

E. Wieschaus One thing that struck me in Cliff's presentation was that it reminded me to what we described as phase transition, not at the molecular level, but at the cellular level during embrionic development. I am thinking, in particular, of experiments by Malcom Steinberg in the 1970's and 80's that involved mixing of ectodermal and endodermal cells from an embryo. They showed that they sorted out into associated aggregates, internalized mesoderm, endoderm and ectoderm, that followed exactly the same rules of surface tension of these aggregates. You could show that cells, as well as molecules, would follow the same physical relationships.

A. Hyman Stefan, do you want to say something?

S. Grill One little comment on that. One thing that always struck me is that in this case and also in special cases in reconstituted systems the objects are very round, which really speaks for a surface-tension effect driving the roundness. Of course, how easily that can be brought back to molecular interactions, is not clear.

A. Hyman Eric, as you mentioned, it is obvious to invoke principles of phase separation and people did it. In developmental biology it didn't go very far. The reason it is so exciting in cell biology is that the more we look, the more phenomena can be explained by phase separation. That is why many papers are now being published on this topic, as mentioned earlier. All the things that seemed to be hard to explain before now seem easy in considering them as phase separation phenomena.

E. Wieschaus But is phase separation really the governing mechanism in the processes whereby these structures actually do form, even in the case of the cell biology you looked at? This is the next question we have to ask.

C. Brangwynne I would like to make a comment. I appreciate the history of the differential adhesion hypothesis by Malcom Steinberg. I think that in that case, the original form of the idea was that there was a free-energy landscape associated with the interaction of individual cells. Maybe, in retrospect, that went too far, because can you use equilibrium concepts at the level of cells? However, I think that your point is a very interesting and important one. I mean: is there a free-energy landscape? Or some interaction energy that we can put in some sort of statistical mechanical framework? Or are there just some effective interactions, effective surface tensions, that are sort of mapping in a analogous way? I think those are kind of critical questions that are to be solved.

N. King First a quick comment and then I want to frame a question. The comment is about how organisms deal with temperature fluctuations and phase transitions. Much of the early animal evolution happened in the oceans where organisms were unlikely to encounter much temperature fluctuations. It is interesting that the warm-body mammals are sort of reverting back to temperature control. I wonder whether in evolutionary perspective you might see a lot of phase transitions happening in marine organisms that then get lost. Another point: one thing that has interested me in biology is modularity and how complexity comes from simple components. As you were describing these phase transitions, I was trying to think what is the modulus. Is it that these different states in going from the uncondensed form to the condensed form to the solid or is it the condensates themselves, can they combine? I wonder where complexity arises and I have no idea about that.

A. Hyman We have now a prepared remark by Stefan Grill that may address one of these things. Does anyone want to say something while we are waiting for the slide projection to come up?

G. Suel I just wanted to follow up on what Daniel and Eric said. I think that this point needs to be made very clear and more general, because I see a divide in this room in the sense that we can think of biological processes and we can have a shelf of physics problems. We can say which physics that has been done explains this process. There can be many of those biological observations that can be explained with simple laws of physics that we already understand. However, I think that will only go so far. There is a need to connect between physics and biology and that will not happen if this is the only type of physics approach that we do. What we need to understand is at what point biology figures out ways to overcome or manipulate those processes in ways that are not just off-the-shelf solutions that have been already identified. If we can think of what side of camp we are on and if we want to connect to biology — because ultimately the goal should be to understand biology — we have to make sure that we look at those sort of problems where biology is manipulating physics, or coming up with unique solutions to overcome limitations or maybe even laws of physics.

A. Hyman This is a perfect introduction to Stefan's remark.

Prepared comment

S. Grill: Evolving active matter: Which materials can evolution generate?

Inside the cell a variety of molecular processes take place, and there is a spectrum of types of organization that arise. Biological matter is characterized by unusual material properties, and the fact that active processes drive the system locally out-of equilibrium. For example, the actomyosin cortical layer underneath the cell membrane in the one-cell stage *Caenorhabditis elegans* embryo, through forces that are generated within this layer, is able to undergo chiral rotatory flows always with the same handedness, see Figure 1 below.[a] At a later stage in development, such rotatory flows are important for determining the left-right body axis of the embryo. What are the material properties of the cortical layer that such movements are based upon? To attempt to answer this question, we make use of approaches that have been developed in the realm of physics, where we think about these

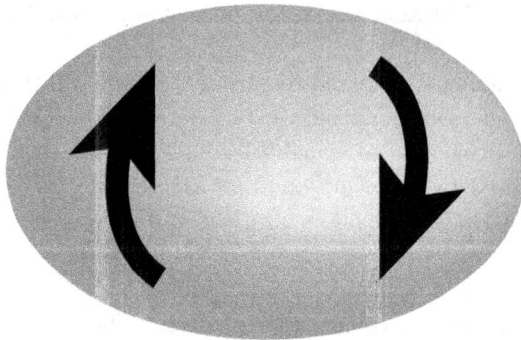

Fig. 1.   The cell surface in the one cell stage *Caenorhabditis elegans* embryo (grey) undergoes rotatory flows (black arrows) of consistent handedness. Rotatory flows are a result of unusual material properties that arise through torque generation via active processes in the actomyosin cortical layer.

---

[a]S. Naganathan, S. Fürthauer, M. Nishikawa, F. Jülicher, S. W. Grill, Active torque generation by the actomyosin cell cortex drives left-right symmetry breaking, *eLife* **3**:e04165, doi:10.7554/eLife. 04165 (2014).

materials as being substances that are near thermodynamic equilibrium but that are constantly kept away from reaching thermodynamic equilibrium through processes that consume a chemical fuel. In the actomyosin cortical layer that undergoes chiral rotatory flows, ATP consumption by the molecular motor myosin gives rise to the generation of molecular-scale forces and, given the helical nature of actin filaments, molecular-scale torques. Broken symmetries and conservation laws can be used to formulate a hydrodynamic theory of the actomyosin cortical layer. We determine a free energy density that depends on hydrodynamic variables, and identify pairs of conjugate generalized thermodynamic fluxes and forces that contribute to the local rate of entropy production.[b] We then make use of an Onsager approach to evaluate how thermodynamic variables interdepend on each other in their quest to relax towards equilibrium. This approach can be utilized to describe how in a thin film of an active chiral fluid, active torque dipoles contribute to fluxes of angular momentum for driving rotatory flows.[c] In such a hydrodynamic theory of active biological matter, generalized hydrodynamic fluxes depend on generalized forces, and physics informs us that all couplings that are allowed by symmetry and the Curie principle exist, and that they can all be important. I would now like to speculate on how the evolutionary process can act on such materials. Evolution acts on molecular mechanisms, and an interesting consequence is that evolution should be able to tune the relative importance of all allowed couplings in this hydrodynamic theory. Evolution can thus generate materials where some of the couplings that are allowed by symmetry are more important and others are unimportant. Hence evolution should be able to explore all material behaviors that are physically possible, and select material behaviors that are needed. An important task for the future is to evaluate which of the physically possible material behaviors have been selected for in the evolutionary process.

---

[b]M. C. Marchetti, J.-F. Joanny, S. Ramaswamy, T. B. Liverpool, J. Prost, M. Rao, R. A. Simha, Hydrodynamics of soft active matter, *Rev. Mod. Phys.* **85**, 1143 (2013); Jülicher, S. W. Grill, G. Salbreux, Hydrodynamic theory of active matter, *Rep. Prog. Phys.* **81**, 076601 (2018).
[c]S. Fürthauer, M. Strempel, S. W. Grill, F. Jülicher, Active chiral processes in thin films, *Phys. Rev. Lett.* **110**, 048103 (2013).

Discussion

A. Hyman Thank you Stefan. Holly Goodson, could you also address that problem?

Prepared comment

H. Goodson: What aspects of biology are predictable? This question is old and contentious, but now is the time to revisit it because of factors including significant recent developments in the understanding of complex systems, synthetic biology, and the molecular diversity of life. Although this question can and should be asked at all possible biological scales, we suggest focusing on the cellular level, as cells are the fundamental unit of life, at least as we presently understand it. Since cells exist at the interface between chemistry and biology, a more tractable way to phrase this question becomes "How do physics and chemistry shape cellular life?"

Here the word "shape" can be considered literally — how do physics and chemistry impact the observed morphologies of cells and organisms, but also more figuratively and more broadly. For example, how do physics and chemistry (including geochemistry) lead to predictable characteristics of metabolic pathways, information processing networks, and structures (physical or ecological) formed from communities of cells?

Answers to these questions are important because they should help identify fundamental principles of biology. And, since physical law is universal, they should also provide insight into what life might look like elsewhere.[a] One obvious approach is to start from the bottom up: take physical principles and look for cases where we can find them at work in biology, as done by D'Arcy Thompson in the early 20th century. Another is to compare divergent biological systems and look for unexpected similarities. Both are fruitful, and can give insights from the deep (e.g., the existence of fundamental design principles for biological networks[b]) to the trivial but informative (e.g. protein polymers are so often observed to be helical because helices are simply the most likely way to form a filament from asymmetric subunits).

In addressing these questions, it is important to recognize that evolutionary biology as a field focuses on how organisms change with time, i.e., how they diversify. Partly as a result, the question "Why are organisms so similar?" has remained under-studied, especially when considered at a cellular scale. Obviously, one explanation for similarities between organisms is the

---

[a] Cockell, Charles C. (2017) The Laws of Life. *Physics Today* **70**, 42–48.
[b] Alon, U. (2006). *An Introduction to Systems Biology: Design Principles of Biological Circuits*, CRC Mathematical and Computational Biology (Chapman and Hall).

common ancestry of life on earth, but physics and (geo)chemistry provide an equally and perhaps even more important additional set of answers.

Significantly, when physics and chemistry are invoked to explain biology, the effect is often phrased in terms of constraints — that physics and chemistry restrict the parameter space in which organisms can live and thus influence what has evolved. However, it is important to remember that physics and chemistry can also be profoundly creative: spontaneous self-organization occurs in dissipative (non-equilibrium) physical systems across scales. Cell biological systems harness this self-organization for processes ranging from the establishment of cell polarity to partitioning of chromosomes.[c] Biological self-organization is observed at scales from the molecular[d] to the ecological.[e] Indeed, as many are aware, there are strong arguments that life itself is a predictable outcome of creative self-organizational processes.[f]

It is sometimes argued that instead of being predictable, biology is inherently contingent, e.g., it is only because of chance events that life on earth existed "the boring billion" and became anything other than a sea of microbial slime. First, it is important to recognize that the sea of microbial slime was (on the basis of presently living organisms that had already diversified by that time) a complex world that almost certainly had many predictable features ranging from aspects of metabolism to ecological structure.[g]

Second, it is critical to observe the parallel between "predictability" as it applies to biology and the predictability of a stochastic chemical system. In the chemical system, free energy differences (i.e., thermodynamics) dictate where the system will end up if given sufficient time, but provide no information about the rate or path of the reaction. If multiple states have similar energies, physics predicts the eventual distribution between them, but not which specific molecules end up where. These uncertainties do not detract from the fundamental predictability of the system. In biology, it is similarly difficult to predict the path or time of an evolutionary transformation, except to say that the process will be constrained by the principles of population genetics. However, given enough organisms and enough time, many aspects of living systems as observed within a biosphere as a whole should be "predictable" (in the sense of the stochastic chemical system

[c]Karsenti, E. (2008). Self-organization in cell biology: a brief history. *Nat. Rev. Mol. Cell. Biol.* **9**, 255–262; Wedlich-Söldner, R., Betz, T. (2018). Self-organization: the fundament of cell biology. *R. Soc. Lond. B Biol. Sci.* **373**, 20170103.
[e]Denny, M., and Benedetti-Cecchi, L. (2012). Scaling up in ecology: mechanistic approaches. *Annual Review of Ecology and Systematics* **43**, 1–22.
[f]Kaufmann, S. A. (1993). The origins of order: Self-organization and selection in evolution (Oxford University Press).
[g]Kaufmann, S. A. (1993). The origins of order: Self-organization and selection in evolution (Oxford University Press); Braakman, R., and Smith, E. (2013). The compositional and evolutionary logic of metabolism. *Phys. Biol.* **10**, 011001.

above) because they are imposed by physics and chemistry as constraints or generated by physics and chemistry through self-organizing processes.

These similarities raise interesting questions about what processes or events play the role of catalysts in evolving biological systems, enabling populations to find new paths to reach previously inaccessible parts of the landscape (defined literally or metaphorically). Equally important is identifying how feedbacks at various scales can alter the fundamental shape of the landscape (e.g., the bio-oxygenation of the atmosphere) and thus the pace and direction of evolution. Answering these questions should improve our understanding of predictability in biology by providing insight into the kinetics of evolutionary transitions.

To paraphrase others, 20th century biology identified most of the parts of biological systems; now it is time to put them together. More precisely, it is time to understand how physics and chemistry lead to predictability in terms of what the biological parts are, how they interact, and how they put themselves together in systems across the range of biological scales. Doing so should enable biology to transcend the detail in which it has been mired and provide a foundation for identifying the fundamental principles of cell biology and of biology, more broadly.

Discussion

A. Hyman Thank you. Those are some excellent summaries of the question that was brought up, which is really the key question: articulate fundamental problems that will inspire future generations. So far, in a way, physics has been reactive to this question. You discover phase separation and you ask to what extent the physics that we know is able to describe this system. But of course we would also like to be predictive, to predict the way biological matter should work and to be able to test ideas. And we need to discuss also to what extent the current physics is not sufficient to describe the phenomena that we are observing. Anyone want to comment on that?

D. Fisher I have a mix of a comment and of a question. Some of the common attributes seem to be from some biology evolving and some other biology adapting, instead of evolving by itself. I mean cyanobacteria and photosynthesis are some obvious examples, and mitochondria and so on. Presumably some of the metabolism that gets traded around...

H. Goodson I fully agree with that, but I do think that there is some underlying structure that is likely to show up again and again. And by predictable I do not necessarily mean that it will happen, I mean it can happen. If you think about the free energy diagram, it is like a sack of sugar sitting on the shelf: it is still predictable that it will eventually go to carbon dioxide. As an example of a metabolism part that may be predictable could be the use of phosphate. You have a certain structure there, maybe not ATP but polyphosphate seems relatively likely to be predictable. Also the use of electron transfer is likely to be predictable, although particular details may not be predictable. The challenge may therefore be to find which aspects are contingent and which aspects are predictable.

D. Fisher Since a lot of evolution is coming from very rare events, the question is which ones are the least unlikely, which end up effectively looking like they are predictable.

A. Hyman Any comments? Steve.

S. Quake I would like to revisit the meaning of the term predictable. Predictable on the basis of what assumptions? The flavor of the comment of things being used across the tree of life seems more that you are asking which aspects are universal... but maybe you have other thoughts about this.

A. Hyman Holly, predictable is an important issue and needs to be clarified. So, Steve, the question you are asking is to what extent you want to be predictable?

S. Quake Predictable on the basis of what assumption? Equations, physical principles, what is it? In what sense do you use that word?

H. Goodson That is a good question. I guess I am using the word in the following sense. I do have in mind a sort of chemical predictability of a free-energy diagram. You may not get to the lower state, because that depends on how high is the hill in between, but if you got something to change the landscape, then you can get to that lower energy state pretty easily.

A. Hyman Andrew?

A. Murray Yes, I want to comment on this too. One of the things we need to do here, as a collective group of people, is to be as rigorous as we possibly can. There is a clear difference between ex-post facto explicable, e.g., "It happens this way and now I can make sense of it" and predictable. Predictable means that you make a prediction that something will happen in the future and then you do some experiments to test whether that happens. I think we really need to be very rigorous about that.

A. Hyman Michael?

M. Elowitz Maybe I can just add that beyond predictable there is also the question of engineerable or buildable. I was just struck in Cliff's talk that there is the possibility of putting domains that are necessary and sufficient to generate phase-separation behavior. One question you can ask from the synthetic biology point of view, is how much of this behavior can we explore by building different systems with different characteristics and doing it in a kind of forward engineering way.

A. Hyman Just in the interest of time, unless someone has a very burning question on what we just discussed, I think I would like to call on our second rapporteur, Jennifer Lippincott-Schwartz, to talk about membranes.

Rapporteur Talk by Jennifer Lippincott-Schwartz: Eukaryotic Membrane Organization: what advanced imaging, quantitative physical analysis and modeling are revealing

Membrane-bound compartments play critical roles within eukaryotic cells, with functions ranging from secretion to endocytosis to energy harvesting. New quantitative approaches that integrate advanced imaging technologies with physical-chemical concepts and computational modeling are making significant contributions to understanding how these compartments form, are maintained, and undergo cross-communication. This integration allows the intricate and changing morphologies of membrane-bound compartments and their modes of protein sorting and retention to be quantitatively studied in living cells. As a direct result, new testable models are emerging for how these compartments operate and are controlled.

1. Introduction

Membrane-enclosed organelles are a ubiquitous feature of eukaryotic cells, occupying approximately one third of cytoplasmic volume (Figure 1). Comprised of an internal lumen and surrounding membrane, these compartments perform specialized but interconnected functions that are essential for proper cell behavior and metabolism. Eukaryotic cells have eight different membrane-bound organelles (nine in the case

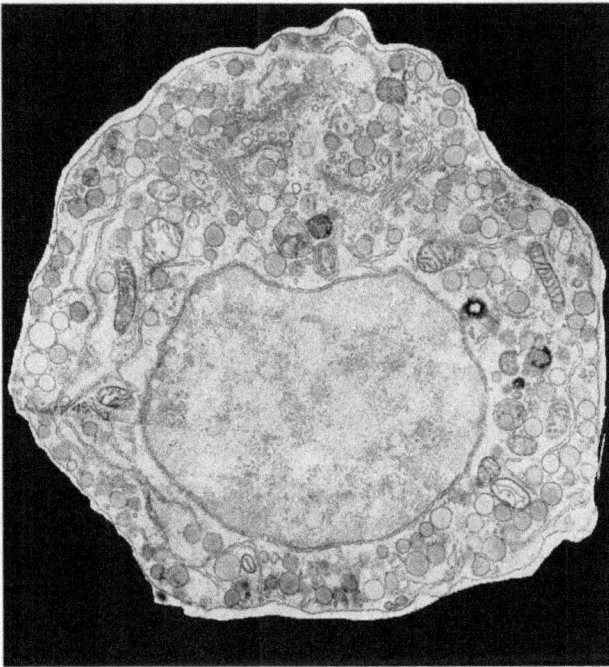

Fig. 1.   Transmission electron micrograph of a fibroblast cell illustrating the variety and complexity of internal compartments within a cell. Obtained from Lydia Yuan, NIH.

of plants and algae), including endoplasmic reticulum, Golgi apparatus, endosome, autophagosome, lysosome, lipid droplets, mitochondria, chloroplast (plants/algae only) and peroxisomes. Eukaryotic cell survivability, morphological diversity and adaptability are intimately linked to these nine organelles' functions, with over 30% of the eukaryotic genome coding for proteins in their membranes or within their lumen. In contrast, simple bacteria and archaeal cells, which lack most of these genes as well as complex, internal membrane-bound compartments, remain small and exhibit limited morphological intricacy.

Three fundamental tasks are performed by eukaryotic membrane-bound compartments. First, they act to take up and digest macromolecules from outside the cell. This is mediated by membrane-enclosed vesicles that bud inward from the plasma membrane carrying extracellular cargo in a process called endocytosis. The endocytic vesicles move through the cytoplasm to fuse with endosomes and lysosomes, where the cargo is redistributed or digested. Autophagosomes intersect this pathway by delivering entrapped substrates to lysosomes for digestion. A second task performed by these eukaryotic compartments is the synthesis, processing and transport of proteins. This occurs through the activities of the ER, Golgi apparatus and transport intermediates, which form the secretory pathway. A final function of eukaryotic organelles is to harness or direct energy-producing pathways, including storing and detoxifying molecules. This is the job of mitochondria, chloroplasts (in plants), peroxisomes and lipid droplets.

The identity and function of eukaryotic membrane-bound compartments have been known for decades, yet many of their properties remain enigmatic. We still do not fully understand, for example, how these compartments are formed or maintained, inter-communicate, or are shaped, nor how they sort proteins and interact with cytoskeletal elements. As discussed here, progress in answering these questions is being notably advanced by new quantitative imaging technologies that integrate physical-chemical concepts with computational modeling.

2. Early Impact of GFP Imaging

The advent of green fluorescent protein (GFP) technology some 20 years ago set the stage for applying physics/chemistry concepts to the analysis of intracellular organelle dynamics.[1,2] Before GFP technology, membrane-bound organelles could only be analyzed after fixation and staining with organelle-identifying antibodies. Since the cells were dead, only a single snap-shot of the organelle's lifetime was captured, with its dynamic attributes inferred from other cells fixed at other time points or conditions. Observing GFP-tagged organelle markers by fluorescence in living cells dramatically changed this as the organelles could now be watched in a single cell over time with minimal photo-damage. Moreover, because the GFP signal could be quantified and correlated to an actual number of molecules, it became possible to measure the mobility and concentrations of proteins in different compartments, as well as to quantify protein exchange rates between compartments.[3]

An early focus of GFP work was on the membrane-bound trafficking interme-diates comprising the secretory pathway. Prior work using in vitro reconstitu-tion approaches assumed these intermediates are small vesicles that diffuse quickly through the cytoplasm, not requiring motor proteins or microtubules to reach their final destinations. This view changed when the transport intermediates were visu-alized using GFP technology which revealed they are elaborate tubule-vesicular structures which translocate along microtubules.[4] This was demonstrated using a GFP-tagged transmembrane cargo called VSVG-GFP, which underwent release from the ER into the secretory pathway after temperature shift from 40°C to 32°C.

In addition to being useful for visualizing transport intermediates, VSVG-GFP's signal could be quantified to determine how levels of the molecule in different com-partments changed after release from the ER by temperature shift. In this way, secretory transport kinetics could be assessed in a single cell with high temporal and spatial resolution.[5] A simple model comprised of a series of linear rate laws con-necting the ER, Golgi and plasma membrane was found to be sufficient to fit the data representing VSVG-GFP flux after its release from the ER5 (Figure 2). More

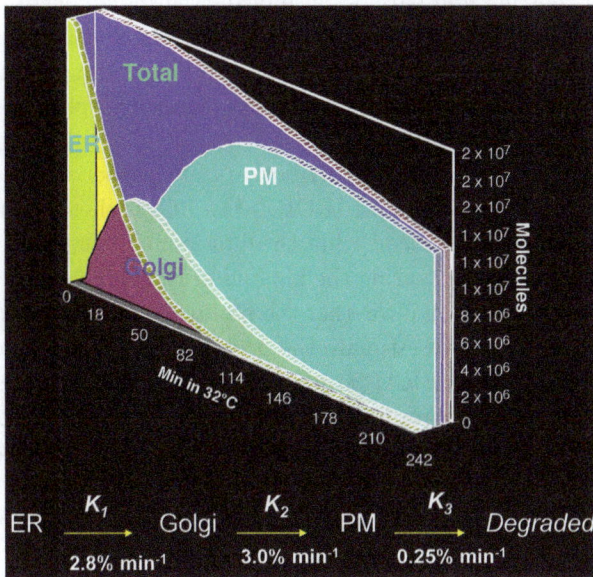

Fig. 2. Kinetic modeling of VSVG-GFP transport through the secretory pathway. Results are from Hirschberg *et al.*, 1998. VSVG-GFP was released from the ER into the secretory pathway by temperature shift. The fluorescent intensities of VSVG-GFP were then measured in the Golgi region and the entire cell over time and used to fit a three-compartment model of the secretory pathway that included ER, Golgi and plasma membrane. The plot shows the change in concentra-tion of VSVG in these three compartments over time, with the number of molecules determined by correlating VSVG-GFP's fluorescent intensity with that of a known concentration of GFP in solution. Distinct rate constants for VSVG-GFP transport out of the three compartments were revealed.

complex nonlinear rate laws (e.g., Michaelis–Menten) involving a changing rate constant (or rate coefficient) as the concentration of VSVG-GFP in different compartments went from high (early in the experiment) to low (late in the experiment) were not required. Rather, at all times VSVG-GFP moved between compartments at a rate equal to a rate constant multiplied by the amount of VSVG-GFP in the donor compartment. The measured rate constants allowed determination of the number of molecules of VSVG-GFP moving into and out of the Golgi at any particular time, but also revealed the average residence time of VSVG-GFP molecules within the Golgi.[5] Various perturbations, including microtubule depolarization, were found to differentially affect the different rate-limiting steps in VSVG-GFP transport.[5] These studies paved the way for using quantitative imaging approaches for investigating multiple aspects of protein transport through the secretory pathway in living cells.[6]

3. Use of FRAP, FLIP and Photoactivation

Over time, new applications of GFP were developed to capture different features of intracellular dynamics. The introduction of confocal fluorescence recovery after photobleaching (FRAP) was a breakthrough in the analysis of protein and organelle dynamics within cells in large part because it allowed researchers to create a transient in the steady-state distribution of fluorescently tagged proteins.[7,8] This was important because otherwise it was impossible to determine whether the fluorescence distribution of a tagged protein within an organelle represented immobile or freely mobile pools of the protein.

FRAP creates a transient in the distribution of fluorescence due to its selective photobleaching of a region-of-interest (ROI).[7] The ROI is then examined over time to assess the return of fluorescence due to diffusional exchange of bleached and unbleached molecules. By measuring the kinetics of recovery into the bleached ROI it becomes possible to determine whether the photobleached population of proteins are mobile or immobile and to estimate the proteins' diffusion coefficient.

One of the first confocal FRAP studies explored whether proteins within the Golgi apparatus are mobile or immobile.[9] The Golgi functions as a sorting and processing station in the secretory pathway, continuously receiving secretory cargo from the ER and exporting it to the plasma membrane. The classical view held that integral membrane enzymes within the Golgi apparatus undergo extensive interactions that "fix" these proteins within particular regions of this organelle. When FRAP was applied to test this model, however, the results showed that GFP-tagged Golgi enzymes experience rapid lateral diffusion, seemingly unhindered by any immobilizing interactions.[9] FRAP methodology applied to the ER soon revealed that ER resident proteins, including chaperones and misfolded proteins, long thought to be part of an immobile ER luminal matrix, are also highly mobile, moving throughout the ER on a timescale of minutes.[9-12] These findings dramatically changed researchers' thinking regarding how resident proteins within different compartments are retained.

Another area impacted by FRAP was in the investigation of machinery underlying the biogenesis of membrane transport intermediates. These intermediates form through the activity of small GTPases that recruit cytosolic 'coat' proteins onto a membrane site that then undergoes shape changes to bud off the donor compartment. The coat proteins were originally thought stable and associated with the vesicle membrane until the vesicle is released from the target membrane. However, FRAP experiments revealed the coat proteins instead undergo continuous and fast cytosol/membrane exchange irrespective of vesicle budding.[13-16] The findings established that budding is a downstream event of multiple binding/release cycles and not directly coupled to coat dissociation as previously thought. Moreover, kinetic modeling of the observed dynamics established specific rates of binding and release for each type of coat protein.[13]

Further insights into organelle dynamics were obtained using variations of FRAP, including fluorescence loss in photobleaching (FLIP) and inverse FRAP (iFRAP).[3,17] FLIP involves repeated photobleaching of one subregion of a cell while visualizing images of the entire cell. Its application revealed that the ER is one continuous tubular membrane system[9] (Figure 3), and that during mitosis both Golgi and nuclear envelope membranes are absorbed into the ER.[18,19] iFRAP involves photobleaching all areas surrounding an ROI to highlight it. This approach revealed the speed and directionality of secretory transport intermediates moving between Golgi and plasma membrane.[5] iFRAP was also used to measure the rate of recycling of proteins between Golgi and ER,[20] and between Golgi and plasma membrane.[21] This was accomplished by measuring changes in fluorescence intensities in each of the compartments during cycling of highlighted pools of the proteins. GPI-GFP molecules, for example, were found to cycle between Golgi and plasma membrane every 70 min.[21] Differences in the rate constants for GPI-GFP leaving the Golgi compared to those leaving the plasma membrane yielded different steady state levels of GPI-GFP, found to be $\sim 10\%$ in the Golgi and $\sim 90\%$ in the plasma membrane.

The introduction of photoactivatable GFP (PA-GFP)[22] allowed even more precise quantification of protein transport between organelles. By selective photo-

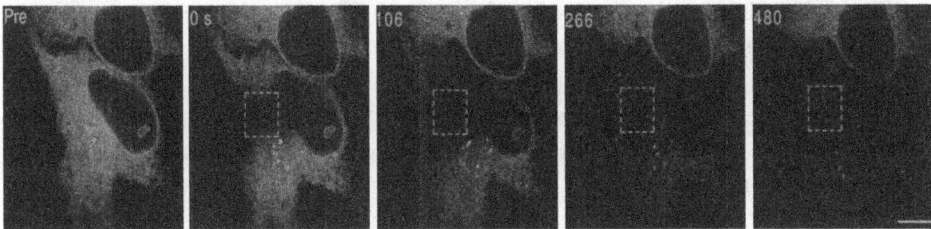

Fig. 3. FLIP experiment showing continuity of ER membranes. Continuous photobleaching through FLIP of a small ROI in the ER is sufficient to remove all of the fluorescent signals from a GFP-tagged, ER-localized membrane protein. Adapted from (Cole *et al.*, 1996).

activation of proteins within an individual organelle, a single structure could now be made visible while the rest remained dark. Similar to iFRAP in enabling a particular ROI to be made visible, photoactivation permitted faster highlighting with no residual fluorescence outside the photactivated region. This was nicely illustrated in an analysis of protein exchange between lysosomes.[22] Photoactivating a single lysosome among hundreds within a cell in less than one second revealed that nearly all lysosomes obtained some fraction of the lysosome's fluorescence within 15 min, through a pathway that was microtubule dependent. The results demonstrated that lysosomes extensively exchange contents through microtubule-dependent trafficking intermediates.

4. Modeling Intra-Golgi Trafficking

Photohighlighting with PA-GFP or iFRAP has had particularly striking results in our understanding of the Golgi. The Golgi apparatus processes and sorts newly synthesized protein and lipid moving through the secretory pathway. Until recently, the most widely accepted model for how the Golgi apparatus accomplishes its diverse and essential trafficking tasks was cisternal progression (or maturation). This model postulates that the stack of Golgi cisternae constitutes a historical record of progression from entry at the cis face to exit at the trans face.[23] Upon arrival at the cis-most cisterna, cargo molecules remain as the cisterna passes, conveyor-belt-like, through an average of seven locations within the Golgi stack on its way to the trans face and exit from the Golgi via transport carriers.

A key prediction of the cisternal progression model was that newly arrived cargo exhibits a lag or transit time before exiting the Golgi. When researchers used photo-highlighting approaches to test this prediction, however, they found that cargo molecules instead exited at an exponential rate proportional to their total Golgi abundance with no lag.[24] Furthermore, incoming cargo molecules rapidly mixed with those already in the system and exited from partitioned domains with no cargo privileged for export based on its time of entry into the system.[24] These contrary results prompted a re-evaluation of the cisternal progression model for Golgi transport. In its place, various proposals have been advanced incorporating physical and biochemical concepts such as phase-partitioning in lipid bilayers.[25]

One well-articulated model, referred to as the Golgi partitioning model, incorporates lipid trafficking pathways and the self-organizing properties of lipids as an integral part of the Golgi[24] (Figure 4). Its key assumption is that the self-associative properties of glycerophospholipids (GPL), sphingolipids (SL) and cholesterol in Golgi membranes lead to phase partitioning of these lipids into two types of domains — one with low SL/cholesterol levels and thin bilayer thickness, and one with high SL/cholesterol levels and thick bilayer thickness. This partitioning, in turn, facilitates the selective lateral segregation of integral membrane proteins residing in or passing through the Golgi because the integral membrane proteins sort by their transmembrane domain thickness.[26] In addition to having two classes

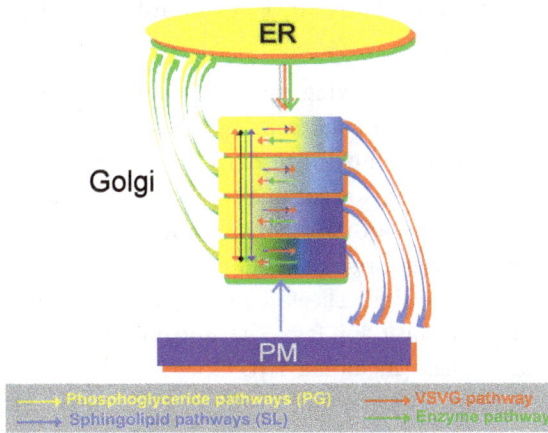

Fig. 4. Diagram of rapid partitioning model. In the Golgi, each cylinder represents one cisterna of an EM-resolved or biochemically resolved Golgi stack. The Golgi membrane lipid environment is modeled as having one component consisting of glycerophospholipids (GPL; yellow) and another component consisting of cholesterol and glycosphingolipids (SL; blue) giving rise to processing domains (left-side, yellow) and export domains (right-side, blue), respectively. Transmembrane cargo proteins (red) move between both lipid environments but concentrate in the export domain, whereas transmembrane Golgi enzymes (green) are excluded from export domains and diffuse within the processing domain.

of membrane domains within every cisterna formed by partitioning, the partitioning model assumes bidirectional trafficking of protein and lipid between Golgi cisternae and that cargo can exit the Golgi from all cisternae. The model also assumes that cargo and enzymes have an optimal lipid environment for preferentially associating within the Golgi.

Simulation and experimental testing of the rapid partitioning model have demonstrated that it can explain many of the major features of the Golgi apparatus.[24] The simulated model generates a gradient in SL/GPL compositions across the stack at steady state, with the ratio lowest in the cis cisternae and highest in the trans cisternae. Resident proteins with different SL/GPL preferences that were simulated showed enrichment in different Golgi cisternae despite their rapid movement between cisternae. Notably, cargo exited the Golgi with exponential kinetics, consistent with the experimental measurements of cargo export in living cells. Finally, a cargo wave pattern across the Golgi stack was observed in response to simulation of a short, low-temperature block and release of membrane traffic, consistent with electron microscopy experiments.[27] These supportive results make a strong case that a self-organizing mechanism involving rapid lipid partitioning plays a major role in controlling intra-Golgi transport. Further work is still needed to see if it possible to incorporate the roles of coats and other membrane trafficking machineries into the model. These additional elements could have a role in partitioning by inducing geometric shape changes (i.e., membrane curvature) that facilitate protein and lipid sorting processes.

5. New Imaging Tools for Studying Endomembrane Organization and Dynamics

Several new imaging tools are playing an increasingly important role in further clarifying eukaryotic membrane organization by allowing visualization of membrane organelles at improved spatio-temporal resolution. One powerful method is multi-spectral imaging, an approach that separates spectra of different fluorescent proteins attached to different organelles-of-interest within a cell[28] (Figure 5). This allows imaging of six or more different organelles simultaneously with no overlap in signal. When combined with lattice light sheet microscopy, a technique using thin light beams to generate a light sheet for fast 3-D isotropic resolution,[29] it is possible to visualize many organelles (including ER, Golgi, mitochondria, lysosome, lipid droplets and peroxisomes) at once, and to do so with unprecedented spatial and temporal resolution and over long time periods.[28] This has permitted new details of organelles to be quantitatively analyzed, including their numbers, velocities, and sizes in the same cell, revealing unexpected interactions and dynamism of these organelles.[28] For example, it was found that in an average COS-7 cell there are ~ 90 lysosomes, 186 peroxisomes, ~ 180 mitochondrial elements and ~ 157 lipid droplets. With respect to dynamism, the ER, the cell's largest organelle comprising about 25% of the cytoplasm (excluding the nucleus), explores over 97% of the cytoplasm every 15 minutes as its elaborate network-like structure is pushed and pulled by cytoskeletal elements.[28] It was also found that each organelle has a characteristic distribution and dispersion pattern in three-dimensional space, impacted by microtubule and cell nutrient status.

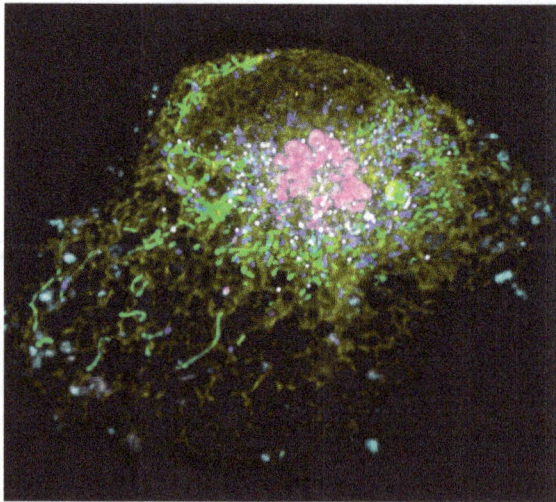

Fig. 5.   Live-cell, six-color, 4D LLS microscopy image showing distributions of ER (yellow), Golgi (pink), mitochondria (green), lysosomes (cyan), lipid droplets (white) and peroxisomes (dark blue). Adapted from (Valm *et al.*, 2017).

Use of multispectral imaging has also provided new insights into organelle-organelle contact sites. These sites are known to play diverse roles in the exchange of metabolites, lipids and proteins between organelles, and are critical for the division and biogenesis of different organelles.[30] Employing multispectral imaging and computational analysis to evaluate organelle contacts among different organelles in the cell, researchers have been able to define an 'organelle interactome' of cells.[28] This organelle interactome changed when cells were exposed to different perturbations, suggesting its importance in regulating cell homeostasis. The ER was found to have the highest frequency of contacts with other organelles, and contacts between ER and mitochondria were the most numerous. Use of spectral unmixing should aid in obtaining new insights into how speicific functions at these contact sites (including lipid, Ca^{++} and ROS exchange) are carried out. Because live cell multispectral unmixing is applicable to any cell system expressing multiple fluorescent probes, whether in normal conditions or when cells are exposed to disturbances such as drugs, pathogens or stress, it offers a powerful new descriptive tool and source for testing hypotheses in the field of cellular organization and dynamics.

Researchers have also recently used an imaging approach that combines grazing incidence of total internal reflection and structured illumination microscopy (GI-TIRF-SIM)[31] to study ER-ER fusion events in live cells. Organelles are known to remodel by fission and fusion events. For most organelles, SNARE proteins drive their membrane fusion processes,[32] while dynamin and dynamin-related proteins control their fission processes.[33] The exception is the ER, which appears to use different mechanisms for these processes. ER fusion has been proposed to be mediated by small GTPases called atlastins that allow ER tubules to come together and fuse.[34] This possibility has been examined using the combinatorial approach of GI-TIRF-SIM because it dramatically improves both spatial and temporal resolution, allowing events occurring on the msec time scale to be imaged at super-resolution.[31] Researchers using this approach have observed that the ER undergoes extreme dynamism, with tubule elements extending peripherally and retracting back continuously.[35] Continued fusion of tubules with each other leads to the generation of a reticular meshwork that is capable of stretching out like a spider-web or retracting back into a tight array within msec. Fission events in this system are infrequently seen, except under conditions of calcium depletion from the ER (i.e., ionomycin). There, the ER quickly fragments. High speed time-lapse imaging of fragmentation revealed the ER lumen responded first, collapsing into concentrated aggregates. This was immediately followed by membrane fission at sites outside the areas of aggregated lumen. This unexpected fragmentation process has yet to be explained but one possibility is that it is driven by luminal aggregation in response to calcium depletion from the ER. In this view, luminal aggregates would draw membranes around themselves (through electrostatic interactions), creating high membrane curvature that leads to membrane fission outside the aggregated zonesGI-TIRF-SIM has also been used, along with lattice light sheet-

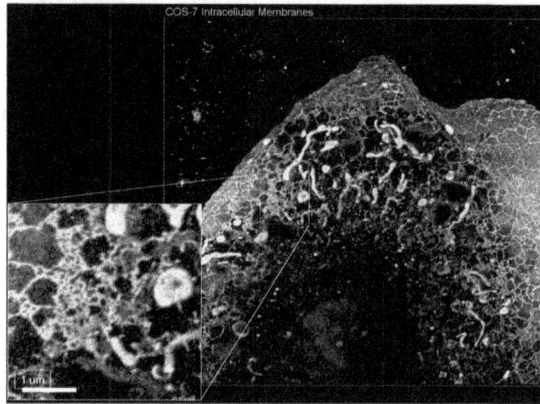

Fig. 6. Lattice light sheet-PAINT imaging using BODIPY-TR methyl ester. An ortho-slice of a COS-7 cell showing internal membrane compartments labeled by BODIPY-TR methyl ester. Clearly visible are the mitochondria and networks of peripheral ER tubules. The zoomed image shows tight matrices of ER. Adapted from (Legant *et al.*, 2016).

PAINT microscopy (LLS-PAINT),[36] to explore the peripheral components of the ER. These have classically been described as comprising both tubules and flat sheets. But researchers employing these new techniques have shown that the peripheral ER system consists almost exclusively of tubules at varying densities; including newly characterized structures termed ER matrices[35,36] (Figure 6). Similar results were obtained using live cell stimulated emission depletion (STED) imaging.[37] The tubular matrices were misidentified as ER sheets with conventional imaging technologies due to the dense clustering of tubular-junctions and a previously uncharacterized novel form of ER motion. The structural conformation of tubular matrices and their ability to quickly transition into looser tubular networks could underlie the ER's ability to rapidly alter its overall organization in response to changing cellular needs. Indeed, the rapid interconversion between loose and tight polygonal arrays of ER tubules likely enables the ER to rapidly reconfigure its spatial footprint in response to intracellular structural rearrangements, cell shape changes, or during cellular migration. Advances in electron microscopy are also providing a more detailed view of the shape and interaction among organelles. A powerful new approach called focused ion beam scanning electron microscopy (FIB-SEM) uses a focused ion beam to collect an image while also milling the specimen surface.[38] As milling can be repeated thousands of times, and can be as thin as 4 nm, organelle distribution can be reconstructed through an entire cell at 4 nm isotropic resolution. FIB-SEM has already revealed unexpected 3-D complexity of the ER,[35] the major site for protein synthesis and the entry portal into the secretory pathway. Regions of the ER in the cell periphery that were previously thought to represent flattened sheets when viewed by light microscopy were found to be tight networks of tubule matrices.[35] Reconstructions of organelles in neuronal processes have revealed the intimate interconnections of organelles with each other.[39] Combining FIB-SEM with correlative light microscopy approaches, such as structural illumination microscopy or PALM/STORM imaging, offers further possibilities for identifying the distribution of specific proteins at high resolution.[35] FIB-SEM is now being used to reconstruct the intracellular organelles of entire volumes of cells at 4 nm isotropic resolution, enabling fine segmentation of organelles throughout the cell. As this can be done in diverse cell types, including those in tissues, FIB-SEM promises to revolutionize our thinking about basic endomembrane organization throughout an entire cell.

All of these technical innovations in imaging, quantitative physical analysis and modeling have made for an exciting time in evolving our understanding of membrane-bound compartments within cells. Not only are new imaging methodologies revealing unexpected morphologies and dynamics of these compartments, but computational methods are providing a rigorous analytical framework for assessing their dynamic properties utilizing physical-chemical concepts. The results are opening up new directions of research for delineating the pathways and mechanisms by which organelles intercommunicate and function within cells, something that is vital to the health of all eukaryotes.

References

1. M. Chalfie, Y. Tu, G. Euskirchen, W. W. Ward, and D. C. Prasher (1994) Green fluorescent protein as a marker for gene expression. *Science* **263**: 802–805.
2. R. Y. Tsien (1998) The green fluorescent protein. *Annu. Rev. Biochem.* **67**: 509–544.
3. J. Lippincott-Schwartz, E. Snapp, and A. Kenworthy (2001) Studying protein dynamics in living cells. *Nat. Rev. Molec. Cell Biol.* **2**: 444–456.
4. J. F. Presley, N. B. Cole, T. A. Schroer, K. Hirschberg, K. J. Zaal, and J. Lippincott-Schwartz (1997) ER to Golgi transport visualized in living cells. *Nature* **389**: 81–85.
5. K. Hirschberg, C. M. Miller, J. F. Presley, J. Ellenberg, K. Zaal, N. B. Cole, E. Siggia, R. D. Phair, and J. Lippincott-Schwartz (1998) Kinetic and morphological analysis of secretory protein traffic in living cells. *J. Cell Biol.* **143**: 1485–1503.
6. J. Lippincott-Schwartz, T. Roberts, and K. Hirschberg (2000) Secretory protein trafficking and organelle dynamics in living cells. *Ann. Rev. Cell and Dev. Biol.* **16**: 557–589.
7. J. Lippincott-Schwartz, N. Altan-Bonnet, and G. Patterson (2003) Photobleaching and photoactivation: following protein dynamics in living cells. *Nature Cell Biol.* **5**: S7–14.
8. J. Lippincott-Schwartz, E. L. Snapp, and R. D. Phair (2018) The development and enhancement of FRAP as a key tool for investigating protein dynamics. *Biophys. J.* **115**: 1146–1155.
9. N. B. Cole, C. L. Smith, N. Sicaky, M. Terasaki, M. Edidin, and J. Lippincott-Schwartz (1996) Diffusional mobility of Golgi proteins in membranes of living cells. *Science* **273**: 797–801.
10. S. Nehls, E. L. Snapp, N. B. Cole, K. J. Zaal, A. K. Kenworthy, T. H. Roberts, J. Ellenberg, J. F. Presley, E. Siggia, and J. Lippincott-Schwartz (2000) Dynamics and retention of misfolded proteins in native ER membranes. *Nat. Cell Biol.* **2**: 288–295.

11. E. L. Snapp, A. Sharma, J. Lippincott-Schwartz, and R. S. Hegde (2006) Monitoring chaperone engagement of substrates in the endoplasmic reticulum of live cells. *Proc. Natl. Acad. Sci. USA* **103**: 6536–6541.

12. D. Marguet, E. T. Spiliotis, T. Pentcheva, M. Lebowitz, J. Schneck, and M. Edidin (1999) Lateral diffusion of GFP-tagged H2Ld molecules and of GFP-TAP1 reports on the assembly and retention of these molecules in the endoplasmic reticulum. *Immunity* **11**: 231–240.

13. J. F. Presley, T. H. Ward, A. C. Pfeifer, E. D. Siggia, R. D. Phair, and J. Lippincott-Schwartz (2002) Dissection of COPI and Arf1 dynamics in vivo and role in Golgi membrane transport. *Nature* **417**: 187–193.

14. T. H. Ward, R. S. Polishchuk, S. Caplan, K. Hirschberg, and J. Lippincott-Schwartz (2001) Maintenance of Golgi structure and function depends on the integrity of ER export. *J. Cell Biol.* **155**: 557–570.

15. W. Liu, K. Moriyama, R. Phair, R. Duden, and J. Lippincott-Schwartz (2005) *In vivo* dynamics of ARFGAP1 and its functional interaction with Arf1 and coatomer on Golgi membranes. *J. Cell Biol.* **168**: 1053–1063.

16. R. Forster, M. Weiss, T. Zimmermann, E. G. Teynaud, F. Verissimo, D. J. Stephens, and R. Pepperkok (2006) Secretory cargo regulates the turnover of COPII subunints at single ER exit sites. *Curr. Biol.* **16**: 173–179.

17. R. D. Phair and T. Misteli (2001). Kinetic modelling approaches to in vivo imaging. *Nat. Rev. Mol. Cell Biol.* **2**: 898–907.

18. K. Zaal, C. L. Smith, R. S. Polishchuk, N. Altan, N. Cole, J. Ellenberg, K. Hirschberg, J. Presley, T. Roberts, E. Siggia, R. Phair, and J. Lippincott-Schwartz (1999) Golgi membranes are absorbed into and re-emerge from the ER during mitosis. *Cell* **99**: 589–601.

19. J. Ellenberg, E. D. Siggia, J. E. Moreira, C. L. Smith, J. F. Presley, H. J. Worman, and J. Lippincott-Schwartz (1997) Nuclear membrane dynamics and reassembly in living cells: targeting of an inner nuclear membrane protein in interphase and mitosis. *J. Cell Biol.* **138**: 1193–1206.

20. P. Sengupta, P. Satpute-Krishnan, A. Y. Seo, D. T. Burnette, G. H. Patterson, and J. Lippincott-Schwartz (2015) ER trapping reveals Golgi enzymes continually revisit the ER through a recycling pathway that controls Golgi organization. *Proc. Natl. Acad. Sci. USA* **112**: E6752–6761.

21. B. J. Nichols, A. K. Kenworthy, R. S. Polishchuk, R. Lodge, T. H. Roberts, K. Hirschberg, R. D. Phair, and J. Lippincott-Schwartz (2001). Rapid cycling of lipid raft markers between the cell surface and Golgi complex. *J. Cell Biol.* **153**: 529–541.

22. G. H. Patterson and J. Lippincott-Schwartz (2002) A photoactivatable GFP for selective photolabeling of proteins and cells. *Science* **297**: 1873–1877.

23. B. S. Glick, T. Elston, and G. Oster (1997) A cisternal maturation mechanism can explain the asymmetry of the Golgi stack. *FEBS Lett.* **414**: 177–181.

24. G. H. Patterson, K. Hirschberg, R. S. Polishchuk, D. Gerlich, R. D. Phair, and J. Lippincott-Schwartz (2008) Transport through the Golgi apparatus by rapid partitioning within a two-phase membrane system. *Cell* **133**: 1055–1067.

25. J. Lippincott-Schwartz and R. Phair (2010) Lipids and cholesterol as regulators of traffic in the endomembrane system. *Annu. Rev. Biophys.* **39**: 559–578.

26. J. C. Holthuis, T. Pomorski, R. J. Raggers, H. Sprong, and G. van Meer (2001) The organizing potential of sphingolipids in intracellular membrane transport. *Physiol. Rev.* **81**: 1689–723.

27. A. Trucco, R. S. Polishchuk, O. Martella, A. Di Pentima, A. Fusella, *et al.* (2004) Secretory traffic triggers the formation of tubular continuities across Golgi subcompartments. *Nat. Cell Biol.* **6**: 1071–1081.

28. A. M. Valm, S. Cohen, W. R. Legant, J. Melunis, U. Hershberg, E. Wait, A. R. Cohen, M. W. Davidson, E. Betzig, and J. Lippincott-Schwartz (2017) Applying systems-level spectral imaging and analysis to reveal the organelle interactome. *Nature* **546**: 162–167. PMID: 28538724.

29. B. C. Chen, W. R. Legant, K. Wang, L. Shao, D. E. Milkie, M. W. Davidson, C. Janetopoulos, X. S. Wu, J. A. Hammer 3rd, Z. Liu, B. P. English, Y. Mimori-Kiyosue, D. P. Romero, A. T. Ritter, J. Lippincott-Schwartz, L. Fritz-Laylin, R. D. Mullins, D. M. Mitchell, J. N. Bembenek, A. C. Reymann, R. Böhme, S. W. Grill, J. T. Wang, G. Seydoux, U. S. Tulu, D. P. Kiehart, and E. Betzig (2014) Lattice light-sheet microscopy: imaging molecules to embryos at high spatiotemporal resolution. *Science* **346**(6208): 1257998. PMID 25342811.

30. S. Cohen, A. M. Valm, and J. Lippincott-Schwartz (2018) Interacting organelles. *Curr. Opin. Cell Biol.* **53**: 84–91. PMID: 30006038.

31. Y. Guo, D. Li, S. Zhang, Y. Yang, J.-J. Liu, X. Wang, C. Liu, R. Price, U. S. Tulu, D. Kiehart, J. Hu, J. Lippincott-Schwartz, E. Betzig, and D. Li. Grazing incidence structured illumination super-resolution imaging of intracellular organelles and cytoskeletons dynamic interactions. *Cell*, in press.

32. D. Ungar and F. M. Hughson (2018) SNARE protein structure and function. *Ann. Rev. Cell Dev. Biol.* **19**: 493–517.

33. R. Ramachandram and S. L. Schmid (2018) The dynamin superfamily **28**: R411–R416.

34. U. Goyal and C. Blackstone (2013) Untangling the web: mechanisms underlying ER network formation. *Biochim. Biophys. Acta* **1833**: 2492–2498.

35. J. Nixon-Abell, C. J. Obara, A. V. Weigel, D. Li, W. R. Legant, C. S. Xu, H. A. Pasolli, K. Harvey, H. F. Hess, E. Betzig, C. Blackstone, and J. Lippincott-Schwartz (2016) Increased spatiotemporal resolution reveals highly dynamic dense tubular matrices in the peripheral ER. *Science* **354**(6311), pii: aaf3928, Epub: 2016, Oct 27, PMID: 27789813.

36. W. R. Legant, L. Shao, J. B. Grimm, T. A. Brown, D. E. Milkie, B. B. Avants, L. D. Lavis, and E. Betzig (2016) High-density three-dimensional localization microscopy across large volumes. *Nat. Methods* **13**(4): 359–365.

37. L. K. Schroeder, A. E. S. Barentine, H. Merta, S. Schweighofer, Y. Zhang, D. Baddeley, J. Bewersdorf, and S. Bahmanyar (2018) Dynamic nanoscale morphology of the ER surveyed by STED microscopy. *J. Cell Biol.* **115**: 951–956.

38. C. S. Xu, K. J. Hayworth, Z. Lu, P. Grob, A. M. Hassan, J. G. García-Cerdán, K. K. Niyogi, E. Nogales, R. J. Weinberg, and H. F. Hess. Enhanced FIB-SEM systems for large-volume 3D imaging. *eLife* 2017 PMID 28500755.

39. Y. Wu, C. Whiteus, C. S. Xu, K. J. Hayworth, R. J. Weinberg, H. F. Hess, and P. De Camilli. Contacts between the endoplasmic reticulum and other membranes in neurons. *Proc. Natl. Acad. Sci. USA* 2017 **114**(24): E4859–E4867. PMID 28559323.

Discussion

A. Hyman Perfect timing Jennifer, thank you. So now, we are going to have some prepared rounds and Frank Jülicher is going to try to link the two different topics.

Prepared comment

F. Jülicher: Membraneless compartments and the emergence of membranes

Membraneless compartments that consist of assemblies of proteins and RNA exist in a large variety of different cells and organisms. Such assemblies can form spherical condensates with liquid-like properties. This suggests that phase separation and droplet formation provide a principle for the organization of chemistry in cells that may have been important already early in evolution and maybe even at the origin of life.

Artificial systems that mimic simple cells are often based on the idea that lipid bilayer vesicles confine and organize biochemistry. Droplets without membranes can also confine chemistry and provide simple models for cell-like systems. However, the spontaneous formation of single lipid bilayers has been a challenge as single or unilamellar lipid bilayers usually form only under special nonphysiological conditions and form typically multilamellar structures. Once formed, it is difficult to construct vesicular systems which can divide. Interestingly, droplets also provide an elegant model system for simple cells. It has been shown recently that such active droplets can undergo cycles of division and growth that are reminiscent of cells. This phenomenon is a consequence of a general physical mechanism.[a] For vesicles it may be harder to construct a minimal system that can divide.

This raises the question of whether active droplets rather than membrane bound vesicles are the best models for simple or early cells that can divide. Membranes are needed to keep different ion concentrations and different values of pH inside and outside a compartment. They also help to avoid the loss of precious molecules. Did membranes arise after simple life forms already existed that initially did not use membranes?

Droplets provide with their surface a two-dimensional compartment with an affinity to certain surface active molecules. This could help the recruitment of molecules to the surface and might under the right conditions facilitate the assembly of a single lipid bilayer at a droplet surface. This may at first appear unlikely, given that a bilayer has two hydrophilic surfaces and a hydrophobic core. However there is a precedent. It has been reported that single lipid bilayers form spontaneousy at an air water interface at certain

---

[a]D. Zwicker, R. Seyboldt *et al.*, *Nature Physics* **13**, 408 (2017).

temperatures.[b] This is a key example of the controlled formation of a single bilayer. Combining all these points, a picture emerges in which droplets act as containers for localized chemistry. Such droplets could guide the assembly of single lipid bilayers on their surface thereby providing a surrounding membrane. Under what conditions this would happen is an open question that should be studied in future in vitro experiments. However, the observed bilayer formation at air-water interfaces demonstrates that such a phenomenon should be expected.

Discussion

A. Hyman We take a comment on that.

E. Siggia To follow up on that comment, a second way to perhaps relate the talks of Jennifer and the phase separation would be to first of all underline the fact that Jennifer has shown that the Golgi is absorbed back to the ER during mitosis and then reforms *de novo* when you go into interphase. Furthermore, she showed with the brefeldin-A experiments that if you depolarize the microtubules or if you block the vesicular trafficking, the Golgi goes back to the ER in a way which is faster than diffusive and plausibly driven by differences in free energy. So I would propose then that the model perhaps is to think of the cell more as a distillation problem, i.e., there is an active process which is consuming free energy to trap it from ER to Golgi to plasma membrane, etc. You could sort that out with these drugs and you would see something like a surface-tension-driven collapse of structures. However, within these structures there is something like phase separation, which involves partitioning of proteins in the lipids, and that protein-lipid mixture is itself a phase, which then perhaps also phase separates within the Golgi and allows a sort of mechanistic way to traffic the enzymes back to the ER and the cargo out of the plasma membrane. So it seems to me that there is a mixture of non-equilibrium-driven processes, which is then allied with equilibrium stuff to accomplish the sorting that was mentioned. As a question I would ask: to what extent are similar processes operating on the endocytic pathway, first as a zoo of vesicles, to those similarly reconstitute and form again as you go through the cell cycle?

A. Hyman So, does anyone else want to comment on these similarities between processes of membranes, membrane-compartment formation and membrane droplets?

D. Fisher I had a question since I only vaguely remember what ER is. Is the process of reorganization — that I have understood, Jennifer, happens on 15 minutes time scales — something which is consuming a lot of free energy

---

[b]N. L. Gershfeld, W. F. Stevens Jr and R. J. Nossal, *Faraday Discussions*, **81**, 19 (1986).

by itself or is that happening spontaneously?

J. Lippincott-Schwartz The jiggling ER dynamics that I showed is actually ATP dependent and is thought to be mediated by the actin-myosin contractile machinery.

D. Fisher So, the 15-minutes time scale is the motion; it is not the reorganization, which is much slower? Is that what you are saying?

J. Lippincott-Schwartz Sorry for the confusion, the 15-minute time scale represents the overall reorganization of the ER over that period. It demonstrates that ER motion is fast, fast enough for this large organelle to explore most of the cytoplasm during this time period.

A. Goldbeter Can you measure also calcium dynamics during these ER movements? And what is the relation with calcium oscillations with the process?

A. Hyman Remember we are not asking questions here, it is the discussion point. So, do you want to elaborate on the point?

A. Goldbeter Because the calcium oscillations are widespread and involve the transport from the ER to the cytoplasm, I was wondering whether you can measure and correlate calcium oscillations if they occur with these oscillations of the ER in the cytoplasm and the organization.

J. Lippincott-Schwartz We are very interested in that question. Right now we are trying to get a calcium probe in the ER that is sensitive enough to do that.

H. Levine So, my naive view, before your talk, of the mitochondria was that it was not a network, that it was really isolated structures of mitochondria. I thought that part of the reason for this was it to be sort of space-filling, to sort of have mitochondria everywhere. Now, the ER does that without breaking up in topologically distinct domains. So, is it just that the mitochondria have a smaller volume, so therefore it just naturally breaks up? Is it understood what the different topologies of those different organelles imply as far as their function?

J. Lippincott-Schwartz This is still being studied by many labs. There are times during the cell cycle where mitochondria actually fuse into a large reticulum where they are interconnected, but most of the time during the cell cycle, mitochrondria are dispersed as separate units in the cell. We think that might have something to do with cellular metabolism and also, for the way that the mitochondria are dispersing their genetic material, also for damaged mitochondria to be autophagized, to be destroyed. You do not want the whole system wrecked by autophagy.

J. Howard I just have a very general point and it relates to the membrane compartments and also to the liquid droplet compartments. How do cells know how big these compartments are? I mean, how do you know that you should put most of your membranes in the ER and not into the Golgi or whatever? Somehow the cell must know how much ER it has got, how much Golgi it has got and how big these things are. I want to know, what kind of feedback or measurement systems do cells have to regulate the size and extent of these various compartments.

A. Hyman Does anyone else want to comment on different compartmental size and feedback?

S. Grill I just have a maybe slightly related comment to the two before, which is of course the question of what sets the mechanism which defines how many little structures there are and what the topology is of the total membrane and surface. The question that comes to my mind here is the following. If we go back to membrane physics, there is a mean and a Gaussian curvature and a mean and a Gaussian rigidity. Has the Gaussian rigidity perhaps a role in determining what the topology of these membrane surface inside the cell is?

A. Hyman Suzanne.

S. Eaton So, I just wanted to comment. I had a comment on both Frank's and Jennifer's talks. So first, Frank, it is interesting actually: I believe it has been proposed that new bacteria and archaebacteria actually might have diverged before there was a membrane, because of the earlier origin of electron-transport proteins compared with lipid biosynthesis inside. So that is interesting from the point of view of the model that you were proposing. And then to Jennifer, I wanted to say the following. So in the model that you proposed for the thickness of the membrane and then helping to direct the secretory pathway where proteins end up, it seems like you are assuming a very static role for this gradient of lipid composition. But of course, there has to be an interplay, because the proteins must also somehow be important in generating that lipid gradient, which is far out of equilibrium, and it has to be happening all the time. So, I think any model you develop for something like that has to take account of the effects in both directions somehow.

J. Lippincott-Schwartz I absolutely agree. It is membrane proteins that help create the lipid gradient across the secretory pathway. This arises because the membrane proteins surround themselves with cholesterol and sphingolipids as they move through the different secretory compartments.

A. Hyman So, now I would like to call Satyajit, because what he is going to say is also going to be very apposite to that point.

Prepared comment

S. Mayor: The active membrane bilayer

Picture a cell surface, where all information from the outside world is parsed and interpreted before it reaches the encoding and decoding machinery of the cell, the nucleus. The cell surface of a typical eukaryotic cell is a bilayer composed of several hundred lipid species and thousands of proteins, including the information transducers, membrane receptors that appear to be merely solubilized in this lipid milieu (see Figure 1). Similar to the condensed systems that Cliff Brangwynne was mentioning in his commentary about the aqueous cytoplasm, the membrane is also a condensed liquid phase but at its core has very little water. Since it is the outer most cover of a living cell, its accessibility provides a tremendous opportunity to study the properties of a living material.

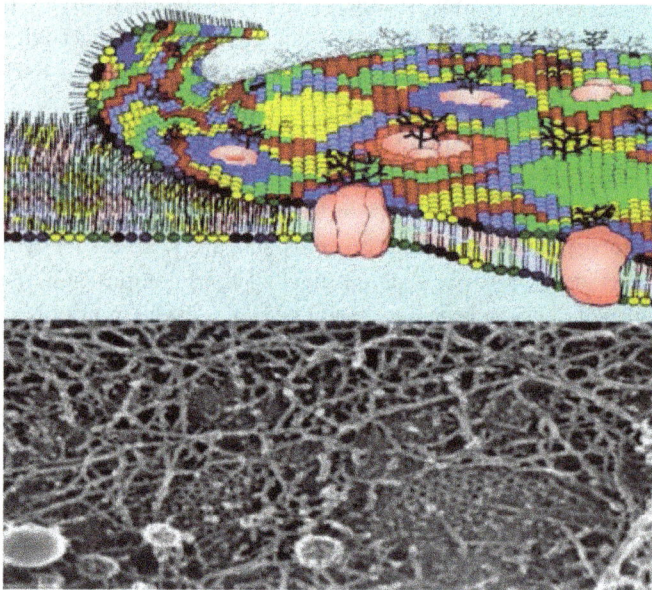

Fig. 1.   The fluid mosaic membrane (top; adapted from Edidin, M. (2003), *Nat. Rev. Mol. Cell Biol.* **4**, 414–418) resting on a cortical actin meshwork (below; adapted from Morone, N., Fujiwara, T., Murase, K., Kasai, R.S., Ike, H., Yuasa, S., Usukura, J., and Kusumi, A. (2006), *J. Cell Biol.* **174**, 851–862) consists of over thousand different lipid and protein species in a eukaryotic cell.

Our understanding of the nature of this membrane bilayer has frequently stopped at considering this lipid membrane as a system in equilibrium wherein its composition gives rise to rich phase behavior. For example, the generation of liquid-ordered and disordered phases in the membrane has been considered important for understanding how this membrane system functions; thermodynamic phase transitions[a] and more recently, critical point fluctuations have been evoked in their generation.[b] Regardless of their origin, these phases have the potential of concentrating molecules in the membrane thereby allowing the sorting or segregation of components for a variety of functions, including membrane traffic and signalling. This has been a driving principle in understanding how this membrane system functions just as Jennifer Lippincott Schwartz has described in the context of the yeast vacuole. However, the plasma membrane of a living cell is hardly at thermodynamic equilibrium.[c] It is subject to many energy consuming processes such as endocytosis, exocytosis, lipid flip-flop, and synthesis, all of which impact its overall composition and shape, contemporaneously. It also does not exhibit typical phase segregation behaviour in the physiological state, unless it is acted upon.

A primary actor at the cell surface is the actin substructure — since it sheaths the cell. The membrane is in contact with the rest of the cytoplasm and in particular, it rests on a dynamic actin substructure, the cortical actin layer (Figure 1). This profoundly influences the properties of this membrane bilayer. The passive element of this mesh impacts the membrane in terms of the diffusional properties of the molecules in the membrane, creating a picket fence-like pattern on the bilayer (Figure 2A). The cortical actin substructure is an active energy consuming visco-elastic mesh powered by motors. The structure and function of this actin mesh is a subject of intense investigation by some bio-physicists and physicists even in this room, and is also bound to influence the shape and composition of the membrane bilayer.

A key feature of this active cortex is the dynamic cortical actin filaments which interact with specific membrane molecules and induce spatial patterns of membrane proteins and lipids (Figure 2B). These patterns are obviously generated by energy consuming mechanisms and are naturally out of equilibrium. It is also becoming apparent that this mechanism provides control on the phase behavior of the components of the membrane. Since signaling receptors can regulate the creation of these dynamic actin filaments, the receptors are able to control and self-organize their local membrane environment. This also impacts their function.

A deeper understanding of this membrane system is emerging from a very lively interaction of soft and active matter theorists and experimentalists who are thinking in terms of active mechanics and hydrodynamics to provide an explanation of the nature of the outer shell of the membrane. The membrane is a prime example of a living material where the bilayer is inextricably intertwined with the energy consuming scaffold, the cortical

---

[a]Simons, K., and Ikonen, E. (1997), *Nature* **387**, 569–572.
[b]Machta, B. B., Veatch, S. L., and Sethna, J. P. (2012), *Phys. Rev. Lett.* **109**, 138101.
[c]Rao, M., and Mayor, S. (2014), *Curr. Opin. Cell Biol.* **29**, 126–132.

actin layer, resulting in its dynamic shape and composition (Figure 3). This is an active actin-membrane composite.[c]

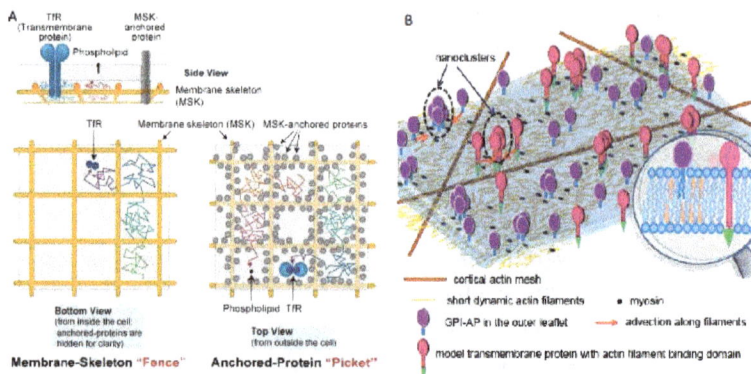

Fig. 2.   (A) Influence of passive mesh adapted from (Kusumi, A., Fujiwara, T.K., Chadda, R., Xie, M., Tsunoyama, T. a., Kalay, Z., Kasai, R.S., and Suzuki, K.G.N. (2012), *Annu. Rev. Cell Dev. Biol.* **28**, 215–250) on the diffusion of a molecule in the membrane, and (adapted from Gowrishankar, K., Ghosh, S., Saha, S., Rumamol, C., Mayor, S., and Rao, M. (2012), *Cell* **149**, 1353–1367). (B) The role of active contractile actin filaments on membrane component organization.

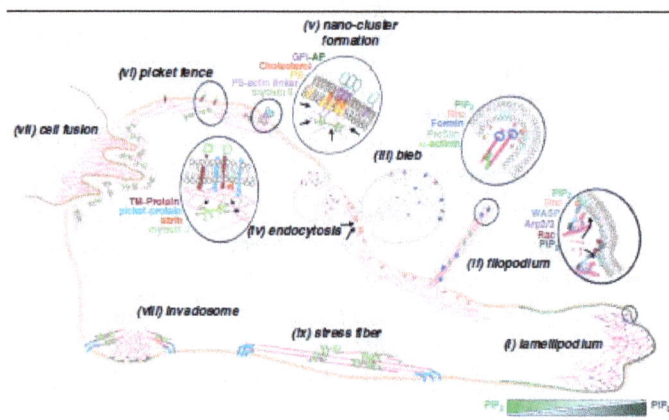

Fig. 3.   Prominent cell actin cortex architectures that influence shape and composition of the membrane. Schematic (anti-clockwise) depicting: the Arp2/3 driven lamellipodium (i); Formin dependent formation of filopodia (ii); Plasma membrane bleb because of a local disconnection from the cytoskeleton and initiation of new actin filaments by Arp2/3 and Formins (iii); formation of endocytic vesicle by Arp2/3 driven actin polymerization propelling the membrane invagination into the cell and promoting scission (iv); myosin driven actin aster formation driving clustering of GPI-anchored proteins (v); influence of the actin mesh on membrane protein diffusion, changes can, amongst others, be induced by action of myosin motors (vi); cell fusion by increased actin in the attacking cell and increased cortical tension by myosin activity in the receiving cell (vii); Arp2/3 driven invadosome formation supported by engagement with integrin based focal adhesions (viii); parallel actin bundles formed by α-actinin and engaging in integrin based focal adhesions (ix). Adapted from (Köster, D. V., and Mayor, S. (2016), *Curr. Opin. Cell Biol.* **38**, 81–89).

All indications are that we may be able to understand how this piece of a living material works since the bare elements of this system may be reconstituted in vitro. These systems should be able to provide both explanation and make predictions about how the biological membrane composite functions in executing its tasks in information flow from the outside to the inside and vice-versa. I end these remarks by saying that the membrane has been a very fertile place for bringing together quantitative methods in imaging in cell biology and non-equilibrium soft matter physics. In my view it represents a very successful model for how physics may inform fundamental questions in biology.

Discussion

A. Hyman Great, so I think one point, Satyajit brought up there, is that now a big step forward is to be able to reconstitute these very complex systems, membrane and compartmental systems, which has not really been possible before and that is what is also very appealing to physicists too. The physicist's approach for simplification and rebuilding is much easier than it was, now that in our biological systems we have much better control on the molecules. So, anyone wants to comment on membranes? Jennifer, you wanted to say one thing about lipid asymmetry before I interrupted you? No, OK. Suzanne?

S. Eaton Now that we have heard about membrane phase separation and the phase separation in the cytoplasm, maybe it would also be worth thinking about how these things might sometimes be coupled and could they interact with each other in some way, generate a microdomain in the membrane that is coupled to a cytoplasmic ...

S. Mayor So, I think, certainly if phase-separating proteins interact with membranes, one will certainly generate patterns of membrane that reflect the propensity of those phase-separating proteins to create those sorts of special environments, but in addition, I think the cortical machinery which is operating under the membrane is creating active patterns. They are not sort of phase-segregating patterns, but they are patterns that are created only because there is dynamics and energy consumption. And there again you see phase segregation of the lipid bilayer, so I think there could be multiple mechanisms by which distinct compositional control can be affected.

A. Hyman More comments?

A. Perelson Jennifer's movies impress me with the complexity of the systems that we are trying to study. I mean, a cell is the basic subunit of much of high-order learning. Here we are looking at interactions probably of thousands, if not of ten thousands, different components and trying to understand how such a system evolved to have all these subcomponents and it is clear that it is not equilibrium. There are so many driven processes, energy dependencies, structural transitions that I think, it will be a challenge for us to envision the physics or the biology of how such a system arose. The second point is really a comment. You made some remark saying that you think you can at least look at the protein-sorting problem by simple mass-action modeling, which is something that we understand. However, I found it confusing because you also showed pictures of the transport protein cargo going down microtubules in a very directed way and we think of mass action as random collisions, just having to do with the density of molecules. And secondly, the targets of the proteins are these organelles, so it is not another molecule that they are colliding with. Can you elaborate a little bit about the sort of structure that you see?

J. Lippincott-Schwartz Yes. So basically, all that we did was, we were calculating the rate at which the cargo was leaving one compartment and arriving and leaving from the other compartments. Again, this is a diffraction-limited image of the whole cell and we just watch how fast are these molecules accumulating within the Golgi. Basically, the rate constant came out very strong, 3% permin, and, it did not change. We were in no way saturating that system and we started with 20 million molecules and through the whole process it was 3% permin that was leaving and arriving at the Golgi, and a similar rate constant — that was a little bit slower — from the Golgi on outward. These are crude rate constants. We believe the microtubule motors are fast on a scale of the sorting event and that is why you are not seeing, for instance, in leaving from the Golgi some delay because the cargo is arriving at a fast enough rate that it is filling up to drive a single rate constant out.

A. Perelson It sounds to me that you are thinking about compartmental models with transport between them and you would be recording rate constants for these transport rates. We can talk about that offline.

W. Bialek I am not sure whether this is the right place to bring this up. As I have been listening to all the discussions: there is an issue about some levels of description. So try not to be too philosophical about it, when you talk about phases. Phases have properties that are not so sensitive to the microscopic details. That is the whole point. I think many of us are hoping for descriptions that are not too sensitive for microscopic details. On the other hand, we know that there is an enormous amount of detail there, and a lot of that detail has been under evolutionary pressure for a long time. So it is not random interchangeable parts, but, on the other hand, you hope that not every detail matters. As Andrew says, are we being critical enough of ourselves about whether when we use coarse-grained descriptions, we have to convince ourselves that the properties of phases that are not sensitive to the details are actually the properties that matter for the cell. That seems to me to be the hard part, right? You can identify something that you recognize as being as "Ah, this is like this thing", but is that thing, the thing that the cell actually cares about? Except for getting some things together or not together, we have not heard so much about that. I think Arup got close. When you get all these things together to organize transcription, can we convince ourselves that it is the phase behavior of this that is actually helping us control transcription? I mean, that is something I certainly have struggled with. That relates also to Daniel's question about whether, even if you give a description in terms of phases, are we in one of the regimes where we are seeing the generic behavior deep inside a phase or has biology found a way to a non-generic point?

A. Hyman Andrew, I think, has a comment on that.

A. Murray This gives the opportunity to make a Brussels-specific comment. So we are at the home of a centralized organization which deals with all sorts of complicated things. And one sort of answer to how much details matter is that cells, like the EU, have a single currency which is ATP, and one of the questions goes back to what Cliff was talking about: sometimes you pay much more than you need to make sure that things go in a completely irreversible direction, like charging tRNAs, and sometimes you are deliberately paying something very close to the cost, cause you want this sort of flexibility that Daniel was talking about. It seems like this might be an interesting idea that sort of percolates amongst us, about how much evolution has selected the cost to match what it desires as processes, cause this is the detail that is fixed by biology.

A. Hyman I think Joe is going to come back to this particular point in the prepared remarks, so we will come back to that later. I wanted to pick out your point, Bill. In a way you can look historically. When people first thought about phase separation in the 1920s, there were no molecules, so that is why they could not go any further. And so, we know as molecular biologists exactly how to do what you want to do. We know how to make mutants that affect phase behavior, that affect the different aspects of it, and put them back in the cell. So, I think that is the great thing of the last 50 years as molecular biology taught us how to manipulate molecules in a very precise way. This means that we can do tests that are predictive, rather than, as you say, simply just describing the system.

W. Bialek The predictive power in traditional physics comes from thinking of phases and phase diagrams which comes precisely from the irrelevance of microscopic details and the fact that macroscopic and microscopic processes have their own rules. So to say, now we can manipulate that. I do not know.

A. Hyman That is a good one. So Steven has not talked for a while.

S. Chu I just want to try a totally different connection. I am looking at these phase changes and they are actually transient phase changes. I want to blur the distinctions between equilibrium and non-equilibrium and just say that, you use these transient phase changes as a multiplier effect. I am just puzzled whether this is what is going on. If you think of a receptor protein on a membrane, it gets a ligand, it has to recruit other things, molecules that are on the membrane (kinases, phosphotases...), so you want this transient phase to bring together those other things. So to me, it is a multiplier effect, because biology is full of multiplier effects that take small signals, make them bigger signals, but not make them so irreversible that you cannot go the opposite way. The Golgi dissolving into the ER and back again is something where, if you have even just a separate ring-oscillator type of mechanism, you want to get that. It is to transport the proteins that are made in the ER out to the Golgi out the cellular membrane and so that makes very good sense to me. You have that. The question is whether you just have a separate ring oscillator border or a modulated one where you have other signals. The actin stuff is the actual signaling for that and so it does make a lot of sense to me that you incorporate all those tools. The desire of physicists to make predictions is overstated at this moment. It is better to just look and see what is going on. The golden standard of physics 'Tell me what is going to happen tomorrow exactly' is overstated at this point. We are still in hunting and gathering mode.

A. Hyman Great! Arup?

A. Chakraborty A very short point. I completely agree with what Steven said. An old example of this is the immune synapse, which is also a phase-separated body and its principle purpose is to amplify down-regulation of these signaling. I completely agree that that is the purpose.

A. Hyman We have two more prepared remarks that I would like to bring in now. So we will get on to them and then we will finish off with some more general discussions. So first Dan Needleman and then Joe Howard.

Prepared comment

D. Needleman: Bioenergetics of the cytoskeleton, embryo development, and infertility

The cytoskeleton governs many cellular behaviors such as growth, division, motility, and response to external stimuli. The organization of the cytoskeleton is dynamically maintained by a constant flow of energy. This energy enters at the molecular level, making the cytoskeleton an intrinsically non-equilibrium material and a paradigmatic example of active matter. While recently developed theories provide a rigorous formalism to quantitatively explain the coarse-grained mechanics and dynamics of active matter, these theories do not properly incorporate dissipation and energy transduction mechanisms. This limitation means that we currently lack an understanding of the energy flows which are responsible for making active matter active. Solving this problem would result in a thermodynamics of active matter which would provide fundamental insight into physics, biophysics, and cell biology. Furthermore, it has been hypothesized that an interplay between energy metabolism and the cytoskeleton may underlie many diseases, including cancer, neurodegenerative disorders, and infertility. Establishing a thermodynamics of active matter might provide a quantitative framework to study this interplay and to improve medical care. My lab is taking an interdisciplinary, multi-pronged approach to investigate these issues. Much of our work is inspired by the goal of understanding, and developing improved treatment for, age related infertility in women. Extensive evidence convincingly shows that the primary cause of age related infertility in women is an increase in chromosome segregation errors in eggs, and subsequent embryos, in older women. However, the cause of this increased rate of chromosome segregation errors remains unknown. Many hypotheses have been proposed. One of which is that a breakdown of mitochondrial energy metabolism in eggs from older women causes those eggs to miss-segregate their chromosomes during meiotic cell divisions. While there is some evidence in support of this hypothesis, it is unclear if it is correct, or how malfunctions in mitochondrial energy metabolism might perturb chromosome segregation. We are attempting to gain a quantitative understanding of mitochondrial energy metabolism in mouse and human eggs and embryos. We are optimizing procedures, and developing improved data analysis, for Fluorescence Lifetime Imaging Microscopy (FLIM) of NADH and FAD autofluorescence. NADH and FAD are central coenzymes in mitochondrial energy metabolism, and FLIM can be used to quantitatively measure their concentration, extent of enzyme engagement, and provides additional information on their local environment. We are constructing and testing coarse-grained models of mitochondrial energy metabolism to allow the molecular information provided by FLIM to be interpreted in terms of cellular energetics. Such FLIM measurements and coarse-grained models may provide a means to investigate correlations between endogenous defects in energy metabolism and chromosome segregation errors, and how experimentally perturbing energy metabolism impacts chromosome segregation errors. In addition to examining the effects of perturbing energy flows, we are also characterizing these flows by directly measuring dissipation with calorimetry. Taken together, we hope that this work will help establish a thermodynamics of active matter and provide new insights into fundamental cell biology. We are also collaborating with clinicians to attempt to improve in vitro fertilization success rates (partially under the auspice of Lumoniva, a startup which Needleman co-founded). Such applied work not only has the potential to improve people's lives, it also helps identify knowledge gaps in related cellular and

developmental biology. Furthermore, the ultimate goal of much of quantitative biology is to develop theories of biological phenomena. If such theories
are really correct, then they should enable predictions that have practical
applications (in analogy to the phenomenal success of physics and applied
mathematics in improving engineering). Conversely, the successful use of
theory in applications is one of the strongest arguments for that theory's
validity.

Discussion

A. Hyman Dan, thank you. Let us go straight to Joe and then we will take
 questions afterwards. Or maybe, Cliff, do you have one comment while we
 are waiting?

C. Brangwynne This issue has come up before in some discussions. You mentioned dissipation of energy and how it is essentially highly inefficient. On
 the molecular motor level, for example, you have highly efficient motor
 energy conversion to work, potentially. And yet, I think somebody said, it
 is like a drunken sailor: cells are essentially wasting all this energy. So what
 is ...

D. Needleman So, ...

A. Hyman Dan, before you go into this, let us go to Joe.

Prepared comment

Jonathon Howard: The cost of signaling

A fundamental question in biology is how much cells pay for information.
In principle, one bit lost costs $kT \ln 2$. But the real cost of biological
information is much, much higher, perhaps 10^3 or even 10^6 times higher.

The high cost of information used in signaling pathways can be appreciated by considering an ion channel, that is used for action potentials and
electrical signaling in neurons. When a sodium channel opens for 1 millisecond about 10^4 ions may pass through it, and this will cost the cell about
5000 ATPs for the sodium-potassium- ATPase protein to pump the sodium
ions back out of the cell. The net cost is $\sim 105kT$, given that the free energy
derived from ATP hydrolysis is $\sim 20kT$. And that is just one ion channel,
yet the opening of thousands to millions of sodium channels are needed
to propagate one action potential. Thus the energy expended by action
potentials in the nervous system is very high. Why is this so "inefficient".

Compare the energetics of the action potential to that of the pump. The
pump has close to 100% efficiency, similar to the efficiency of ATP synthase,
which makes ATP. Even motor proteins, that are responsible for heart contraction, locomotion and intracellular transport are quite efficient, being

able to generate mechanical work with an efficiency of 10–50% depending on the load. Thus, signaling appears to be energetically costly compared to "housekeeping" processes that make ATP, maintain the ionic gradients across cell membranes or such power muscles.

Other signaling processes — such as G-protein or phosphorylation pathways — are also expensive. Even the central dogma is incredibly expensive when we take into account the splicing of RNA, which remove on average 95% of the RNA transcribed from human genes. Even motility can be expensive: a sperm might consume 1,000,000 ATP/second to move, yet a single kinesin could, in principle, carry a sperm at the same velocity and dissipation at a greatly reduced consumption. Of course in this case the cargo (male genome) is valuable; but what sets the market rate for its transport? We also see this inefficiency with the depolymerizing kinesin-8: it spends thousands of ATPs walking to the end of a microtubule, then just takes off a single tubulin dimer from the end. In this case, we think that the cell is paying for information about the lengths of the microtubules.[a]

In summary, cells pay a large price for information yet we have no idea about what sets this price. In my view, understanding the cost of signaling will go a long way towards understanding Schroedinger's "order-from-order" principle espoused in his well-known book What is Life: how much does it cost to keep disorder at bay?

Discussion

A. Hyman Thanks, Joe and Dan. Any questions on those or any comments?

A. Walczak I have a naive question. What is the scale on which we can now measure energy? I understand that for molecular motors it may be easy, but more for chemical reactions. Calorimetry seems like a large-scale measurement. So can we measure the energy expenditure of one chemical reaction and what are the scales to which we have access and how well can we measure these energies? What is the precision of these measurements and also what is the variability in the measurements that we do?

J. Howard You know, I think it is hard. You can measure ATP being hydrolyzed, because you can use chemical probes to be able to see at a molecule-by-molecule basis when they are hydrolyzing γ-phosphate, for example. But then, you kind of jumped to the whole cell measurements and calorimeters, and the problem there is that there is a limit to how precise you can measure temperature, which is what you need in order to measure heat. So I think you are hinting at: we really need new techniques for being able to measure heat and energy at very small level.

---

[a]Vladimir Varga *et al.* (2009), Kinesin-8 motors act cooperatively to mediate length-dependent microtubule depolymerization, *Cell* **138**: 1174–83.

A. Walczak Can you see variability from presumably identical cells?

J. Howard It may require a level of precision which is difficult to obtain.

A. Hyman Sorry, we do not have much time left.

D. Fisher So I just like to give a simpler example in some ways of Joe's general point (which is a very important one), which is just phosphorilation cascades for signaling within cells in response to an external input in the bacterial chemotaxis. It is one of the examples where there are typically many more phosphorilations and dephosphorilations done than or has biology found a way to a non-generic point, in order to do the transmission of the information, and they can either make it more nonlinear or make it better in some other way, but they are certainly making it more inefficient. So again, it is an example where the cell wants to control things well by burning through a lot of extra energy.

A. Hyman Steve?

S. Chu It is probably a speculative answer to those questions. The short answer is: time is money. For communication, electron-communication over long distance is the fastest thing you have, so you might want to pay a premium of lots of ATP to get signals across. I would not call that a theory, I would call that a speculation. As for a theory for 'time is money', for diffusion versus directed motion, you can get more quantitative: how much do you pay in order to get it? Because diffusion goes as a square root, but if you want to get something from here to there over a 100 microns, you better pay for some real directed motion. Finally, the spermatid is a totally different thing. You are not going along microtubule, you are swimming. And, time is money here too, because you want to get to the egg first. So, it is all "time is money". Now, this is Steve Quake's point, I would not call it theory, there is not enough math in it.

A. Hyman Nicole?

N. King I will continue a conversation I was having with Joe over coffee, which is that I was surprised to see that there are some phenomena that are close to being 100% efficient and there are many that are not, and I think that maybe there is some information in that. Are they sloppy? So for instance, these 50 000 motors: is that a physical imperative or is that the accumulation of evolutionary noise? There may be some value in thinking: here is something that has evolved under strong selection and here is something for which there is a lot of opportunity to be messy. I do not know if there is information in that. It is just the way it is.

A. Hyman Ottoline, do you have a comment?

O. Leyser Yes. It is just a quick comment. There are quite a lot of organisms that are photosynthetic and make ATP out of light and so are not in principle energy-limited in the same way. I think it would be very useful in addressing some of these questions to compare a system like that with a system that is generating ATP by burning sugar and therefore having to collect the sugar, which is a more difficult job than collecting light. And I do not think that there have been enough comparisons of that sort.

A. Hyman First Steve, then Holly.

S. Quake I want to return to what Nicole and Steven were sort of nibbling around and what Joe touched on. It has always bothered me, when people appeal to optimality in biology, especially in a context of evolution. It has been very successful in physics as extremum principles, but I wonder under what conditions is it safe to assume that something has been optimized in the evolutionary sense? You can kind of argue: well, maybe some intuitive sense that a particular protein, like a pump, will be optimized in some sense. The question of the motor running the sperm, the one Nicole raised, I think. Is it reasonable to assume that some optimum is reached or not and under what conditions can you make these assumptions. I think there is a real gap conceptually, about how we think about these problems.

H. Goodson A quick thing I was going to add on the question of why there is so much energy spent in the cell cycling in the embryo: it may be because it is multi-cellular. So I am very curious to see for a yeast, or something like that, do you get a lot less energy? Because for a multi-cellular organism you really need to regulate that so much more.

J. Skotheim I just want to push back a little bit against the importance or pre-dominance of ATP or focus on ATP in all processes. I certainly appreciate that it is important and so on, but there are a lot of cases, think of cell growth, where the cell is really not making a whole lot of ATP, not nearly as much as it could given some molecule of glucose. It is using input molecules to build more cells. Certainly, there are contexts where ATP can be limit-ing, but there are a lot of contexts where it is not and I do not think it is the only unit of energy that we should think about when we think of cells.

A. Hyman Uri, and that is probably the last question. We will finish off with Uri Alon.

U. Alon Just to talk about optimality and about "time is money". So for example, if you try to optimize something uni-dimensional, like a capitalist organiza-tion optimizing only money, you can get to some conclusions. For example, if you make and break something and spend energy, you can get great response time. You just need to stop breaking it and concentrations shoot up. That is a tremendous use of energy. But response time is only one objective and biological systems have multiple objectives. And some of the problems when we look at optimality is considering a single objective. When we look at multiple objectives, there are tools for multiple objectives optimization. The point of making optimality assumptions is not to pat ourselves on the back, to say how optimal and wonderful biology is. Rather, it is a theoreti-cal stance what you say: OK, let us assume what is going to be optimal and what the conditions are. Then you can compute, compare to the proteins, it does not work, you change if in the end you are prepared to discover the possibility that it is optimized. But if you keep that in mind, you can do experiments you would not do otherwise and discover more about biological systems. And by now, there are spectacular examples coming from animal behavior and ecology, to proteins, to wasps laying eggs in figs, to balancing specificity in catalysis, where multi-objective optimality is enlightening to show how biological systems work.

S. Quake But, it is not plausible to assume that everything is optimized! That is like saying evolution is over. Some things must be in the process of still being optimized.

U. Alon There is a difference...

A. Hyman I think we will miss our lunch. Let us then not discuss ATP now, we will have to discuss that afterwards. There is a lot of ATP drop going around here. I just like to thank our two rapporteurs and the prepared comments. It was an excellent session and thank you very much.

Session 2

Cell Behavior and Control

Chair: *Andrew Murray*, Harvard, USA
Rapporteurs: *Michael Elowitz*, Caltech, USA and *Terence Hwa*, UCSD, USA
Scientific secretaries: *Geneviève Dupont*, ULB, Belgium and *Didier Gonze*, ULB, Belgium

A. Murray Good afternoon! I would again like to thank the Solvay family for their generosity and making this event possible, and to curse the gentlemen sitting next to me for making me chair of this session. I have a very important reminder from the staff. If you are giving a prepared remark, if you are making an unscripted remark, please slowly give your name because the meeting is recorded and transcribed. The most difficult thing for those recording and transcribing the meeting is to decipher your name. If you mumble or say it quickly they will not know who you are. The organisation of this afternoon session is slightly different from this morning. Our rapporteurs will speak one before and one after the break. Immediately afterwards we will have the prepared remarks. And then we will have the general discussion. I would encourage you to make that discussion as general as it can possibly be. One of the things that Boris and I have talked about is that what could possibly emerge from this meeting would be a set of questions that might at least pay homage to master Hilbert and the mathematical questions he asked over one century ago, to instruct our successors what might be interesting problems to consider. Daniel Fisher was mentioning to me over lunch that mathematics apparently today is very short of conjecture.

I will very briefly introduce the session. This is about cellular behaviour and control. I am just going to make a couple of points. The first is that if you measure the density of components and if you take a very sophisticated device like a Boeing 787 Dreamliner and a budding yeast cell, the density of parts inside the budding yeast cell exceeds that of the airplane by a factor of 10^{19} which is a rather large number. The second point to make is that, after Monod, I would like to remind you is that the dream of every cell is to become two and one of the things that is interesting and may represent one of the best mechanisms for controlling the composition of cells because every cell halves the errors of his parents and quarters the errors of its grand parents in terms of its effort to maintain a particular composition. We will hear about regulatory mechanisms. And the last point I want to make is a point against the hegemony of Jacob and Monod, about gene expression. I am going to describe an experiment. The experiment is the following: You fertilize a starfish egg. Having fertilized it, you remove the DNA by sucking out the nucleus and now you ask the question: How many times does that fertilized egg divide successively, not growing, just making the cells smaller and smaller. How many people think it does not divide at all? Divide once? 5 times or fewer? 10 times? More that 10? So the answer is 10 times. The point is this is a cell with no DNA. There is no gene expression being controlled. As far as we know all mRNA's are translated constantly. The only thing that is being regulated is the destruction of a small number of proteins like cyclins and a cycle of post-translational modification. This produces a lot of joy and excitement. We know about cell division, the regulation of the structure of endoplasmic reticulum, etc. And it reveals rather clearly that changing the expression of genes and the whole transcription machinery is not required for any of these events. And with that, I give the floor to Terry.

Rapporteur Talk by T. Hwa: Cell Behavior and Control

A cell is the smallest unit of a freely living system. Our understanding of a cell is measured by our ability to predict and manipulate cellular behaviors in response to environmental and genetic changes. This has been a primary area of activities for physicists entering biology since the renaissance of quantitative biology in the late 1990s. Much of the research effort in this area can be categorized as "bottom-up", in the sense that researchers strive to gain insights into cellular behaviors through characterizing the underlying molecular control systems. Substantial progress has been made in this direction, particularly in few-component, localized control systems with direct links to clear cell behaviors. More recently, there has been increasing efforts in a complementary "top-down" approach, in which one aims to gain insight on control mechanisms through characterizing cell behaviors. This report will briefly cover both approaches, with examples taken primarily from studies of microbial systems—the simplicity of the latter at both the levels of molecular interactions and cellular behaviors has made them favorite subjects of quantitative studies. Progress in higher eukaryotic is summarized in the report by Michael Elowitz, with exemplary systems (e.g., circadian oscillation, and determinations of cell size, cell fate) discussed by Albert Goldbeter, Jan Skotheim, and Ben Simons in this meeting.

1. Molecular Circuits and the Bottom-Up Approach

For about ten years starting in the late 1990s, a dominant direction in quantitative biology has been the studies of "molecular circuit". These are interacting regulatory or signaling systems, believed to act as *modules*[1] that drive different classes of cell behaviors, such as switching, adaptation, and oscillation. Adaptation of chemotaxing bacteria to the concentration of attractants (so that they can respond to the concentration gradient instead of the concentration itself), was one of the first endogenous control circuits dissected in a quantitative manner.[2] The regulatory strategy was examined in the context of control theory[3]; later, connection to molecular interactions was established,[4–6] and uniqueness of the regulatory strategies for adaptation was investigated.[7] The phenomenon of bistability was characterized quantitatively for the *lac* promoter,[8] the best-known bacterial sugar uptake system. In parallel, synthetic approaches were used to establish simple genetic circuits amendable to quantitative characterization *in vivo*, including the toggle switch[9] and the repressilator,[10] with more robust versions achieved later.[11,12] Further down at the molecular level, characteristics of gene regulation were quantified for exemplary bacterial systems and found to be in agreement with predictions of mechanistic models.[13–15] *De novo* development of transcriptional and post- transcriptional control were also demonstrated using synthetic and evolutionary approaches.[16,17]

Stochasticity in gene expression is another area extensively investigated by quantitative biologists. Starting with characterizing the sources of stochasticity in bacteria,[18] a growing number of researchers went on to establish the molecular determinants of stochasticity,[19–22] characterize manifestations in stochastic cell behaviors, e.g., in metabolic switching,[23–25] swimming[26,27] and differentiation,[28,29] and explore the origin and consequences of stochasticity.[30,31] Together, this collection of studies established beyond doubt that predictive, quantitative description starting with known molecular characteristics is possible in living cells, at least in bacteria.

It has turned out to be much more difficult to achieve similar level of characterization and predictive understanding for eukaryotes, even for simple eukaryotes like the budding yeast. One of the difficulties is that some components of the eukaryotic gene regulatory systems are characterized only at a qualitative level, including nuclear transport, mRNA splicing, and chromatin modification, none of which have counterparts in bacteria. Another difficulty is that a high degree of eukaryotic control circuits involves multiple layers of phosphorylation cascades, for which quantitative knowledge is, again, severely lacking.

2. Cell-level Behavior: Difficulty of the Bottom-Up Approach

Even for bacteria, extending the existing quantitative description beyond small-scale circuits with well-characterized components has been unexpectedly difficult. Take, for example, the bacterial stress response system in *E. coli*. The master sigma factor RpoS driving the expression of the general stress regulon has numerous factors regulating its transcription,[32] the synthesized rpoS mRNA interacts with many small RNA regulators,[33] and the sigma factor itself is further regulated post-translationally.[34] Figuring out all the underlying interaction parameters is a daunting challenge well beyond the existing technology. Even more challenging is to identify the sources of the inputs, i.e., the factors that trigger the activities of the dozens of regulators that affect the levels of functional RpoS proteins, without the knowledge of which the study of stress response cannot be connected to environmental perturbations that the system is designed to respond to. Equally difficult challenges lie on the output end: As a sigma factor, RpoS drives the transcription of many mRNA encoding mostly stress-response proteins. However, in stressful environments where there is often very little net protein synthesis, the availability of resources (e.g., nutrients or energy during starvation) needed to synthesize these proteins is perhaps the biggest unknown among all the factors listed above.[35,36]

Valiant attempts have been made in synthetic genetic systems to minimize the unknown coupling of circuit elements to the environment and general machineries in bacteria.[37,38] But even there, after 20 years of trials and experiences in synthetic biology, it is amazing how difficult it is to get synthetic circuits to work as designed. Even adapting a working circuit from one growth condition to another is a formidable challenge, a common knowledge that is rarely admitted however.[39–41]

What is it that makes us so limited and powerless in understanding complex cellular control and behavior? I suggest that the problem may lie in one of the fundamental tenets of molecular biology, that the study of molecular interactions will ultimately lead to an understanding of the underlying functions, may not be effective generically in capturing cell-level behaviors. This tenet, which has been the foundation of the bottom-up paradigm, has given biology many rewards since the 1950s: The knowledge of the double-helix structure of DNA led directly to the universal mechanism of genetic inheritance,[42] the study of *lac* operon led to the general scheme of genetic control circuits.[43] However, in general, biomolecular systems are not as modular as what one might naively expect it to be. In many cases, cellular behaviors involve the interaction of many molecular components. Thus, as the focus of biomedical research shifts towards the more integrated, higher-order (i.e., 'system-level') behaviors in recent years, the shortcomings of the bottom-up paradigm become more and more apparent.

3. Bacterial Growth Control: Dimensional Reduction and the Top-Down Approach

As a concrete example, I will describe in some detail below the phenomenon of bacterial growth which my lab has devoted some efforts investigating in the past decade. Let us try to describe how the rate of biomass growth λ, a basic measure of cell behavior, is affected by changes in the external level of a key nutrient $n(t)$, say a carbon source. In a bottom-up approach, we want to link the growth rate to molecular interactions and processes that give rise to biomass growth. As depicted in Figure 1, one might start this description with the flux of nutrient uptake J_C, given by Eq. (1) where k_C is the specific rate of uptake and C is the concentration of catabolic proteins. Next, one needs to describe the set of biochemical reactions which convert the external nutrient into the thousands of internal metabolites. Let the concentration of the ith metabolite be m_i; its net synthesis flux ν_i generally depends on the concentrations of multiple metabolites as well as proteins (enzymes and/or regulators), with the concentrations of the latter collectively denoted as $\{p_j\}$; see Eq. (2). The protein concentrations themselves change according to Eq. (3), with the first term on the right-hand side being the synthesis flux for the ith protein, χ_i being the fraction of total protein synthesis flux J_P directed towards protein i, and the second term being the dilution rate due to cell growth. Importantly, χ_i is

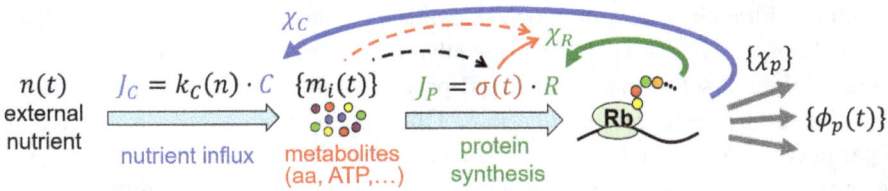

Fig. 1.   **Schematic model of bacterial growth control.** External nutrients (n), brought into the cell at the flux J_C by catabolic proteins C, drives the synthesis of many internal metabolites $\{m_i\}$. Some of these metabolites (including amino acids, ATP and GTP) fuel the ribosome for protein synthesis and affect the translational speed σ (dashed black arrow), and some affect the regulation of ribosome synthesis χ_R (dashed red arrow). We hypothesize that the regulation of ribosome synthesis χ_R is actually determined according to the translation speed σ, i.e., the dashed red arrow is implemented via the solid red arrow. This would result in a huge dimensional reduction, with the mathematical effect of replacing two complex (and unknown) sets of reactions represented by the two dashed arrows by a simple function represented by the solid arrow.

specified by the gene regulatory function g_i, which in turn depends on the set of metabolites and proteins as indicated by Eq. (4). The protein synthesis flux itself is given by Eq. (5), with R being the concentration of ribosomes, and σ being the average rate of protein synthesis by a ribosome; the latter again depends on the set of metabolites and proteins, via the function ν_σ as indicated by Eq. (6).

$$J_C(t) = k_C(n(t)) \cdot C(t) \tag{1}$$

$$\frac{d}{dt} m_i = \nu_i\Big(\{m_j(t)\}, \{p_j(t)\}; n(t)\Big) \tag{2}$$

$$\frac{d}{dt} p_i = \chi_i(t) \cdot J_P - \lambda(t) \cdot p_i \tag{3}$$

$$\chi_i(t) = g_i\Big(\{m_j(t)\}, \{p_j(t)\}; n(t)\Big) \tag{4}$$

$$J_P(t) = \sigma(t) \cdot R(t) \tag{5}$$

$$\sigma(t) = \nu_\sigma(\{m_j(t)\}, \{p_j(t)\}) \tag{6}$$

$$J_P(t) = \lambda(t) \sum_i p_i \tag{7}$$

$$J_C(t) = J_P(t) \tag{8}$$

Equations (1)–(6) are supplemented by several other relations: Because the biomass density is nearly invariant across different growth conditions[44] and protein is the major component of biomass,[45,46] we can regard the total cellular protein concentration $P_0 \equiv \sum_i p_i$ to be a constant. Applying this to Eq. (3) (after summing over all i), we obtain Eq. (7). Also, we assume that cells do not allow substantial leakage of internal metabolites, so that the nutrient influx J_C is converted completely[a] into protein synthesis flux as indicated by Eq. (8). Together, Eqs. (1)–(8) provide a minimal mathematical description of how changes in the nutrient level $n(t)$ drives changes in the growth rate $\lambda(t)$ in the bottom-up approach. We see that this approach requires the full knowledge of the biochemical reactions $\{\nu_j, \nu_\sigma\}$ and the regulatory functions $\{g_j\}$, which involve not only thousands of variables comprising of the full set of metabolites and proteins, but also many more parameters needed to specify each of these functions. Obtaining the forms of each reaction and its parameters (for the relevant *in vivo* conditions) is, needless to say, a daunting challenge. One may even wonder whether this full knowledge is too difficult for the cells themselves to 'master'. In the following, I suggest that the cell uses regulatory functions to implement *dimensional reduction*, such that the dynamics of bacterial growth is actually determined by only a few parameters despite the formal involvement of a large number of reactions and parameters. We will see that the use of

---

[a] Again, this is under the simplifying assumption that protein is the dominant component of cellular biomass. Other biomass components (nucleotides, lipids, etc) can be incorporated by slightly generalizing the formulation.

phenomenological top-down approach is effective in revealing the existence of this dimensional reduction as well as in pin-pointing its mechanistic origin.

To see what behavior is encoded in the system described by Eqs. (1)–(8), let us consider first the steady-state limit, where the nutrient n is provided much in surplus and cells reach the state of balanced exponential growth. In this limit, all concentrations become time independent, i.e., $dm_i/dt = 0$ and $dp_i/dt = 0$. We denote all the steady-state values of the time-dependent variables by asterisks. Combining Eqs. (3) and (7) we see that in steady state the fractional protein synthesis flux $\chi_i^* \equiv g_i(\{m_i^*\}, \{p_i^*\})$ is simply given by the absolute abundance of that protein, as a fraction of the total proteome, i.e., $\chi_i^* = p_i^* / \sum_i p_i$. This relation, together with Eqs. (5) and (7), gives a key relation between the ribosomal fraction ($\chi_R^* = R^* / \sum_i p_i^*$) and the growth rate ($\lambda^*$), Eq. (9), while a similar relation between the catabolic fraction ($\chi_C^* = C^* / \sum_i p_i^*$) and the growth rate, Eq. (10), results from Eqs. (1), (7) and (8).

$$\lambda^* = \sigma^* \cdot \chi_R^* = \nu_\sigma(\{m_i^*\}, \{p_i^*\}) \cdot g_R(\{m_i^*\}, \{p_i^*\}) \tag{9}$$

$$\lambda^* = k_C^* \cdot \chi_C^* = k_C^* \cdot g_C(\{m_i^*\}, \{p_i^*\}) \tag{10}$$

Equations (9) and (10) describe how the growth rate λ^* is determined by concentrations of the metabolites $\{m_i^*\}$ and proteins $\{p_i^*\}$, which set the levels of the regulatory functions g_R, g_C, as well as the translation rate of the ribosomes, ν_σ. On the other hand, a simple linear relation (Figure 2(a), green symbols) is known to exist between the ribosome content and the growth rate, for the growth of *E. coli* and various other microbes growing on a wide variety of nutrient sources.[45,47,48] The empirical relation is captured by Eq. (11). In recent years, the abundances of catabolic proteins were also shown to follow a simple relation, a negative linear relation, with the growth rate; see Figure 2(a), blue symbols for growth of E. coli in minimal medium with different carbon sources.[49,50] This is captured by Eq. (12).

$$\chi_R^* = \chi_{R,0} + \lambda^*/\sigma_0 \tag{11}$$

$$\chi_C^* = \chi_{\max} \cdot (1 - \lambda^*/\lambda_C) \tag{12}$$

How is it possible that different nutrient sources involving different sets of enzymes and metabolites belonging to distinct metabolic pathways, e.g., glucose and succinate which use opposing pathways for central carbon metabolism, follow the same relations, Eqs. (11) and (12)? One might argue that perhaps these relations exist due to some underlying optimization principles, and cells have, through evolution, adjusted the numerous molecular parameters appearing in Eqs. (9) and (10) to ensure that these relations are followed. However, even invoking the evolution argument cannot explain why the same two relations are followed also by mutants with various defects in carbon catabolism or in protein synthesis.[48,49]

$$g_R(\{m_i^*\}, \{p_i^*\}) = \hat{g}_R(\nu_\sigma(\{m_i^*\}, \{p_i^*\})) \tag{13}$$

$$\hat{g}_R(\nu_\sigma) = \chi_{R,0}/(1 - \nu_\sigma/\sigma_0) \tag{14}$$

$$g_C(\{m_i^*\}, \{p_i^*\}) = \hat{g}_C(\nu_\sigma(\{m_i^*\}, \{p_i^*\})) \tag{15}$$

$$\hat{g}_C(\nu_\sigma) = \chi_{max} \cdot (1 - \nu_\sigma \hat{g}_R(\nu_\sigma)/\lambda_C) \tag{16}$$

An alternative possibility is that the forms of the regulatory functions g_R, g_C and the translation rate ν_σ contain some special structure that yield the simple relations Eqs. (11) and (12), describing the striking effect of dimensional reduction seen in Figure 2(a), as *generic* solutions. A hint of this structure is revealed by simply taking both Eqs. (9), (10) and Eqs. (11), (12) seriously: Eqs. (9) and (11) yield a relation between χ_R^* and σ^*, $\chi_R^* = \chi_{R,0}/(1 - \sigma^*/\sigma_0)$, as depicted by the green line in Figure 2(b). This correlation implies an underlying *causal* relation between the regulatory function g_R which sets the value χ_R^* and the translation rate ν_σ. This causal relation is depicted by Eq. (13): It states that the regulatory function for ribosome biogenesis, g_R, depends on the metabolite and protein levels not directly, but indirectly through its dependence on the translation rate ν_σ, which itself depends directly on the many metabolites and proteins (e.g., all of the amino acids, ATP, GTP, tRNAs, tRNA synthases, elongation factors, etc). The dependence of the regulatory function itself, $\hat{g}_R(\nu_\sigma)$, is simply given by the empirical correlation shown in Figure 2(b) (green curve) and contains only two parameters, $\chi_{R,0}$ and σ_0, whose values are fixed by the growth law Eq. (11) (green symbols in Figure 2(a)). It describes the regulation of ribosome biogenesis as an increasing function of the translation rate.

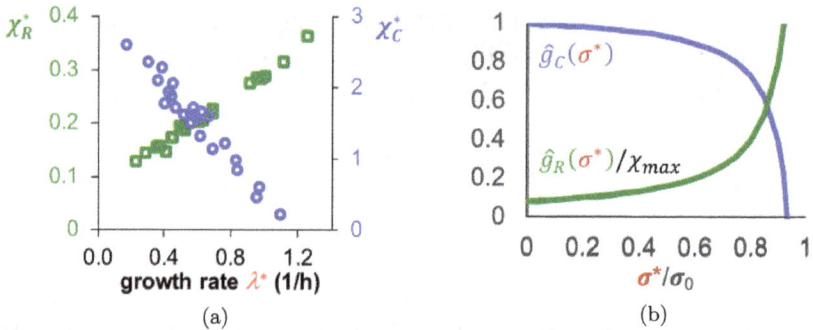

Fig. 2.   **Growth laws and regulation functions.** (a) For exponentially growing bacterial cultures, there are a number of "growth laws" characterizing robust correlations between the expression levels of various proteins and the growth rate. For growth in minimal medium with varying quality of carbon sources, the ribosome content exhibits a positive linear relation with the growth rate (green symbols and Eq. (11)),[48] while the level of catabolic proteins exhibits a negative linear relation (blue symbols and Eq. (12)).[49] (b) The regulatory functions controlling the expression of ribosomal and catabolic proteins, $\hat{g}_R(\sigma)$ and $\hat{g}_C(\sigma)$, respectively, can be deduced from the growth laws.[51] Their forms, as given by Eqs. (14) and (16), are shown as the green and blue lines, respectively.

Applying the same analysis to the regulation of catabolic proteins, we find that Eqs. (10) and (12) yield a relation between χ_C^* and σ^*, as depicted by the blue line in Figure 2(b). This correlation implies another underlying *causal* relation between the regulatory function g_C which sets the value χ_C^* and the translation rate ν_σ, as depicted by Eq. (15), with a regulatory function $\hat{g}_C(\nu_\sigma)$ given by Eq. (16). Again, there are only two parameters which are fixed by the growth law Eq. (12).

The above analysis shows that uninterpretable expressions for the growth rate λ^* (Eqs. (9) and (10)) derived from an explicit bottom-up approach (Eqs. (1)–(8)), when combined with simple phenomenology (Eqs. (11) and (12)) obtained from a top-down approach, produces a crucial hypothesis on the regulatory strategy underlying dimensional reduction observed for bacterial growth: Eqs. (13) and (15) suggest that the regulation functions sense the activity of a cellular process, the average translation rate, which appears to play the role of an internal measure of the state of cell growth. Consistent to this expectation, a robust relation between translation rate and growth rate has indeed been established recently for a broad variety of growth conditions.[52] The suggested regulatory strategy is also quite plausible molecularly. It has been long established for *E. coli* that the direct regulator of ribosome biogenesis is the signaling molecule ppGpp.[53-55] The trigger of ppGpp synthesis has been a subject of debate; however recent cryoEM studies are clarifying[56,57]: RelA, the main enzyme which synthesizes ppGpp, is normally autoinhibited; but the inhibition is relieved when RelA is caught in a ribosome whose A-site is occupied by an uncharged-tRNA. This picture suggests that RelA activity (hence ppGpp level) reports the duration that a ribosome is not engaged in translation, which can be taken as a direct measure of ribosome (in)activity.

So far, the proposed regulatory strategy Eqs. (13)–(16) merely provide a plausible rationalization of how dimensional reduction may take place mechanistically. But if it indeed captures the underlying regulatory strategy, then there are many consequences that can be explicitly tested. One such tests reported recently is on the kinetics of growth transitions.[51] Given the explicitly form of the regulatory functions Eqs. (13)–(16), one can use them in the general kinetic Eqs. (1)–(8) to *derive* an explicit dynamical equation for the translational activity $\sigma(t)$:

$$\frac{d}{dt}\sigma = \sigma \cdot [k_C(t)\hat{g}_C(\sigma) - \sigma\hat{g}_R(\sigma)]. \tag{17}$$

This equation, supplemented by the regulatory functions Eqs. (14) and (16), completely specifies the growth dynamics in response to changes in external nutrients, modeled here by a time-dependent uptake rate, $k_C(t)$. The solution, $\sigma(t)$, can be further integrated to obtain the instantaneous growth rate $\lambda(t)$, as well as the protein concentrations $R(t)$, $C(t)$, etc. The predictions made by this model are verified quantitatively for a dozen of nutrient shift experiments in *E. coli* using at most one free parameter.[51]

4. Summary and Outlook

The subject of "Cellular behavior and control" is dominated by a dichotomy between the simplicity of behavior at the cell level and the apparent complexity of control at the molecular level. Despite limited early success in our understanding of small-scale circuits, the bottom-up approach alone has not been effective in helping us to dissect and understand more complex circuits due to the large number of poorly characterized variables and parameters involved. On the subject of cell growth, phenomenological laws derived from top-down studies have been useful in guiding bottom-up modeling. We describe here an explicit example how simple cell-level behavior (namely, the growth laws) can arise from a novel regulatory strategy whose inputs are not the concentrations of regulators but the rate of translational elongation, which itself depends on many variables and parameters but whose details do not affect growth control.

What this regulatory strategy accomplishes, as captured by Eq. (14), is analogous to Boltzmann's fundamental postulate of statistical mechanics, that the equilibrium distribution of a system with many degrees of freedom depends only on conserved quantities such as the total energy and not directly on the individual degrees of freedom themselves. For a classical gas, this simplicity ultimately arises from the nature of the Hamiltonian dynamics and the occurrence of molecular chaos. In bacterial growth, we suggested above that this may be arranged as a result of a clever molecular trick that allows RelA to sense translational activity and synthesize the signaling molecule ppGpp in response to changes in this activity.

Eukaryotes do not use the ppGpp pathway to regulate ribosome biogenesis. However, the budding yeast *S. cerevisae* exhibits a similar relation as Eq. (11) between ribosomal content and the growth rate.[58] It is possible that another molecular system has been employed to detect the rate of translational activity. Moreover, it is quite plausible that cells employ different molecular tricks to sense a number of key internal processes and use the outcome to control various aspects of cell behaviors. Discovering additional regulatory systems of this type, and most importantly, identifying the source of the true signals they sense, may be key to gaining a comprehensive view of how cells navigate through the complex web of molecular interactions to implement a set of coherent, coordinated behaviors.

It should be noted that despite progress made in understanding the control of biomass growth which is conveniently studied at the bulk level, substantial gaps still exist in understanding the growth of *individual* cells: The instantaneous growth rates of cells (e.g., the rate of cell elongation for rod-shaped bacterium such as *E. coli*) has a rather broad distribution and can typically deviate by as much as 20–30% from the average bulk growth rate for several generations.[59,60] Striking correlations have been reported between fluctuations in the growth rate and in the expression of various genes[61,62]; however, causality among the observed fluctuations is not clear,[63] and the origin of such large, prolonged fluctuations is unknown. Another topic of intense recent interest is that of cell-size control. Phenomenological characterization established the lack of correlation between the sizes of mother and daughter of exponential growing *E. coli* cells,[59,64] suggesting an "adder" rule by which cell division occurs after a cell accumulates a certain amount of mass (length) after birth.[65–67] However, molecular mechanisms underlying such a rule are still elusive. Nor is it known how cell-size is set differently for different growth conditions. In a related study, an important element of cell size control in yeast has been revealed recently. The key regulator Whi5 is shown to be synthesized in a growth-rate independent manner, such that its concentration, set by dilution due to cell volume increase, can be used to sense cell volume and hence control cell size.[68] However, it is again not known how this system can be used to set cell size differently in accordance with the growth condition. The pursuit of these important regulatory processes in their physiological context for simple model organisms, as well as extensions to higher eukaryotes including cancer, will likely dominate the study of cell behaviors and control in the next 5–10 years.

References

1. L. H. Hartwell, J. J. Hopfield, S. Leibler and A. W. Murray, From molecular to modular cell biology, *Nature* **402**, C47 EP (1999).
2. U. Alon, M. G. Surette, N. Barkai and S. Leibler, Robustness in bacterial chemotaxis, *Nature* **397**, 168 (1999).
3. T.-M. Yi, Y. Huang, M. I. Simon and J. Doyle, Robust perfect adaptation in bacterial chemotaxis through integral feedback control, *Proc. Natl. Acad. Sci.* **97**, 4649 (2000).
4. A. Celani, T. S. Shimizu and M. Vergassola, Molecular and functional aspects of bacterial chemotaxis, *J. Stat. Phys.* **144**, p. 219 (2011).
5. V. Sourjik and N. S. Wingreen, Responding to chemical gradients: bacterial chemotaxis, *Current Opinion in Cell Biology* **24**, 262 (2012).
6. Y. Tu, Quantitative modeling of bacterial chemotaxis: Signal amplification and accurate adaptation, *Annu. Rev. Biophys.* **42**, 337 (2013).
7. W. Ma, A. Trusina, H. El-Samad, W. A. Lim and C. Tang, Defining network topologies that can achieve biochemical adaptation, *Cell* **138**, 760 (2009).
8. E. M. Ozbudak, M. Thattai, H. N. Lim, B. I. Shraiman and A. van Oudenaarden, Multistability in the lactose utilization network of escherichia coli, *Nature* **427**, 737 (2004).
9. T. S. Gardner, C. R. Cantor and J. J. Collins, Construction of a genetic toggle switch in escherichia coli, *Nature* **403**, 339 (2000).
10. M. B. Elowitz and S. Leibler, A synthetic oscillatory network of transcriptional regulators, *Nature* **403**, 335 (2000).
11. M. R. Atkinson, M. A. Savageau, J. T. Myers and A. J. Ninfa, Development of genetic circuitry exhibiting toggle switch or oscillatory behavior in escherichia coli, *Cell* **113**, 597 (2003).
12. J. Stricker, S. Cookson, M. R. Bennett, W. H. Mather, L. S. Tsimring and J. Hasty, A fast, robust and tunable synthetic gene oscillator, *Nature* **456**, 516 EP (2008).
13. T. Kuhlman, Z. Zhang, M. H. Saier and T. Hwa, Combinatorial transcriptional control of the lactose operon of escherichia coli, *Proc. Natl. Acad. Sci.* **104**, 6043 (2007).
14. E. Levine, Z. Zhang, T. Kuhlman and T. Hwa, Quantitative characteristics of gene regulation by small rna, *PLOS Biol.* **5**, 1 (08 2007).
15. R. Brewster, F. Weinert, H. Garcia, D. Song, M. Rydenfelt and R. Phillips, The

transcription factor titration effect dictates level of gene expression, *Cell* **156**, 1312 (2014).

16. H. M. Salis, E. A. Mirsky and C. A. Voigt, Automated design of synthetic ribosome binding sites to control protein expression, *Nature Biotechnol.* **27**, 946 EP (2009).

17. F. Poelwijk, M. de Vos and S. Tans, Tradeoffs and optimality in the evolution of gene regulation, *Cell* **146**, 462 (2011).

18. M. B. Elowitz, A. J. Levine, E. D. Siggia and P. S. Swain, Stochastic gene expression in a single cell, *Science* **297**, 1183 (2002).

19. A. Becskei, B. B. Kaufmann and A. van Oudenaarden, Contributions of low molecule number and chromosomal positioning to stochastic gene expression, *Nature Genet.* **37**, 937 (2005).

20. M. Kærn, T. C. Elston, W. J. Blake and J. J. Collins, Stochasticity in gene expression: from theories to phenotypes, *Nature Rev. Genet.* **6**, 451 (2005).

21. L. Cai, N. Friedman and X. S. Xie, Stochastic protein expression in individual cells at the single molecule level, *Nature* **440**, 358 (2006).

22. A. Sanchez and I. Golding, Genetic determinants and cellular constraints in noisy gene expression, *Science* **342**, 1188 (2013).

23. M. Acar, J. T. Mettetal and A. van Oudenaarden, Stochastic switching as a survival strategy in fluctuating environments, *Nature Genet.* **40**, 471 EP (2008).

24. J. A. Megerle, G. Fritz, U. Gerland, K. Jung and J. O. Rdler, Timing and dynamics of single cell gene expression in the arabinose utilization system, *Biophys. J.* **95**, 2103 (2008).

25. G. Lambert and E. Kussell, Memory and fitness optimization of bacteria under fluctuating environments, *PLOS Genet.* **10**, 1 (09 2014).

26. E. Korobkova, T. Emonet, J. M. G. Vilar, T. S. Shimizu and P. Cluzel, From molecular noise to behavioural variability in a single bacterium, *Nature* **428**, 574 (2004).

27. A. J. Waite, N. W. Frankel and T. Emonet, Behavioral variability and phenotypic diversity in bacterial chemotaxis, *Annu. Rev. Biophys.* **47**, 595 (2018).

28. G. M. Süel, J. Garcia-Ojalvo, L. M. Liberman and M. B. Elowitz, An excitable gene regulatory circuit induces transient cellular differentiation, *Nature* **440**, 545 (2006).

29. J. R. Russell, M. T. Cabeen, P. A. Wiggins, J. Paulsson and R. Losick, Noise in a phosphorelay drives stochastic entry into sporulation in bacillus subtilis, *The EMBO J.* **36**, 2856 (2017).

30. E. Kussell, R. Kishony, N. Q. Balaban and S. Leibler, Bacterial persistence, *Genetics* **169**, 1807 (2005).

31. A. Raj and A. van Oudenaarden, Nature, nurture, or chance: Stochastic gene expression and its consequences, *Cell* **135**, 216 (2008).

32. P. Landini, T. Egli, J. Wolf and S. Lacour, sigmas, a major player in the response to environmental stresses in escherichia coli: role, regulation and mechanisms of promoter recognition, *Environ. Microbiol. Rep.* **6**, 1 (2014).

33. A. Battesti, N. Majdalani and S. Gottesman, The rpos-mediated general stress response in escherichia coli, *Annu. Rev. Microbiol.* **65**, 189 (2011).

34. R. Hengge, Proteolysis of σS (RpoS) and the general stress response in escherichia coli, *Res. Microbiol.* **160**, 667 (2009), Special issue on proteolysis in prokaryotes: protein quality control and regulatory principles.

35. J. Mandelstam, The free amino acids in growing and non-growing populations of escherichia coli, *Biochem. J.* **69**, 103 (1958).

36. H. Link, T. Fuhrer, L. Gerosa, N. Zamboni and U. Sauer, Real-time metabolome profiling of the metabolic switch between starvation and growth, *Nature Methods* **12**, 1091 EP (2015).

37. B. Wang, R. I. Kitney, N. Joly and M. Buck, Engineering modular and orthogonal genetic logic gates for robust digital-like synthetic biology, *Nature Commun.* **2**, 508 EP (2011).

38. C. Lou, B. Stanton, Y.-J. Chen, B. Munsky and C. A. Voigt, Ribozyme-based insulator parts buffer synthetic circuits from genetic context, *Nature Biotechnol.* **30**, 1137 EP (2012).

39. S. Klumpp, Z. Zhang and T. Hwa, Growth rate-dependent global effects on gene expression in bacteria, *Cell* **139**, 1366 (2019/05/17 2009).

40. J. R. Kelly, A. J. Rubin, J. H. Davis, C. M. Ajo-Franklin, J. Cumbers, M. J. Czar, K. de Mora, A. L. Glieberman, D. D. Monie and D. Endy, Measuring the activity of biobrick promoters using an in vivo reference standard, *J. Biol. Engin.* **3**, p. 4 (2009).

41. H. M. Zhang, S. Chen, H. Shi, W. Ji, Y. Zong, Q. Ouyang and C. Lou, Measurements of gene expression at steady state improve the predictability of part assembly, *ACS Synth. Biol.* **5**, 269 (2016).

42. J. D. WATSON and F. H. C. CRICK, Genetical implications of the structure of deoxyribonucleic acid, *Nature* **171**, 964 (1953).

43. F. Jacob and J. Monod, Genetic regulatory mechanisms in the synthesis of proteins, *J. Molecul. Biol.* **3**, 318 (1961).

44. N. Nanninga and C. Woldringh, *Molecular Cytology of Escherichia coli (ed N. Nanninga)* (Academic Press, 1985).

45. H. Bremer and P. Dennis, *Escherichia coli and Salmonella typhimurium: Cellular and Molecular Biology (eds F.C. Neidhardt and R. Curtiss)* (ASM Press, 1996).

46. M. Basan, M. Zhu, X. Dai, M. Warren, D. Sévin, Y.-P. Wang and T. Hwa, Inflating bacterial cells by increased protein synthesis, *Molec. Syst. Biol.* **11**, 836 (2015).

47. M. Schaechter, O. Maale and N. O. Kjeldgaard, Dependency on medium and temperature of cell size and chemical composition during balanced growth of salmonella typhimurium, *Microbiology* **19**, 592 (1958).

48. M. Scott, C. W. Gunderson, E. M. Mateescu, Z. Zhang and T. Hwa, Interdependence of cell growth and gene expression: Origins and consequences, *Science* **330**, 1099 (2010).

49. C. You, H. Okano, S. Hui, Z. Zhang, M. Kim, C. W. Gunderson, Y.-P. Wang, P. Lenz, D. Yan and T. Hwa, Coordination of bacterial proteome with metabolism by cyclic amp signalling, *Nature* **500**, 301 EP (2013).

50. S. Hui, J. M. Silverman, S. S. Chen, D. W. Erickson, M. Basan, J. Wang, T. Hwa and J. R. Williamson, Quantitative proteomic analysis reveals a simple strategy of global resource allocation in bacteria, *Molec. Syst. Biol.* **11** (2015).

51. D. W. Erickson, S. J. Schink, V. Patsalo, J. R. Williamson, U. Gerland and T. Hwa, A global resource allocation strategy governs growth transition kinetics of escherichia coli, *Nature* **551**, 119 EP (2017).

52. X. Dai, M. Zhu, M. Warren, R. Balakrishnan, V. Patsalo, H. Okano, J. R. Williamson, K. Fredrick, Y.-P. Wang and T. Hwa, Reduction of translating ribosomes enables escherichia coli to maintain elongation rates during slow growth, *Nature Microbiol.* **2**, 16231 EP (2016).

53. M. Cashel, The control of ribonucleic acid synthesis in escherichia coli : Iv. relevance of unusual phosphorylated compounds from amino acid-starved stringent strains, *J. Biol. Chem.* **244**, 3133 (1969).

54. J. Ryals, R. Little and H. Bremer, Control of rrna and trna syntheses in escherichia coli by guanosine tetraphosphate, *J. Bacteriol.* **151**, 1261 (1982).

55. K. Potrykus, H. Murphy, N. Philippe and M. Cashel, ppgpp is the major source of growth rate control in e. coli, *Environ. Microbiol.* **13**, 563 (2011).

56. A. Brown, I. S. Fernández, Y. Gordiyenko and V. Ramakrishnan, Ribosome-dependent activation of stringent control, *Nature* **534**, 277 EP (2016).
57. A. B. Loveland, E. Bah, R. Madireddy, Y. Zhang, A. F. Brilot, N. Grigorieff and A. A. Korostelev, Ribosomerela structures reveal the mechanism of stringent response activation, *eLife* **5**, p. e17029 (2016).
58. E. Metzl-Raz, M. Kafri, G. Yaakov, I. Soifer, Y. Gurvich and N. Barkai, Principles of cellular resource allocation revealed by condition-dependent proteome profiling, *eLife* **6**, p. e28034 (2017).
59. S. Taheri-Araghi, S. Bradde, J. T. Sauls, N. S. Hill, P. A. Levin, J. Paulsson, M. Vergassola and S. Jun, Cell-size control and homeostasis in bacteria, *Current Biol.* **25**, 385 (2019/05/17 2015).
60. A. S. Kennard, M. Osella, A. Javer, J. Grilli, P. Nghe, S. J. Tans, P. Cicuta and M. Cosentino Lagomarsino, Individuality and universality in the growth-division laws of single e. coli cells, *Phys. Rev. E* **93**, p. 012408 (2016).
61. H. Salman, N. Brenner, C.-k. Tung, N. Elyahu, E. Stolovicki, L. Moore, A. Libchaber and E. Braun, Universal protein fluctuations in populations of microorganisms, *Phys. Rev. Lett.* **108**, p. 238105 (2012).
62. L. Susman, M. Kohram, H. Vashistha, J. T. Nechleba, H. Salman and N. Brenner, Individuality and slow dynamics in bacterial growth homeostasis, *Proc. Natl. Acad. Sci.* **115**, E5679 (2018).
63. D. J. Kiviet, P. Nghe, N. Walker, S. Boulineau, V. Sunderlikova and S. J. Tans, Stochasticity of metabolism and growth at the single-cell level, *Nature* **514**, 376 EP (2014).
64. M. Campos, I. V. Surovtsev, S. Kato, A. Paintdakhi, B. Beltran, S. E. Ebmeier and C. Jacobs-Wagner, A constant size extension drives bacterial cell size homeostasis, *Cell* **159**, 1433 (2019/05/17 2014).
65. A. Amir, Cell size regulation in bacteria, *Phys. Rev. Lett.* **112**, p. 208102 (May 2014).
66. S. Jun and S. Taheri-Araghi, Cell-size maintenance: universal strategy revealed, *Trends in Microbiol.* **23**, 4 (2019/05/17 2015).
67. S. Jun, F. Si, R. Pugatch and M. Scott, Fundamental principles in bacterial physiology—history, recent progress, and the future with focus on cell size control: a review, *Rep. Progr. Phys.* **81**, p. 056601 (2018).
68. K. M. Schmoller, J. J. Turner, M. Kõivomägi and J. M. Skotheim, Dilution of the cell cycle inhibitor whi5 controls budding-yeast cell size, *Nature* **526**, 268 EP (2015).

Prepared comments

S. Eaton: How do Biological Systems Cope with Temperature?

The question I would like to bring up today is the fascinating question of how, in particular cold-blooded, animals cope with changes in temperature. In very extreme temperatures they can enter dormant states, so simply shut down. But even more interesting is their ability to develop and reproduce over a very wide range of temperatures.[a] Why is that surprising? Because for one thing we know that the physical properties of matter, including living matter, change with temperature. For example, raising temperature increases the creep compliance of cells — that is in the upper corner of the slide (see Figure 1). Thinking about what we heard in the last session, we know that temperature strongly affects phase transitions both in membranes and the cytoplasm and this is really important for the dynamic spatial organization of cells.

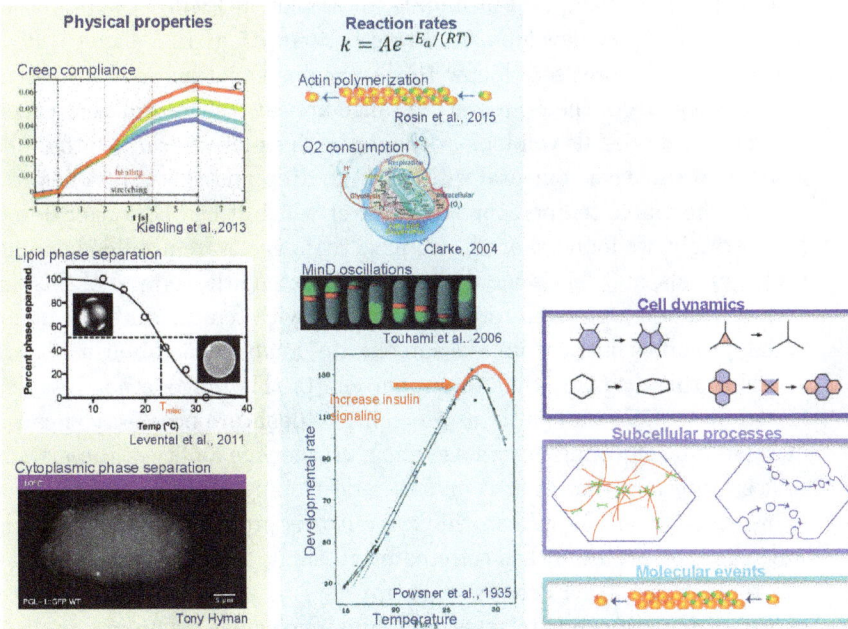

Fig. 1.   The slide presented by S. Eaton at the conference.

[a]Brankatschk M., Gutmann T., Grzybek M., Brankatschk B., Coskun U. and Suzanne Eaton, A temperature-dependent shift in dietary preference alters the viable temperature range of Drosophila, *bioRxiv* 059923, 2016/1/1; Brankatschk M. *et al.*, A temperature dependent switch in feeding preference improves drosophila development and survival in the cold, *Developmental Cell* **46**(6):781–793.e4 (2018).

I do not want to so much focus on these problems, which are big problems, but more on the direct effect of temperature on chemical reaction rates. So temperature exponentially increases chemical reaction rates in a way that depends on the activation energy of that reaction, in a way that is described by the Arrhenius equation. If you look at actin polymerization in a test tube, it is temperature dependent and it follows this equation (see Figure 1). Somehow even more surprising is the fact that quite complex biological processes that depend on networks of reactions also qualitatively follow this kind of relationship. For example, oxygen consumption in plants and animals seems to follow this and different kinds of cellular oscillations, including for example MinD oscillations in bacteria.[b] So this raises a couple of really interesting questions. One, how the temperature dependence of these reaction networks emerges from the temperature dependencies of all the different reactions that constitute them. But then it also poses the problem: how does an organism actually ensure that the many different networks it is made up of actually respond in the same way to temperature. And we do know that even at an organismal level temperature can speed things up. For example, the middle lower plot in Figure 1 shows a four-fold increase of the developmental rate between 15 and 27 degrees. It does not exactly and precisely follow the Arrhenius equation or if it does then different regions of the temperature range appear to have different effective activation energies. If you look at the top of this plot you can see that above 27 degrees the developmental rate starts to drop and this is just before the edge of the viable temperature range after which it can no longer develop. Interestingly, we found that all you have to do is elevate insulin signaling in these animals, and this is enough to actually extend the top end of this range over which developmental rate can speed up with temperature. Insulin signaling promotes nutrient uptake and elevates anabolic metabolism. So what we think this means is that the upper end of this range is not just determined by the fact that proteins are going to denature because you can just increase it by elevating insulin signaling. Rather the ability of these animals to keep reactions coordinated and keep developing faster with temperature is just not trivial. So we can think about this problem in several different ways. Maybe evolution has selected every single molecular reaction in the cell to have the same activation energy and to respond similarly to temperature. Or you can think that individual molecular reactions can diverge with temperature but that they are connected in networks by checkpoints and feedbacks that somehow make the output of the whole sensible. But then you need to come back and ask how an animal would ensure that its different networks respond in the same range of temperature.

---

[b] Jacob Halatek, Erwin Frey, Highly canalized MinD transfer and MinE sequestration explain the origin of robust MinCDE-protein dynamics, *Cell Reports* **1**(6):741–752 (2012).

Let me finish this up quickly by saying that I think that now with the quantitative approach we are developing in developmental biology we can get at these questions by looking essentially at developmental rates at multiple scales. For example, rates of tissue shape change, we can quantify rates of tissue shape change, we can decompose them into the rate of shape change conferred by different cell dynamics. And then at levels underlying that at the rates of different sub-cellular processes that contribute to this cell dynamics and finally all the way to the rates of molecular reactions in a test tube. I think that this will eventually lead to answering this interesting unsolved problem.

U. Alon: Bacterial Growth Laws out of Steady State[c]

When a complex phenomenon shows simple patterns, theoretical research can make progress. A prime example of such simple patterns in biology are bacterial growth laws. Despite the fact that bacterial growth depends on thousands of reactions, certain cell components show nearly linear relations with the growth rate. These linear relations hold in different growth conditions. For example, the fraction of ribosomes in the proteome goes as $R = a + b\, mu$, where mu is the growth rate. These growth laws helped to form theories of resource allocation in bacteria, and to quantitatively explain diverse physiological processes.

Growth laws so far were measured in bacteria growing exponentially for many generations. This steady-state situation is relevant to only some natural circumstances. Often, bacteria face changing conditions in which growth rate changes. Thus, it is important to test the possibility of growth laws out of steady state. As in physics, testing systems out of steady state can reveal new aspects of their dynamics and structure.

An out-of-steady-state situation which may be simple enough to understand is a step-like nutritional change: an improvement called upshift or a downgrade called downshift. Bacteria have been known for decades to change their growth rate within minutes following such shifts. This means that ribosomes and transport systems must have spare capacity — otherwise changes in growth rate would take far longer as new ribosomes and pumps are made. This spare capacity cannot be revealed by steady-state measurements alone — indeed, many theories that used the steady-state growth laws assumed that ribosomes work at full capacity.

Spare capacity means that cells are not optimized for instantaneous maximal growth rate. This opens the question what multi-objective optimization is at play. There are at least two possibilities: spare capacity can provide an advantage when conditions change often enough, because they increase growth rate immediately after the change. Spare capacity can also

---

[c]E. Metzl-Raz *et al.*, *eLife* **6**:e28034 (2017); M. Mori *et al.*, *Nature Commun.* **8**, 1225 (2017).

help avoid toxic levels of intermediates following an upshift that occurs when capacity is full.

It would be interesting to explore growth laws out of steady state, in order to reveal additional attributes of cell resource allocation and additional objectives at play during evolution.

J. Skotheim: On the Biosynthetic Mechanisms Linking Cell Growth and Division

My prepared remark relates to Terry Hwa's grand challenge number three "To what extent is the phenomenological theory of bacterial growth applicable to describe the growth of higher eukaryotic cells?"

But, before I go on to my remark I'd like to object to the term 'higher eukaryotic cell' from the standpoint of cell growth and nearly all cell biological functions 'lower' eukaryotic cells, such as yeast, perform just as well, or perhaps even better. Certainly, they can grow faster. And, studies in yeast have more often than not been in the vanguard of cell biological and genetic studies found later to also hold true for animal cells. Therefore, I would suggest, for the purposes of generalizing the phenomenological or physiological understanding of growth that Terry and others have made so much progress on bacteria, we ought to first look at yeast. The main advantage of studying the 'higher' eukaryotic cells for cell biological purposes is that they are 'larger' and therefore more amenable to microscopy studies. I suggest for the remainder of the meeting we term them 'larger' eukaryotes.

In any case, the key question that we have been investigating in my group is the question of how cell growth triggers cell division — which we study in both smaller and larger eukaryotic cells. It is known that for a large variety of eukaryotic cells there is a growth requirement, meaning that in order for a cell to divide, it needs to grow a specified amount first. What we want to understand is how that works? What is biochemically different about this larger cell than this smaller cell that then results in triggering division? This is where I think our work may intersect the phenomenological work on bacterial growth.

What we found in yeast was that larger cells triggered division because of a differential size-dependency of the expression of certain genes. In general, larger cells are able to make more protein faster and to grow faster than smaller cells because they simply have more biosynthetic machinery. They have more RNA polymerase, and more ribosomes, and, in many cases, make protein at a rate in direct proportion to their size.

Importantly, cell division activators follow the general rule that larger cells make more. On the other hand, important cell division inhibitors do not. We found that large and small cells make the same amount of division inhibitors, in yeast, this includes the important division inhibitor Whi5. Thus, larger cells have proportionally more division activators than division inhibitors and readily divide, while smaller cells have proportionally more inhibitors than activators, and do not divide. Instead, they grow for a while before reaching the size of the larger cells and then dividing.

Thus, the central question facing us now is to identify and understand the mechanisms through which this differential size dependency in protein synthesis happens. How do some genes take advantage of the increased biosynthetic capacity of larger cells, while others do not? What happens when you change cell growth rate? Is that the same as changing the biosynthetic capacity of the cell? Are the same genes limited, or are there different genes?

This is where I think a phenomenological understanding that Terry hinted at in his remarks could really help. Especially if the theory of bacterial growth were developed to be more granular and to account for what growth and synthesis phenomenologies exist at the level of single genes.

In any case, it is clear to me now that we really need a much better quantitative understanding of the central dogma of biology. We need a much better quantitative understanding of protein synthesis and how this interacts with the fundamental mode of cell geometry, which is cell size.

R. Phillips: Allostery and the Molecular Switch: Supporting Actors with Leading Roles

Adaptation to the world around us is one of the signature features of living organisms of all types, from humans to the bacteria that colonize us. A central conduit to such adaptation is provided by molecular events in which receptors, for example, detect some external signal which results in a variety of signaling cascades and concomitant cellular responses. Our everyday experience reveals this familiar phenomenon when changes in the light level in our surroundings leads to adaptations in our eyes that make it possible for us to see in conditions with light intensities differing by more than ten orders of magnitude.

At the single-molecule level, perhaps the most ubiquitous mechanism of implementing responses to the external world comes in the form of allosteric proteins, molecules that change their structure between inactive and active conformations depending upon whether there is a specific regulatory ligand present or not. For example, in the case of "light as ligand", the allosteric molecule that begins the visual signal cascade is rhodopsin while in the classic case of bacterial chemotaxis, it is the chemoreceptors themselves that respond allosterically to the presence of chemoattractants. Already in the 1960s, it was understood how to turn this broadly acting mechanism into statistical mechanical language that is easily recognized by physicists and that reveals a great unity to many different biological phenomena ranging from oxygen transport by hemoglobin to the action of ion channels in our muscles to the signaling behind bacterial quorum sensing, all of which are mediated by allosteric proteins.[d,e]

One of the great challenges we face however in this era of high-throughput experimentation is how the various signaling and regulatory networks are linked in turn to the ligands that control them.[f] It is now common-place to see diagrams of the regulatory networks that control metabolism in microbes or embryonic development in multicellular organisms that give the impression that we understand the "wiring diagram" of these processes. And yet, even for what many consider biology's best understood organism (*E. coli*), we are still completely ignorant of how more than half of its genes are regulated, with the question of how the regulatory networks are controlled in turn by their corresponding regulatory ligands being even more enigmatic. Even in those reasonably well understood cases such as the famed wiring diagram for the control of early development in the sea urchin,[g] there is a huge gap in our understanding of these regulatory networks because many of the arrows linking the various nodes are not static and are instead dynamically controlled by a deluge of small molecule regulators. Though at first cut these regulatory diagrams make it seem as though regulatory ligands have supporting roles, in fact, by way of contrast, much of what makes living organisms what they are is their ability to react to the external world, and it is often through the actions of these poorly known molecular mediators that these reactions occur. In the context of the 2017 Solvay Congress, my "prepared remarks" call for a predictive and quantitative theory-experiment dialogue both at the level of individual allosteric molecules and the entirety of the networks they are part of. This effort requires not only "fact-based" discovery of the "allosterome" (the suite of regulatory ligands and the allosteric molecules they regulate), but also of the full quantitative dissection of how specific molecules and pathways give rise to both physiological and evolutionary adaptation.[h]

This call to action in the specific case of signaling, regulation and adaptation is part of a larger appeal for the future involvement of physics with our study of the living world. Writ large, the time has come for an increasingly demanding role for predictive theory in our study of living organisms. In his autobiography, Darwin famously spoke of mathematics serving as a

---

[d]Monod, J., Wyman, J. and Changeux, J. P., On the nature of allosteric transitions: a plausible model, *J. Mol. Biol.* **12**, 88–118 (1965).

[e]Kirschner, M. and Gerhart, J., *Cells, Embryos and Evolution*, (Blackwell Science, Inc., 1997).

[f]Lindsley, J. E. and Rutter, J., Whence commeth the allosterome? *Proc. Natl. Acad. Sci.* **103**, 10533–10535 (2006).

[g]Davidson, E. H. *et al.*, A genomic regulatory network for development. *Science* **295**(5560):1669–78 (2002).

[h]Milo, R., Hou, J. H., Springer, M., Brenner, M. P., and Kirschner, M. W., The relationship between evolutionary and physiological variation in hemoglobin, *Proc. Natl. Acad. Sci.* **104**, 16998–17003 (2007).

scientific sixth sense, an idea that has been a truism in physics for centuries. For example, after Newton's formulation of the law of universal gravitation, more than a hundred years of effort was put into drilling down into the rigorous applicability of this law to the solar system with figures such as Euler, Clairaut and d'Alembert focusing on the Sun-Earth-Moon system and luminaries such as Lagrange and Laplace focusing on the stability of the solar system with special reference to the orbits of Jupiter and Saturn.[i] The goal of this work was a definitive and rigorous demonstration of the validity of the law of universal gravitation to the solar system as a whole and it is this style of thinking that physics offers in biology, despite the protestations of many that "biology is different" and hence defies such quantitative dissection. The study of allostery provides a preliminary but incomplete glimpse of this kind of theory-experiment dialogue in action.

My biggest hope for the future of biology is that the style of thinking familiar with physics in which there is a rich interplay between theory and experiment characterized by committed experimental programs designed specifically to test theoretical ideas will lead in the coming century to the rise of a principled description of the living world. Further, one of the great gifts that biology can give to physics is to force us to deepen our understanding of non-equilibrium phenomena and how to describe them, making this a truly wonderful time for scientists of all types to be studying the seductive and mysterious world of living organisms.

Discussion

D. Fisher So we heard a lot about evolution and some part of that includes function but also proteins evolving on long time scales in ways which we would think they were very conserved, like residues, pairs of residues that are positive or negative and people use that to get structure and so on. Why do they have to change at all? If it happens at a pair level, why does the whole pair change? Somehow the proteins are changing under evolution in a way that it is not clear what the selection is, what makes them change in that way. And one of the things that people have thought about — but I do not know what they have done — is whether or not there is a primary selection on the proteins not interacting with others. And really every protein is always a little bit under selection, both in avoiding interactions with other proteins and then some other protein changes in some functional ways but everything seems to adjust. So the real question is whether there is an overall pushing towards trying not to interfere with each other too much and often one sees what people call neutral evolution in protein is really

---

[i]Laskar, J., *Is the Solar System Stable*, arXiv:1209.5996v1 (2012).

associated with that.

J. Lippincott-Schwartz I just want to pursue Terence's really interesting question related to bacteria that are going to stationary phase. Presumably in the system that you are modeling, there is no degradation or turnover of the ribosomal system. Presumably when bacteria are going to stationary phase you are now having ribosome turnover at some level and I think that's important because in the eukaryotic kingdom, ribosomes rapidly turn over and it is absolutely critical for the eukaryotic cell to be able to adapt to starvation. The ribosomes are huge pools of free amino acids and there are in fact two different translational pathways in eukaryotic cells, one that is tuned to anabolic growth versus catabolism. So you could tell us how you might be thinking about integrating that in your model.

T. Hwa Basically, if you switch on your genes suddenly for bacteria, then of course you still observe protein synthesis and quite some comes from ribosomes. Qualitatively this was already known in the 60s. But quantitatively it is very difficult to characterize because that turnover production is only a tiny fraction. So the problem to understand stationary phase is a methodological problem. You can use special methods to try to pull out the newly synthesized proteins but when you talk about such a small fraction it is not working very well. A lot of people study stationary phase by looking at which genes are turned on, which genes are turned off, using RNA-seq. But this means nothing. I mean every RNA is turning over quickly and there is only a subset turning into protein and we have no idea about the rules under which certain RNA is turning into proteins. This is at the molecular mechanism level. These technical questions will be solved certainly in 5–10 years. But more fundamentally, if we want to develop a quantitative understanding of stationary phase, then we need to say something about survival and death. To me that's really fundamental, but because it is a stochastic process, what does it mean? When can a cell no longer be reviving to growth? That's a much harder question. This is a long-term job.

E. Siggia Firstly I respond to Daniel's question. Wendell Lim has data showing that signaling pathways during evolution do not bind things and bind other things. A positive comment vis-a-vis the work of Suzanne Eaton: The problem of temperature compensation is of course a defining feature of the circadian cycle, that is to say the period should be 24h, independent of the temperature, yet the circadian cycle has to be temperature-entrainable, so it cannot ignore temperature. There is some theory for how the cell is capable of that based on just other phenomenological things that the circadian cycle has to do and then some recent experiments from Michael Young's laboratory and collaborators who showed for Drosophila that this is actually realized. But this is an enduring question, certainly.

J. Howard I would like to ask a question to Rob and maybe related to other talks as well. The allosteric system he wrote down is sort of statistical mechanical equilibrium equation. It is remarkable to me that this level of control would be approachable from a thermodynamic equilibrium point of view, considering that it is controlling something which is so much more dissipative. I wonder why does it work so well that formalism.

R. Philips The best answer is: I don't know. First thing is separation of time scales. I am not going to claim that it is universally valid. Think about ion-gated channels. The binding and unbinding of those ligands is so fast compared to the time that they open and close, they are effectively at equilibrium because of that separation of time scales. Another interesting thing that I do not fully understand, is that you can write down the chemical master equations for a lot of these things, solve the probability distribution, compute things and you end up getting more or less the same mathematical structure but with different effective parameters. I find that really fascinating. But why the allosteric model works so well, I do not know about that one, but the same model is used beautifully for chemotaxis and quorum sensing and I think it is separation of time scales that explains why it is ok.

A. Murray I may just amplify that by saying that one of the things we understand least well are the things at the most immediate level of control. The slowest thing is gene expression. It is easy for us to monitor that because of the ability of nucleic acids to hybridize each other. We do somewhat a worse job looking at changes in the modification of proteins. We have less general methods but the methods are the same. And what we do by far the worst on is measuring the fluxes of molecules through things like allostery and almost certainly because they are the most abundant molecules and therefore stochasticity is the least and it is by far the most controllable system. There is a huge lacuna in our knowledge about how biology works.

A. Hyman Just coming back to the point on temperature compensation. It is true that the circadian clock has to respond to temperature variation in cold-blood animals like Drosophila, but in the mammalian cells there is no reason to do that. Does it get a huge advantage in terms of its ability to be a circadian clock?

E. Siggia I think my friends told me that some of the synchronization among the organs in mammals comes about from temperature fluctuation. They are indeed very entrainable by temperature.

A. Hyman My general question is when you move towards warm-blood animals, does it give an incredibly evolutionary advantage, because the protein could start to evolve and I was wondering about its ability to operate in 15–20 degrees temperature.

H. Goodson Just a quick thing to follow up to Rob. That's a beautiful example of the type of convergent evolution I was trying to suggest in terms of predictability in cell biology.

W. Bialek I also want to pick up on Rob's talk, in response to this issue on why equilibrium models work. It gets back to what we talked in the morning. You can also think about the example of ion channels, the voltage-gated ion channels. The open and closed states are at equilibrium with each other despite the fact that there is current flowing through the channel which is dissipating an enormous amount of power. This is not true I guess for most enzymes. If they are busily catalyzing some reactions and if you ask about the population of different states, that's affected by whether the reactions they are catalyzing are near or far from equilibrium. But apparently the ion channel is insulating itself so that open means that the current can pass but the power that is then dissipated does not feed back. And so it suggests that equilibrium is not even an issue of whether the system is near at equilibrium or is affected by the driving force. It is something that has evolved too, somehow. In different systems the answer actually comes out differently, presumably for some good reason.

D. Fisher How do you know it is an equilibrium, rather than just a stable steady state? What do you mean by equilibrium?

W. Bialek Take away the ions that carry the current but measure the gating charges so you know whether it is open or closed. You can count which state the channel is in independently of whether the current is flowing.

S. Quake Just to return to Tony and Suzanne question about temperature change. I just want to point out that some years ago Rustem Ismagilov did some experiments with Drosophila embryos and they were able to have different halves of the embryo at different temperatures and follow the developmental clock change on slower side and things became distorted. So the temperature still matters and if you create gradients, it really mixes the things up.

A. Murray Perhaps I can ask a question of the audience, just for a second. Eugene's question strikes me as a very interesting question and I wonder if someone wants to comment in response to that. I want to make one preliminary comment. If you are looking at the evolutionary history of the catalogue of genes in organisms, genes and what we call pathways appear and disappear as a coherent unit. So in that evolutionary sense, it may not completely be a mindless abstraction of humans but it would be interesting to know if there are other comments on this topic.

E. Koonin It is actually not entirely correct what you just stated. When you look empirically at gain and loss of genes, at least in prokaryotic genomes, pathways are not particularly coherent, which is maybe explained by communal biochemistry and common goods. So one needs to be extremely careful about that, and coming back to Daniel's comment, which was not about pathways but completely pertinent. I think a key issue in many aspects of biology is their relevant level of hierarchy or, speaking in genetic terms, relevant level of epistasis. How does the strength of intra-gene epistasis relate to the strength of inter-gene epistasis, etc.

H. Levine The part of the reason we all think of pathways is that it is at least possible to imagine how those evolved. You start writing out something very simple and then you accrete other elements that make the picture very blurry from the point of view of what we see. But at least we can imagine someway in which each accretion did not dramatically change the basic function. When you start the other way, when you say the whole structure, the whole network is doing these complicated things, it is very hard to see how that emerged. I think that's the excuse we have. That does not mean that it is true but that is at least the excuse that we can use when we do that.

P. Rainey I would like to make a general comment that is relevant to some ideas expressed this morning and also relevant to this discussion. The idea that pathways, organelles, cells and so forth might appear to lack coherency, or might be better "designed", but there is no particular reason that pathways and so on should be perfect from an engineering point of view. Natural selection optimises reproductive success and that might be achieved via the evolution of pathways that perhaps do not make a lot of sense from a strictly engineering perspective.

N. King I would like to come back to the question of pathways. I should say I have a second rate background in both chemistry and physics but I have always thought of these being peaks of activity from noisy proteins and I like the idea that some enzymes are promiscuous and I suspect that these proteins are primarily doing one thing but there are a lot of other things they do poorly and transcription factors is one example. The biological consequences of knocking out a particular transcription factor can often be defined in clear, scorable, quantitative traits but now we know these transcription factors are setting down all over the genome in ways that probably do not matter. And getting back to the question of gene evolution, maybe you have done these yourself but for instance, if we look across all animals we find only 37 genes that are conserved in all animals. So that means that is evolutionarily defined groups of organisms that have traits that are unifying are built upon genomes that are very dynamic and have clearly a lot of redundancy and pleiotropy. And when it comes down to about thinking what is a pathway, I think that it is interesting.

J. Howard For me what the important thing is how does one answer this question, how does one actually define a pathway. If you have the interaction matrix of all genes or all proteins or whatever, how do you interpret that? What kind of measure do you have on that to say this is a pathway or nearly a pathway or whatever? Measurement is important.

E. Koonin A very quick return to this. Perhaps different approaches might be workable here. If you take the interaction matrix, there are specific algorithms to detect communities or modules in such matrices, that might be more relevant, at least a little more relevant than arbitrary definitions of pathways.

M. Elowitz Staying on the pathway question, I think when you go to developmental signalling pathways, things like Bmp and Notch and Wnt, there is an issue of cross-talk which is to say, are these pathways really pathways or are they all interacting with each other. One important issue is time scale, in other words, many of these pathways — if there are pathways — activate ligands for other pathways. So when we use techniques that measure only after very long times we can see one enzyme activating others but it could be that on faster time scales they are more tightly defined.

U. Alon The question of how pathways and modules evolve actually presents us with a challenge because when you try to evolve a system on a computer with some inputs and outputs you get a distributed network that distributes the function among components of the network and you cannot understand how that works. This is how neural network deep learning works. It can learn but it is very difficult to see how it works. So you can ask how it is that modules and pathways evolve inside a distributed system and there has been some progress on that in the last 10 years. One way you can get modules is by driving the system crazy by letting it evolve for thousands of generations for one problem and then switching for thousands of generations for another problem and then back for thousands of generations for the first problem and that doesn't work unless those problems share sub-problems. For a cell, eating, moving, killing itself, etc are shared sub-problems, each time with a different set of signals and combinations. And if you do that you get evolvable networks that have modules. Each module does one of the functions and can rapidly rewire each other to meet the new circumstances. Also when you have cross-rewiring like in the brain you also get those modules. There have been advances in understanding evolution, in fact along the historic line over millions of years we have learned many different environments but they have shared sub-problems and that gives the basis for why there are pathways and modules and without that, it is extremely difficult for human being to fully understand the cell.

M. Desai One thing that, I think, is implicit to a lot of discussions, for instance in Terry's comments and in many others, is that evolution acts as a sort of reason why we can think about optimality in many of these systems and in particularly in regulatory systems and also, as we were talking about, constrains how pathways and modules can be organized. And I wonder if it is worth also thinking about ways in which evolution can act in opposition to optimality in the sense, for example, if we think about regulation as a response in a short time scale to environmental changes and often it seems like evolution should act in a way which makes these kinds of regulatory responses less optimal depending on the time scales of change that organisms are facing where long time scale changes and environmental challenges can drive systems through evolution to places where regulatory responses over short times scales are less optimal and less well organized.

S. Eaton I am also going to the question of optimization. It is interesting to think about proteins that have moonlighting functions, so for example, one of the subunits of ATP synthase, which is also, in the cytoplasmic membrane, a lipoprotein receptor, which seems to be unrelated to its function. Or metabolic enzymes like GAPDH that are important in glycolysis but they do completely unrelated things. It seems too easy to solve these problems by duplication and divergence — why does this hang around and then also what does it imply about optimizing one function over another function. Can you optimize both at the same time?

A. Murray It is now time for coffee, but I will exercise chairman's dubious prerogative of making one last remark which is one possible response to what Uri was talking about. One of the things that preoccupies many people, is how complicated and messy eukaryotic wiring is compared to bacterial wiring. And particularly for bigger eukaryotes and so on, one possibility is what bigger eukaryotes have done is manage to keep selective conditions more or less the same because they can move around and change what they eat and therefore they have evolved neural nets as responses which makes it impossible for us to understand them even though we are them. And with that, coffee!

Rapporteur Talk by Michael B. Elowitz: Cells as Devices: information processing by biological circuits

Living cells are extraordinary devices: they proliferate, sense, communicate, remember, compute, and develop into complex tissues and organisms. These capabilities arise from molecular circuits operating within cells. A central challenge is that even when we know key components and interactions within these circuits, it is still difficult or impossible to predict their response to perturbations. More fundamentally, we lack a conceptual understanding of the overall design logic of most genetic circuits. That is, we don't understand why their components are wired together in specific ways. These issues are not only critical for explaining and predicting cell behaviors, but increasingly urgent for opening up the ability to modify cellular functions and program entirely new ones using synthetic biology approaches.

1. Introduction

Living cells can be analyzed from different perspectives. On the one hand, they are built out of molecules, which interact according to the laws of physics and chemistry. On the other hand, cells can equally well be thought of as complex molecular 'devices' that execute a diverse repertoire of specific programs. This view evokes ideas from engineering and computer science and stresses the programmability and information processing roles of cells and their molecular components. There is no fundamental conflict between these views, both are essential for a complete understanding of the cell, and both form critical components of the Physics of Living Matter, the topic we are discussing at this Solvay conference. However, because Living Matter is distinguished from ordinary materials in large part by the internal information contained in its structures and the programmability of its dynamics, the device view of the cell provides a useful and powerful framework for addressing fundamental biological questions.

The view of the cell as a programmable device is not new. We've known for decades that cells use circuits, or 'pathways', of interacting genes and proteins to control their own behavior and coordinate with other cells in multicellular contexts. Within the cell, they process, transmit, and store information. They act as modules that actuate complex processes such as cell division or death. They enable immune functions that eliminate disease. And, perhaps most astonishingly, they enable single cells to generate complex, precisely organized multicellular embryos and organisms. While our interest here is fundamental, there is also an ulterior motive: the ability to program useful new behaviors in cells. These synthetic biology approaches should enable us to use cells as therapeutic devices that respond to specific conditions within a host, and coordinate with other engineered or endogenous cells to cure disease in ways that no drug, no matter how specific it is for its molecular target, could do. Achieving this vision will both require and contribute to understanding the fundamentals of cellular circuit design.[1]

However, despite enormous progress, the art and science of gene circuit design remain nascent. Even when we know the molecular components and biochemical interactions that comprise cellular circuits, our ability to answer basic questions about their underlying design remains limited. We often cannot say what function a given circuit performs, why it uses a particular circuit architecture, and how it will respond to perturbations. And we similarly lack the ability to rationally and predictively program new synthetic circuits from scratch.

Here, I would like to start from the view of the cell as a programmable device, and ask to what extent the design principles of cellular programs resemble, or differ from, the more familiar paradigms of electronics. Are cells simply encoding familiar algorithms in a new material substrate, or are they using different kinds of programs altogether?

Electronic circuit design depends on multiple levels of physics, from the underlying condensed matter physics of semiconductors to the higher level physical principles of computation and information theory. While living cells implement their circuitry with genes and proteins rather than silicon, cells, like electronic devices, are based on programmed dynamical behaviors, internal states, stochastic noise, and sophisticated information processing capabilities. As a result, many of the physical principles of electronic circuit design apply directly to biological circuits. At the same time, it is equally true that living matter has evolved to perform very different functions than electronics, such as self-replication and multicellular development, and it does so using different circuit architectures that require new theory and physical principles to understand. The application of existing theory will not be sufficient. To make progress, we will need to allow for the discovery of new design principles specific to biology.

This Solvay conference report provides an opportunity to consider more *specifically biological* principles of circuit design.[2,3] Because the space of genetic circuitry is vast, I will describe a few emerging biological circuit design principles that differ fundamentally from those we are most familiar with in other fields and therefore exemplify the kinds of new paradigms that we need to develop. I will discuss examples where biological systems work with component combinations rather than individual signals, control processes in time rather than through concentrations or levels, and regulate the fraction of cells that respond rather than the level of response in each individual cell. In each of these paradigms, it increasingly appears not only that the biological circuit design works differently than we would have imagined or designed, but also that there are principles that we can begin to understand in the natural context and potentially apply to synthetic circuit design in the future.

2. Combinatorics and Computation in Cellular Communication Systems

Many genetic circuits utilize multiple seemingly redundant components. Transcription factors, chromatin regulators, and signaling proteins typically occur in families of similar, but non-identical, proteins that operate in concert. Just as neural information is distributed across many neurons, cellular communication and signal processing systems may distribute information and regulation across these sets of closely related, co-expressed, and interacting components. An appealing idea is that cells utilize these factors combinatorially, representing states as "vectors" of concentrations of different proteins. Understanding the design principles of combinatorial systems could help us to make sense of pathways that otherwise seem bafflingly, and unnecessarily, complex.

Most developmental communication pathways use multiple, partly redundant ligands and receptors in the same process. For example, in the Bone Morphogenetic Protein (BMP) signaling pathway, $\sim$15–20 different ligands activate a combinatorial set of heteromeric receptors, each composed of 2 type I and 2 type II subunits. Those subunits in turn come from a set of $\sim$7 different type I and type II receptor proteins. It is not clear whether or how the cell can retain any information about which ligand contacted which receptor, since diverse signaling complexes activate the same or similar downstream targets. The prevalence of similar "promiscuous ligand-receptor" architectures[4] across pathways such as FGF, Wnt, Notch, and others provokes the question of why cells would evolve a system with so much apparent

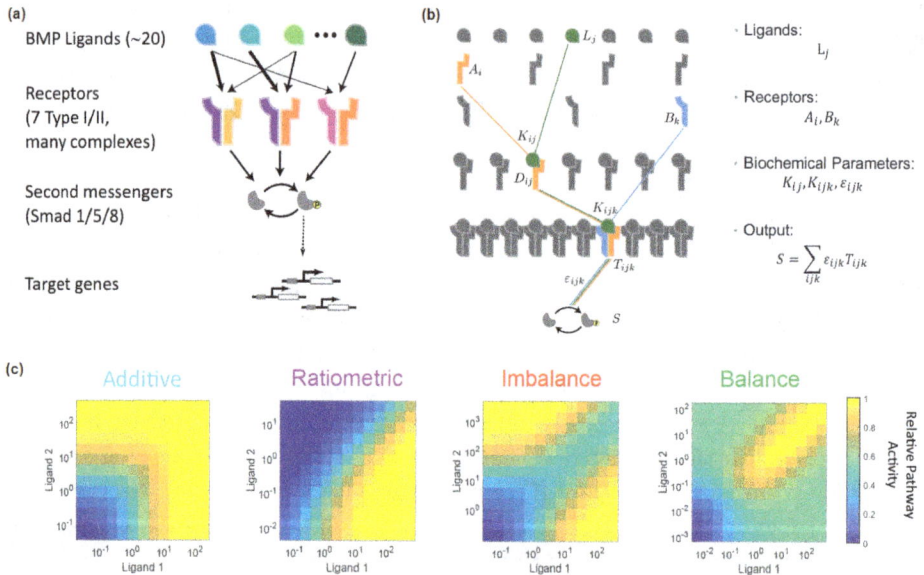

Fig. 1.   Promiscuous ligand-receptor interactions compute complex functions of multi-ligand inputs. (a) In promiscuous ligand-receptor architectures, a family of ligands (top) can each interact with multiple receptor variants (middle). The resulting ligand-receptor complexes activate downstream effectors, such as Smad1/5/8 for the BMP pathway by phosphorylation. These regulators in turn activate downstream target genes. (b) Simple mathematical models can represent formation of competing ligand-receptor complexes, each of which can have a distinct specific activities. (c) This simple architecture can compute a variety of functions of ligand combinations, including the examples shown here. Each matrix represents relative activity of the pathway in response to the indicated combination of two ligands. Different functions result from different choices of the biochemical parameters, e.g. K_{ijk} and ε_{ijk}, in the model.

redundancy rather than a seemingly simpler architecture based on orthogonal inter-actions between specific ligand-receptor pairs. In recent work we quantitatively ana-lyzed the response of the BMP pathway across two-dimensional ligand concentration spaces.[3] These simple measurements revealed that the promiscuous architecture can perform computations on inputs of ligand combinations (Figure 1). These functions range from trivial additivity, in which pathway output is proportional to the sum of the concentrations of two ligands, to more complex functions such as 'imbalance detectors' that respond strongly when the ratio of two ligands is either very high or very low, but respond weakly at an intermediate ligand concentration ratio.

Mathematical modeling revealed that these functions arise because the parame-ters that control the concentrations of different signaling complexes (i.e. affinities) are independent of the parameters that control the enzymatic activity of the result-ing complexes. The result is a kind of biochemical computer that uses protein-protein interactions to directly compute complex functions of the signals it is sens-ing, without requiring downstream transcriptional networks. In other words, the detection of a signal and the processing of that signal occur in a single process. More generally, these results suggest that promiscuous protein systems could be providing computational functions in a manner loosely analogous to the way simple computational neural networks process information through weighted connections among nodes.

3. Epigenetic Regulation and Probabilistic Control of Cell States

Multicellular organisms must control the distribution of cell types or states: the abundances of different immune cell types in the blood, the relative proportions of absorptive and secretory cells in the intestinal crypt, and so on. This points to another fundamental difference between biological circuits and their electronic coun-terparts: cells often translate continuous instructional inputs, such as the concentra-tions of signaling molecules, into probabilistic rather than deterministic responses. They can thereby be thought of as controlling the fraction of cells in an otherwise equivalent population that take on a particular fate.

One example of such probabilistic, or 'fractional' control, occurs in the bacteria *Bacillus subtilis*, where signaling and nutritional inputs control the fraction of cells that activate a genetic competence pathway allowing them to take up DNA from other cells. The underlying control circuitry uses principles of excitability and takes advantage of stochastic 'noise' in the expression of cellular components.[5,6] Similarly, cells can also stochastically switch between a repertoire of different states controlled by sigma factors (specialized transcription factors) which activate in a stochastic, but coordinated, manner.[7] Stochastic state-switching strategies have been suggested to provide 'bet-hedging' functions, which allow populations to effectively anticipate potential future conditions, and distribute distinct or incompatible functions across a population of cells.[8]

Probabilistic, fractional control also occurs in multicellular organisms, but is

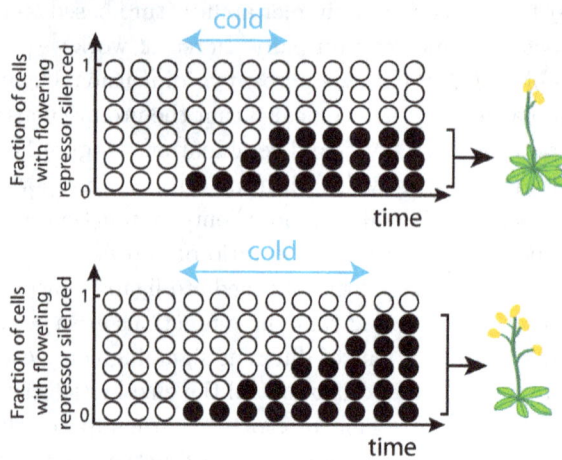

Fig. 2. Vernalization is a cold-induced fractional control process. Cold temperatures cause a monotonic increase in the fraction of cells in which the FLC repressor of flowering is silenced (black circles). This fraction in turn controls the amount of flowering, which more silencing leading to more flowering. FLC silencing occurs through the PRC2 (Polycomb 2) chromatin modification system.

implemented quite differently. Beautiful studies from the labs of Martin Howard, Caroline Dean and others have elucidated the system that plants use to control flowering in response to cold temperature, a process called vernalization.[9,10] In this system, periods of cold temperature increase the fraction of cells in the organism in which the FLC gene, which encodes a repressor of flowering, is epigenetically silenced. Critically, FLC is silenced in a probabilistic, all-or-none fashion at a relatively slow rate, such that the fraction of cells in which FLC is silenced accumulates over extended periods of weeks to months. Silencing occurs through histone methylation by the PRC2 polycomb system. Once silenced, the FLC gene remains silenced as the cell proliferates. Thus, the integrated amount of cold is effectively recorded in the monotonically increased proportion of silenced cells. Later, this cell fraction will control the amount of flowering (Figure 2). The plant thus converts an analog input (temperature over time) into a digital memory (the number of silenced FLC alleles) and then later converts it back to an analog output. Analog-digital conversion is of course familiar in electronics. But the principle of representing digital information in the size of different cell subpopulations, and the way in which this is implemented molecularly, are more uniquely biological. More work will be necessary to understand how analog and digital representations are used in other biological contexts, and to develop the physical theory that will allow us to understand how analog-digital conversions play out in more complex contexts.

The fundamental role of epigenetic regulation in vernalization provokes the broader questions of what functional roles epigenetic regulatory pathways provide for cellular natural or synthetic circuits. Chromatin regulators such as DNA methyltransferases and histone methylases can modify histones or DNA bases near a gene, affecting their expression state, sometimes in a heritable manner. Chromatin regulation involves a large family of different modifications and protein components that read, write, and erase them. As with signaling pathways, much is known about the molecular componentry and biochemical interactions within these pathways, but the functional capabilities they offer often remain more obscure. In particular, it has remained largely unclear why so many different modifications of corresponding readers, writers, and erasers are necessary.

Recently, dynamic single cell studies and the use of synthetic constructs that enable direct control of the recruitment of chromatin regulatory factors, also known as synthetic epigenetics, has begun to enable bottom up methods for addressing these questions.[11] These studies have shown that recruitment of diverse chromatin regulators to a target promoter can cause silencing to accumulate slowly across a population but in an all-or-none manner at the single cell level, similar to what happens naturally in the vernalization system.[12] They further suggest how we might begin to map chemical modifications to dynamic properties such as the rates of silencing and the duration or reversibility of transitions between actively expressing and silent gene expression states. Being able to understand the dynamic device properties of chromatin regulators independently of the specific processes in which they are used might allow us to begin to understand epigenetic regulatory modules as ready-to-use fractional control modules. In this view, different types of regulators and modifications would operate in different regimes, or offer distinct capabilities, such as different degrees of epigenetic stability. They might also couple differently to cell states, with the durability of some modifications more or less sensitive to the global state of the cell than others. A major challenge is to develop a physical theory that would allow us to interpret the full repertoire of chromatin regulators in terms of a specific set of capabilities or modes.

4. Dynamics: From Concentrations and States to Frequencies, Durations, and Schedules

While epigenetics integrates inputs over timescales of days or weeks, many cellular systems make extensive use of dynamics on much faster timescales. Even as single-cell methods are opening up a far more direct snapshots of single-cell states, some circuit features may be understood more simply in terms of timing than in terms of concentrations. Just as engineered systems make extensive use of oscillatory signals, cells similarly use dynamics to encode information.

Environmental inputs have been shown to control the fraction of time that a transcription factor is active, by modulating the frequency or duration of stochastic activity pulses, rather than controlling the precise time series. This type of behavior occurs when transcription factors such as Crz1 and Msn2 in yeast activate in stochastic pulses, whose frequency, duration, or other characteristics are modulated by various inputs.[13-18] In other cases, different inputs can activate distinct genomic targets and cellular responses by generating different dynamic patterns or temporal durations of transcription factor activity. The relative timing (or phasing) with which different factors activate can also play a key role in controlling cellular processes.[19] In signaling pathways, different ligands can activate the same receptor with different dynamics, which in turn are decoded to selectively activate distinct cellular targets. This paradigm has been observed in growth factor signaling,[20,21] Notch,[22] and multiple other contexts, suggesting that it is a general principle used by cells to transmit information.

Beyond signaling, time-based regulation increasingly appears to function ubiquitously in diverse processes, including some of the most central pathways of the cell, such as p53.[23] Pioneering studies by Kageyama and others has revealed pervasive oscillations of key regulators in diverse stem cell types.[24,25] Recently, Cai and co-workers identified unsynchronized oscillations of global gene expression levels in stem cells.[2] Together, these results suggest that pulsing, oscillation, and other time-based control mechanisms are pervasive.

Nevertheless, most cellular dynamics remain difficult to detect. Observing them requires temporal measurements of time in individual cells. New methods that can track single-cell dynamics are likely to reveal a pervasive use of dynamics to encode and represent information in diverse cellular processes. Most importantly, recognizing the central role of time-based regulation can help us identify the most relevant and informative variables in the system, and thereby lead to alternative representations for key processes.

5. Conclusions and Open Questions

Thanks to astonishing technological progress and many emerging new conceptual insights, it is becoming possible to imagine a longer term future in which we will understand cellular components, circuitry, and behavior in terms of a set of fundamental principles of living matter. These principles would explain what is currently a bewildering variety of different components and biochemical interactions in terms of a smaller number of underlying principles. They would also enable us to use these components in a more rational way to engineer new cellular capabilities. However, many fundamental questions and challenges remain.

One exciting challenge is to understand how genetic circuits specify cell types and their potential transitions. Single cell profiling methods are now allowing increasingly dense sampling of the space of cell states.[2,26,27] Emerging approaches based on synthetic biology approaches that enable cells to actively record their own cellular histories within their genome promise to complement these snapshots with individual cell dynamic histories.[28–31] It will thus become possible to ask questions not only about what cell types exist, but also about how they are organized molecularly and phenotypically, how they are dynamically established and stably maintained by underlying circuits, and what other, non-natural, states are possible. Are cellular states in a multicellular organism organized hierarchically, or mosaically (or, most likely, both)? Beyond the use of positive feedback loops and epigenetic memory, what circuit architectures enable cells to remain in one state, or differentiate to a constrained set of possible others? How many different types of cell fate control circuits exist, and how do they differ in their ability to control states and transitions? From a synthetic biology point of view, it is critical to understand what fundamental principles govern the kinds of cellular properties that can be created and combined. More broadly, cells rarely function alone. Circuits extend across multiple cells and cell types, both in development and normal physiology. Different design principles may apply to circuits at this multicellular level than to the circuits operating within the cell.[32,33] Answering these questions will require combining single cell analysis with emerging genome engineering approaches to begin systematically designing, building, and exploring the space of potential cell states. Beyond understanding per se, payoffs can also include the ability to engineer biomedically useful new cell types, and more predictably manipulate tissue and organ levels properties.

A deeper understanding of physical theory of biology will likely require not only existing principles from physics and engineering but also new principles specific to the biological context. How will we find these principles? Historically, transformations in scientific understanding sometimes emerge from alternative ways to represent existing knowledge. In biology, at the level of proteins, we routinely translate between nucleotide sequence, protein sequence, domain organization, and three-dimensional structures, choosing appropriate representations for the problem at hand. By contrast, genetic circuit analysis has been dominated by diagrams of molecular components connected by arrows. While useful, this representation doesn't scale well to large circuits, fails to directly represent the roles of temporal dynamics and spatial dimensions, and doesn't, by itself, provide a logical rationale for circuit architectures. Further progress in genetic circuits may hinge on the discovery of alternative representations that enable new and different ways of conceptualizing, analyzing, and designing genetic circuits. These representations might be based on specific dynamic operating "modes" or component combinations, and should ideally be able to map back to standard representations.

References

1. M. Elowitz, W. A. Lim, Build life to understand it. *Nature* **468**, 889–890 (2010).
2. E. Lubeck *et al.*, Dynamics and spatial genomics of the nascent transcriptome in single mESCs by intron seqFISH. *bioRxiv* (2018) (available at https://www.biorxiv.org/content/early/2018/06/05/339234.abstract).
3. Y. E. Antebi, N. Nandagopal, M. B. Elowitz, An operational view of intercellular signaling pathways. *Curr. Opin. Syst. Biol.* **1**, 16–24 (2017).
4. T. D. Mueller and J. Nickel, Promiscuity and specificity in BMP receptor activation, *FEBS Letters* **586**(14):1846–59, (2012).
5. G. M. Süel, J. Garcia-Ojalvo, L. M. Liberman, M. B. Elowitz, An excitable gene regulatory circuit induces transient cellular differentiation. *Nature* **440**, 545–550 (2006).
6. G. M. Süel, R. P. Kulkarni, J. Dworkin, J. Garcia-Ojalvo, M. B. Elowitz, Tunability and noise dependence in differentiation dynamics. *Science* **315**, 1716–1719 (2007).

7. J. Park *et al.*, Molecular time sharing through dynamic pulsing in single cells. *Cell Syst.* **6**, 216–229.e15 (2018).

8. M. Ackermann, A functional perspective on phenotypic heterogeneity in microorganisms. *Nat. Rev. Microbiol.* **13**, 497–508 (2015).

9. J. Song, A. Angel, M. Howard, C. Dean, Vernalization – a cold-induced epigenetic switch. *J. Cell Sci.* **125**, 3723–3731 (2012).

10. H. Yang *et al.*, Distinct phases of Polycomb silencing to hold epigenetic memory of cold in. *Science* **357**, 1142–1145 (2017).

11. A. J. Keung, J. K. Joung, A. S. Khalil, J. J. Collins, Chromatin regulation at the frontier of synthetic biology. *Nat. Rev. Genet.* **16**, 159–171 (2015).

12. L. Bintu *et al.*, Dynamics of epigenetic regulation at the single-cell level. *Science* **351**, 720–724 (2016).

13. L. Cai, C. K. Dalal, M. B. Elowitz, Frequency-modulated nuclear localization bursts coordinate gene regulation. *Nature* **455**, 485–490 (2008).

14. N. Hao, E. K. O'Shea, Signal-dependent dynamics of transcription factor translocation controls gene expression. *Nat. Struct. Mol. Biol.* **19**, 31–39 (2011).

15. N. Hao, B. A. Budnik, J. Gunawardena, E. K. O'Shea, Tunable signal processing through modular control of transcription factor translocation. *Science* **339**, 460–464 (2013).

16. D. Gonze, M. Jacquet, A. Goldbeter, Stochastic modelling of nucleocytoplasmic oscillations of the transcription factor Msn2 in yeast. *J. R. Soc. Interface* **5**, Suppl 1, S95–109 (2008).

17. C. Garmendia-Torres, A. Goldbeter, M. Jacquet, Nucleocytoplasmic oscillations of the yeast transcription factor Msn2: evidence for periodic PKA activation. *Curr. Biol.* **17**, 1044–1049 (2007).

18. R. Martinez-Corral, E. Raimundez, Y. Lin, M. B. Elowitz, J. Garcia-Ojalvo, Self-amplifying pulsatile protein dynamics without positive feedback. *Cell Syst.* **7**, 453–462.e1 (2018).

19. Y. Lin, C. H. Sohn, C. K. Dalal, L. Cai, M. B. Elowitz, Combinatorial gene regulation by modulation of relative pulse timing. *Nature* **527**, 54–58 (2015).

20. S. D. M. Santos, P. J. Verveer, P. I. H. Bastiaens, Growth factor-induced MAPK network topology shapes Erk response determining PC-12 cell fate. *Nat. Cell Biol.* **9**, 324–330 (2007).

21. J. G. Albeck, G. B. Mills, J. S. Brugge, Frequency-modulated pulses of ERK activity transmit quantitative proliferation signals. *Mol. Cell* **49**, 249–261 (2013).

22. N. Nandagopal *et al.*, Dynamic ligand discrimination in the Notch signaling pathway. *Cell* **172**, 869–880.e19 (2018).

23. J. E. Purvis *et al.*, p53 dynamics control cell fate. *Science* **336**, 1440–1444 (2012).

24. H. Shimojo, T. Ohtsuka, R. Kageyama, Oscillations in notch signaling regulate maintenance of neural progenitors. *Neuron* **58**, 52–64 (2008).

25. I. Imayoshi, F. Ishidate, R. Kageyama, Real-time imaging of bHLH transcription factors reveals their dynamic control in the multipotency and fate choice of neural stem cells. *Front. Cell. Neurosci.* **9**, 288 (2015).

26. E. Lubeck, A. F. Coskun, T. Zhiyentayev, M. Ahmad, L. Cai, Single-cell in situ RNA profiling by sequential hybridization. *Nat. Methods* **11**, 360–361 (2014).

27. K. H. Chen, A. N. Boettiger, J. R. Moffitt, S. Wang, X. Zhuang, Spatially resolved, highly multiplexed RNA profiling in single cells. *Science* **348**, aaa6090 (2015).

28. K. L. Frieda *et al.*, Synthetic recording and in situ readout of lineage information in single cells. *Nature* **541**, 107–111 (2017).
29. A. McKenna *et al.*, Whole-organism lineage tracing by combinatorial and cumulative genome editing. *Science* **353**, aaf7907 (2016).
30. F. Farzadfard, T. K. Lu, Emerging applications for DNA writers and molecular recorders. *Science* **361**, 870–875 (2018).
31. M. Chan *et al.*, Molecular recording of mammalian embryogenesis. *bioRxiv* (2018), p. 384925.
32. Y. Hart *et al.*, Paradoxical signaling by a secreted molecule leads to homeostasis of cell levels. *Cell* **158**, 1022–1032 (2014).
33. O. Karin, U. Alon, Biphasic response as a mechanism against mutant takeover in tissue homeostasis circuits. *Mol. Syst. Biol.* **13**, 933 (2017).

Prepared comments

O. Leyser: Is it time for a classification of biological switches?

A major challenge in biology is to understand how the behaviors of biological systems emerge from the dynamic interactions within and between the levels of organization that constitute the system. Over recent years, there have been major advances in defining the molecular level components that deliver these behaviors. Categorization of these molecules into classes that encapsulate their biochemical activity, such as kinases, transcription factors, receptors, etc has made a significant contribution to conceptual understanding of cellular function.

Our increasing understanding of the action of these system components has catalyzed efforts to provide a similar categorization for networks of these components.[a] This provides an important tool to understand dynamical systems properties, since the behavior of these network motifs can be analyzed in vivo, in silico and using synthetic biology approaches. For example, the properties of coherent feed-forward in comparison to a simple linear unbranched pathway provide insight into the circumstances under which this network motif might be deployed in biological systems.[b] This approach can be considered as the next step in building understanding of biological systems bottom up.

An interesting question is whether we have sufficient understanding to allow a parallel top down approach, characterizing mesoscopic dynamic behaviors, and inferring the likely regulatory architectures that could deliver them. An illustrative example is switching behaviors. A biological switch can be defined as a system with two alternative stable states, with the possibility to transition between them in response to one or more stimuli. The properties of this transition in response to a stimulus can be characterized experimentally. For example, the transition can be sharp or gradual, it can be linearly or non-linearly sensitive to the strength of the stimulus, and it can be more or less easy to reverse, with or without hysteresis. In the same way that it is possible to infer network behaviors from the regulatory architecture of the component parts, it should be possible to infer regulatory architectures of component parts from the characteristics of higher level behaviors, such as switching. Building a vocabulary that encapsulates mechanistic understanding of systems at this mesoscopic scale could be an important step in efforts to develop intuition about emergent system properties.

---

[a]Milo *et al.* (2002) *Science* **298**:824–827.
[b]Mangan and Alon (2003) *PNAS* **100**:11980–11985.

A. Goldbeter: The cell cycle and the circadian clock: Dynamics of two coupled cellular rhythms

The cell cycle and the circadian clock provide two exquisite examples that illustrate how regulatory networks control the dynamics of cellular processes. Both networks display autonomous oscillations; moreover, they appear to be coupled. Before discussing the consequences of such coupling let us recall the salient properties of each of these networks, which underlie two major cellular rhythms.

The mammalian cell cycle is controlled by a network of cyclin-dependent kinases (CDKs). Different complexes between cyclins and CDKs control the transitions between the successive phases G1, S (DNA replication), G2 and M (mitosis) of the cell cycle. The network consists of four CDK modules: cyclin D/CDK4-6 controls progression in G1, cyclin E/CDK2 controls progression in G1/S, cyclin A/CDK2 controls progression in S/G2, and cyclin B/CDK1 controls the G2/M transition. The CDK modules are coupled through multiple modes of regulation involving, a.o., cyclin synthesis and degradation, CDK inhibitors such as p21, and CDK regulation through phosphorylation-dephosphorylation.

A model for the CDK network driving the mammalian cell cycle has been proposed by Gérard and Goldbeter.[c] The time evolution of this model is governed by a set of 39 nonlinear, ordinary differential equations. To capture the dynamics of the model it is useful to build a bifurcation diagram showing the qualitative behavior of Cyclin B/CDK1 (one of the key variables of the CDK network, taken as representative of the whole CDK network) as a function of the level of growth factor (GF) which induces the transition from the quiescent state (G0) into the G1 phase of the cell cycle. Below a critical GF value, the network evolves to a stable steady state. Above the critical GF level the CDK network undergoes sustained oscillations associated with cell proliferation. Sustained oscillations in the various cyclin/CDK complexes correspond to the transient, ordered, repetitive activation of the four CDK modules controlling the successive phases of the cell cycle. Therefore they can be associated with cell proliferation, while the evolution to a stable steady state corresponds to cell cycle arrest.[c]

The CDK network is designed in such a way that each module activates the next module(s) and inhibits the previous module(s). Such a regulation allows for the temporal self-organization of the network in the form of sustained CDK oscillations in which each CDK module is activated in turn, in a transient, repetitive manner. The oscillations observed in the model for the CDK network are of the limit cycle type. This type of oscillations is particularly robust with respect to perturbations. Indeed, for a given set

[c]C. Gérard, A. Goldbeter (2009) *Proc. Natl. Acad. Sci. USA* **106**, 21643–48; C. Gérard, A. Goldbeter (2014) *Interface Focus* **4**: 20130075.

of parameter values, the same limit cycle is reached, regardless of initial conditions.

The second network considered pertains to the mammalian circadian clock. The molecular mechanism of this clock is based on the induction of the Per and Cry genes by the activators CLOCK and BMAL1, and on the inhibition of CLOCK-BMAL1 by the PER-CRY complex. A second loop of negative autoregulation involves the control of Clock and Bmal1 expression by CLOCK-BM1AL1 via REV-ERB. The model based on this regulatory mechanism predicts the spontaneous occurrence of circadian oscillations with a period close to 24h.[d]

The dynamics of the model for the mammalian circadian clock is similar to that of the model for the mammalian cell cycle. Sustained oscillations again correspond to the evolution to a limit cycle trajectory. This behavior is also similar to that obtained in the model for the Drosophila circadian clock.[e] The latter model was based on the experimental observations of Hall, Rosbash and Young, who received the 2017 Nobel Prize in Physiology or Medicine for their pioneering work on the molecular regulatory mechanism of the circadian clock in Drosophila.

In the last decade, experiments have shown that the mammalian cell cycle is coupled to the circadian clock. The first mode of coupling uncovered involves the induction by the circadian transcription factor BMAL1 of the kinase Wee 1, which inhibits CDK1 (aka Cdc2) in the cell cycle clock network. Coupling of the mammalian cell cycle to the circadian clock through Wee1 allows for entrainment of the cell cycle by the circadian clock to a period of 24h or 48h.[f] The domains of entrainment take the form of "Arnold tongues" as a function of the coupling strength (measured by the rate of Wee1 mRNA synthesis controlled by BMAL1) and of the autonomous period of the cell cycle prior to its coupling to the circadian clock. Outside the domains of entrainment to 24h or 48h, complex oscillations of various waveforms can occur, including endoreplication (CDK2 then oscillates in the absence of CDK1 oscillations), bursting oscillations and chaos. The question arises as to the physiological significance of the two latter modes of dynamic behavior. Current work pertains to the effect of bidirectional coupling, given that the mammalian circadian clock appears to be coupled to the cell cycle. We are currently exploring the dynamical consequences of such bidirectional coupling and how it affects the synchronization of these two major cellular rhythms.

---

[d] J.-C. Leloup, A. Goldbeter (2003) *Proc. Natl. Acad. Sci. USA* **100**, 7051–56.

[e] A. Goldbeter (1995) *Proc. R. Soc. Lond. B* **261**, 319–24; J.-C. Leloup, A. Goldbeter (1998) *J. Biol. Rhythms* **13**, 70–87.

[f] C. Gérard, A. Goldbeter (2012) *PLoS Comput. Biol.* **8**(5): e1002516.

B. Simons: **Genome-scale oscillations of DNA methylation in embryonic stem cells**

A key challenge in biology is to understand the molecular basis for cell fate decision making. Progress in this area has been confounded by the fact that information on cell fate cannot be easily extracted from the molecular profiling of fixed samples. However, the development of single-cell profiling methods that, with current technology, can access the transcriptome, DNA methylome and chromatin accessibility of individual cells offers the potential to target the molecular basis of cell state heterogeneity, fate stochasticity and flexibility, ubiquitous features that have emerged from the application of functional cell lineage tracing assays. To illustrate the potential opportunities in this area, Simons presented a case study, based on work carried out with Steffen Rulands, Wolf Reik and colleagues, revealing evidence for coherent genome-scale oscillations of DNA methylation in embryonic stem cells.

Epigenetic alterations, such as histone modifications and DNA methylation, are thought to play a critical role in the regulation of transcriptional programmes. Following fertilization, the paternal and maternal genomes of mammals undergo global demethylation, resetting the epigenome for naive pluripotency. Then, during exit from pluripotency, epiblast cells in the inner cell mass move through a phase of epigenomic reprogamming, where the average levels of DNA methylation increase. Single-cell profiling of DNA methylation shows that, during this phase, mouse ESCs grown in serum conditions show large-scale cell-to-cell variability in DNA methylation. In some cells, the distribution of DNA methylation levels is tilted towards low values, in others high, and in others it is somewhere in-between. However, remarkably, this heterogeneity is not static, but is associated with coherent genome-scale 2-3 hour oscillations of DNA methylation. These autonomous oscillations are driven by the paradoxical co-expression of DNMTs and Tets, enzymes that methylate and demethylate the DNA. The autocatalytic binding activity of DNMTs combined with the time-delayed action of Tets drive local oscillations of DNA methylation, that become synchronized across the entire genome.

This example is instructive as it shows evidence for emergent cooperative phenomena appearing at the subcellular scale, and it emphasizes the potential for single-cell approaches to reveal dynamic information. It also raises interesting questions: How do oscillations of DNA methylation become synchronized across the genome? Why is the amplitude of oscillations correlated only with CpG density? Do these oscillations drive dynamic changes in transcription? And do these changes contribute to lineage priming and symmetry breaking in the developing embryo?

Discussion

D. Fisher I guess this is a general comment and question, picking up on some other things said earlier. Terry made a comment about GFP being not the only thing that the cell is putting out. Rob made comments about the importance of the allostery, but it seems that most of the things that Michael talked about happen on very slow time scales. Proteins do an awful lot of things on short time scales. I had an interesting discussion with Paul Nurse and Fred Cross on whether yeast can even run the whole cell cycle without any gene regulation at all. So, I just wonder what the potential is for the future of really trying to get much more at the protein modification level, dynamics, pathway, circuitry in the cell rather than being limited to the very slow processes.

M. Elowitz There is a wide range of time scales, as you said. If you want to look at fast phosphorylation time scales and things like that, nuclear localisation has become a really effective reporter for that and there is recent work from Markus Covert's lab, trying to make new, more flexible nuclear localisation-based reporters for phosphorylation states. That is one way.

E. Koonin I would like to bring up a very general point as we are supposed to do. The unifying theme that came out of the presentations of both esteemed rapporteurs is dimensionality reduction or the availability of rather simple phenomenological models to account for cellular behaviour. So, the question that I want to submit is this: How does that happen? Is there some kind of master integrator within the cell, some kind of central processing unit that integrates all these signals, or alternatively, any search for such master integrating unit would be futile and we should rather think about these processes in terms of self-organized criticality.

T. Hwa Well, I gave you an example of how it works for this particular case. First, we assumed something simple, that there is dimension reduction because the cell is doing dimension reduction. In the particular case I described, it is that the cell picks some core variables, each variable picking up a process, and is detecting the fluxes rather than the concentrations. And the flux itself is an integration. It is a kind of a global measure of how the cell is doing. Actually, we established two such cases: both forward and backward.

J. Lippincott-Schwartz This is a comment and question about the cycling, especially the methylation cycling you are talking about and also a sort of broader transcriptional cycling. I am wondering whether you have considered the possibility that it is related to metabolic cycles of glycolysis versus respiratory oxidative phosphorylation and shifts between those two systems. For instance, when you are running glycolysis, the amount of ATP that is being generated but also NADPH, NAD levels are altering and you are mainly doing anabolism. You are building up acetyl CoA, fatty acids, etc. At some point the cell then starts potentially methylating and shifting the phosphorylation state and that is going to shut off glycolysis and the anabolic pathway of the cell. The cell will shift to more catabolism where basically what you are driving is oxidative phosphorylation. My question is whether any of you have looked at the metabolic readout in terms of what type of metabolism these cells are doing and whether that is coupled to the cell cycle, the circadian rhythm or the other types of cycling that you are seeing.

B. Simons All I can say is that it is a great question and we do not know. One thing that we did check is that the oscillations are not driven by a single oscillator like S1, which is also oscillating. But whether downstream it has an effect on metabolism or any feedback from metabolism is a great question that we did not give out.

A. Murray As far as I know, there are no such oscillations in bacteria.

A. Goldbeter Back with Msn2 oscillations, there has been evidence from the lab of Michel Jacquet in Paris that they are linked to cAMP oscillations. cAMP oscillations would control PKA, the cAMP-dependent protein kinase, which would phosphorylate Msn2 and control the in and out movement between the cytoplasm and the nucleus. And the mechanism of the cAMP oscillations seems to be involving a negative feedback through phosphodiesterase. A mechanism envisaged by Bill Loomis at UCSD for cAMP oscillations in Dictyostelium where they occur also with a period of about 5 minutes.

U. Alon I want to address Ottoline's question about classification of switches and say that generally it touches on viewing biology as a reverse engineering problem. So, we need to face the phenomenon and try to understand the underlying mechanism. A key question is whether the problem is one to one, or many to one. How many solutions there are that are going to give an oscillation, a switch, etc? For switches, for example, you can count on one hand basic classes of mechanisms: Goldbeter-Koshland zero-order ultra-sensitivity (proposed in 1981) and then, there is whether you have hysteresis or not, something you can test. Cell cycle for example has a hysteretic switch, which is very great because then you are not sensitive to noise around the threshold. And then, you have switches with positive feedback loops, double positive feedback loops, etc. It will probably be a handful to one, not millions to one, not one to one. Then, we can ask when to choose one of these mechanisms and when not. This is related to each one having multiple engineering properties: robustness, response time, epigenetic memory, etc. Again, there is a handful of possibilities. We can think of a program where we can understand, the trade-offs in each of these handful of mechanisms. This is something that can help biologists like you because we have these "acid-tests". Recently, there is a paper by Fred Cross, Rahi and collaborators where you can tell the differences between a feed-forward and a feed-backward mechanism for adaptation. And we know that adaptation can happen in only two ways: incoherent feed-forward or feed-back. There is a way to tell the difference if you give two pulses of input. The feed-back loop can sometimes skip the second pulse. The feed-forward would never do that. This type of an acid-test can be tested by experimentalists because one only has to do dynamic inputs and outputs. So, this is a simple feasible score for reverse engineering on the mesoscale.

E. Wieschaus I was struck in Michael's talk by the example of vernalization and the cell culture example where decisions are stochastic in a population and then following up on that, Ottoline raising the question of switches and I was trying in my mind to decide what the switch was. How do we define a switch? Intuitively, one element is non-linearity, meaning that you have an increase in signal and at some point, that signal flips you into another state and that can be stochastic, and so that is fine. The other is a certain permanence of that state and that is generally so. When we think of a switch being thrown, it is not reversible. Two questions are coming to my mind based on these descriptions. The first is how non-linear does a response have to be for you to call that a switch or for us to regard it as a switch? How non-reversible does it have to be? Cells respond to environment. So, if we want to classify switch architectures and circuits, how is a switch different from a response? The other question I was raising as a developmental biologist is that all these descriptions were stochastic and in terms of individual cells that make choices in a population. Some cells make it, some cells do not. Very often, you see that development depends on groups of cells making uniform commitments to particular states. You can achieve that groupness by two ways: waiting long enough, or by talking to each other. Is there any other ways of thinking about those processes? It is true that there are examples where stochastic choices in single cells lead to meaningful states in development. But, from my perspective, the most common phenomena are polyclonal, i.e. groups of cells which make decisions which are then uniformly inherited in the group.

C. Marchetti I actually wanted to ask a question about this classification of switches, and this question was partly answered because my question was about how exactly a switch is defined, and what are the criteria that you have in mind for classification. Physicists love classifications, so I think this is a great point to raise. It seems from what I heard that they are classified according to non-linearity, or reversibility versus irreversibility and so on. Those are classifications that are certainly useful from the point of view of engineering, but I wonder whether a most useful approach might be to think in terms of their functional significance, rather than a more mathematical type of classification.

S. Quake I want to return on what Eric, Ben, Michael all have been talking about, which is a very deep question, about cell fate choices and cellular identity that are very right now, both from the experimental and theoretical perspective. These are things that have been very dramatically engineered artificially. Shinya Yamanaka showed that if you express four transcription factors, you can make a cell go back in time, which is astounding. And Marius Wernig that with three transcription factors, you can make artificial fate decisions from fibroblasts to neurons. These are incredibly interesting and those are done in artificial cell culture conditions. Many of these questions, most interesting biologically, take place in the organism, in the context of a tissue in the organism. Hematopoiesis is a classic example. We have this combination of great experimental tools now, Ben alluded to those, the ability to basically enumerate essentially every RNA molecule in a cell that provides a beautiful portrait of what is going on, and many complementary tools, the lineage tools that Michael talked about, and variations of those. This is a really fundamental question about metastability, plasticity, whether these decisions are reversible or if they happen in a coordinated way. But there are many examples that happen in an uncoordinated way as well, in the immune system. It is thus great for this intersection of physics and biology, both theoretically and experimentally, in health and disease. Many important aspects of cancer for example are recapitulated in cell lines in artificial in vitro experiments, but you need the full complexity of the organism and take it apart to understand how these decisions are going wrong.

E. Siggia To quickly follow up on Steve's remark, in the field of vertebrate signaling, one can do a great deal more with cells in a dish than actually takes place in an embryo. So specifically, there are old experiments for the BMP perturbation on isolated Xenopus cells which show that you basically over a factor of 20 in BMP level recapitulate all the various gene markers that you see in the embryo, which would argue that the relevant concentration range in the embryo is a factor of 20, as defined by getting cell states. Of course, Michael Elowitz will show you a vast range of concentrations in vitro, cells might do other things but it may not be relevant to the embryo.

N. Wingreen I want to poll the audience: tap into the collective wisdom here on a question that has puzzled me for a while. I have a number of colleagues who showed me data occasionally from eukaryotic cells exposed to some signals, and I have seen data from Thomas Gregor's lab on Dictyostelium, and Jared Toettcher lab on eukaryotic cell lines, and one of the things I see universally is that the cells have very different magnitude of responses. And the cells seem to be quite individual. So, you hit them with the signal, and some of the cells will respond well, and you hit them again and they respond well, some of the same cells apparently identical given exactly the same signal will barely respond at all. Of course, there is a mechanistic question, but my puzzle is also: Is that meaningful in biology, is that important or we are just giving the wrong signal and perhaps they are not expecting to have some intermediate level of signals, they are always going to see high or low signals? These different responses would thus be a laboratory artefact and I really do not know how to think about that and I would appreciate guidance.

A. Hyman I actually want to return to this discussion before the break about what is the definition of a pathway because for cell biologists, things are spatially segregated. If there are several molecules together in a compound, we tend to think of that as a pathway. So, in other words, the definition of a pathway is a definition which includes spatial segregation. If they are 1 micron apart, molecules would diffuse between them in a second, but for 10 microns it can take 10 seconds. So, that is how we tend to see it as cell biologists. It is a sort of compartmentalization process, which really ends up as a time scale issue. If the time to communicate between the different compartments is a lot longer than the time to get through the signal transduction pathway then you would tend to think they are separate pathways. I did not hear anyone talk about that this morning, so I thought to link the two together.

N. King One key question is how do we represent and think about what is going on in the cell, or across cells, then in tissues. I thought this question of classification was also very interesting. How do we define different types of responses? A tension that exists in biology that I am not sure actually occurs in physics is the fact that when we represent something, we are oversimplifying it in many cases. I often worry about specific words and about classifying (Steve, I think your group has been looking, for instance, at flow cytometry of different cell types) because the closer you look the more diversity and heterogeneity you see. I suspect that the more switches you look at, you will see them as a continuum. So, in your representations, you are really applying to the highest signal but there are long shoulders. Maybe the pathway issue is similar, which is that there is a primary activity that you can flow information but there is a cloud of activity that is going on around it. I would be interested to hear from physicists if I am making things more complicated than they need to be. My impression is that when a physicist represents something, that representation is true across phenomena, whereas for biology it is often inaccurate, not only for the things that we are representing but also for the things that are related to it.

A. Goldbeter I would like to come back to one aspect which was mentioned by Michael in his talk about pulsatile signalling by transcription factors. I think that this extends to other examples in physiology, which are very important and pertain to hormone secretion. Most hormones are secreted in pulses. For example, the gonadotropin-releasing hormone (GNRH) is released by the hypothalamus every hour. The frequency is very important because another frequency does not succeed in inducing the release of the hormones LH and FSH, which control ovulation. It is a clinical condition, which is known, that some women are unfertile because of lack of pulsatility of GnrRH. They can be implanted with a pump delivering the hormone at the physiological frequency that restores the levels of LH and FSH. Growth hormone is the same and the underlying principle is probably receptor desensitization, which is involved in adaptation because a cell will adapt to a constant stimulation and will not respond. That is, by the way, used in cancer therapy by hormones to saturate the receptors and desensitize them. So, there is an optimal frequency of stimulation, which relies on kinetics of receptor desensitisation and resensitisation.

M. Elowitz I completely agree and this point connects a bit to the issue raised earlier about integrating metabolic dynamics. This was raised with cAMP, and PKA. I think we have a problem really, which is kind of cool, which is dynamics across many different levels, from the organismal level down to the intracellular level and I guess that the challenge is how do we read out dynamically all of those levels at the same time to understand dynamic relationships to one another. This is a big challenge.

W. Bialek Actually, it would be fun to talk about Nicole's question "What are we doing when we are simplifying the representations?" Lots of people could say something about that. I want to pick up on something that Michael said casually about systems with many ligands and many signaling pathways: "This is not the way we would have designed it". This relates also to, if I may, to the casual use of words like "optimal" earlier today. To give an example related to communication, you know that you can communicate without error over a communication channel that has noise. This is a great theorem due to Shannon in 1948. If you ask: "How did he prove the theorem?", the answer is that he considered codes in which you take the discrete signals you want to transmit and you map them into completely random signals. It took fifty years until anybody managed to produce a code that actually achieved the Shannon bound. You can argue that the reason is that they were very carefully designing particular transformations that were designed to correct certain types of errors but they never achieved the bound that would have been gotten by doing things truly random, which of course you do not want to do as a practical matter because it is hard to decode. So, very simple optimisation problems can have very surprising solutions. So, when we say "That is not the way I would have done it, that is not what's optimal", we should be very careful.

S. Chu Let me go back to a much higher level, something Terry said. If you think what is important in people — birth, life, death, sex, love, taxes, and there are equivalents of that in cells as well. Your point was that say these were the important things for cells so let us go backward and figure out what the cell does to do this. The questions like "five things mapped to the one", and all the cell cycles are wrapped up into this. Thirty years ago we thought that DNA was the heart of everything. We then realized that there was epigenetics, and that maybe RNA is closer to the truth. But then I heard a comment somewhere along the line that said that there are a lot of messenger RNA around the world and not all of it gets translated. So, RNA may not be the proxy for proteins, which are what ultimately decides how the cells respond to the things that are important to cells. I just want to throw into the mix one experimental thing that has not been approached yet: single cell mass spectrometry, which is amazing. You can go down five orders of magnitude in protein expression of the single cell. Let us look at the mapping of the epigenetic modifications, the RNA, the potential interference of that RNA and how it maps onto the single proteins where you can get down to copies of 5–10 proteins in single cells measured by this dramatic dynamic range. That is another tool that we should pay attention to. We should concentrate on what is very important to the cell and the organism, and all the other stuff is a map to the mechanism for how the cell does this. But do not forget that there are new technologies that have not been utilized yet, especially the mass-spec stuff.

T. Lecuit It strikes me that the way we are discussing cellular control in time and space in this session does not consider much the things that were discussed this morning and beyond, mainly the fact that cells are amazingly organized. There is dynamics in this organization. To give you an example, there are oscillatory dynamics of sub-cellular pathways which cannot be properly understood if we do not actually capture the dynamics of cytoskeleton components that exist in the cell, which themselves step into motion. This is a question and encouragement to actually articulate better dynamics of molecular interactions that we can model with on/off kinetics, explain sort of bistability and dynamics and to incorporate this with the dynamic organization of compartments, of cytoskeletal components and all the things that we have discussed this morning. I do not think we can understand any signalling and computation in any meaningful way if we do not do that. I would like to see how people react to this general comment.

T. Hwa One common theme of discussion is molecular interactions, the definition of a pathway, and this is one way of looking at interactions: this is a bottom-up approach. I would like to share something about my own experience in studying signalling systems in *E. coli* and some systems that were thought to be understood for 30–40 years as catabolic repression: how glucose comes in and shuts off the transport of other things. When you look at it physiologically, you realize the glucose is nothing special and it has to be a different reason. Eventually we established the real mechanism, as some sort of global coordination. My own limited experience is that if you only follow the molecular interactions, there are too many directions to go. Most of the time, it is an inefficient way because too many things are happening. I personally find that it is much more effective and efficient if you know that there is this effect already and then, you know there has to be a reason. And then, it is a matter of finding this effect. Talking back to the pathway and the modularity, when I was getting into biology, learning it in the late nineties, there was a very influential article that our Chair was co-author of. "Modularity" was in the title. I went back to read that article before coming here. There are two views of modularity. Back when I first read this paper, I understood modularity as a type of insulation of molecular inter-actions and so forth, and it was a bottom-up approach. But rereading it now I saw that the physicist authors were also emphasizing the importance of the phenomenology defining modularity. A top-down view of looking at modularity is that the modules are functional response units. You have a physiological perturbation, and you have a physiological response and the response is what defines the module.

J. Lippincott-Schwartz I am going to speak to support what Thomas was saying in terms of the need to integrate our spatial understanding of cells with the modelling, computational perspective. Because I think that there are a lot of things that you really do not understand unless you actually understand these intracellular pathways. For example, we just heard about this cycling of secretion, and you think this is related to desensitization of the receptors. But, well, when a growth factor binds to a receptor, it gets internalized. There is a receptor-mediated endocytic process that down-regulates that receptor and shuts off that process. So, it is that pathway that is actually controlling. And it is how that pathway gets reset that drives the cycle back to return. This is particularly important in cancer, K-Ras mutants trigger micropinocytosis. That down-regulates glucose receptors, so the cells are now glucose-starved. It is massively trying to respond to that and many of the phenotypes, the response that you see in that cancer cell, can be described based simply on that down-regulation of growth factors, receptors, etc and a shift in metabolism in response to that. I keep going on with examples. For instance, transcription factors like Yap respond to beta catenin and in response to mechanical stress, beta catenin gets internalized. This triggers the Yap release and translocation into the nucleus, to drive the transcription pathway. These are all connected to larger scale mechanical forces and intracellular pathways in a big way. We are going to try to understand how to control this system. We have to understand what is happening at the cellular level.

A. Murray Space matters.

D. Fisher Returning to the points made by Ned, Nicole, Eric, there was a long debate about switches, states, how many states there are in cells and how they are responding. But, we basically do not know what the cells care about. And this is coupled to Terry's topic as well. There are maybe things that we are categorizing that cells do not particularly care about. We do not know if we are barking up the right tree. The other point, coming back to this modularity and the way you are separating things, is the timescale. If you just look at going from basic molecular timescales in nanoseconds up to the division time of a bacterium, it is twelve orders of magnitude in time. Just that factor by itself tends to mean that any given process, say for ribosome to go through its cycle once, almost everything is really fast compared to that or very slow compared to that. And at any given timescale, there is not much going on. So that already gives you quite a lot of separation between different things going on. One of the things that I hope will come out of this meeting is that anyone knowing anything vaguely to do with physics will ban from drawing arrows and bars without timescales on. Numbers we do not know those but at least we know something about the timescales. Without that, in some ways, one is throwing out so much and claiming that this is a function in any loose sense of the explanation.

A. Murray It is now time to close. Many great things have been discussed. People have emphasized the importance of space. People have emphasized the importance of time. Questions have been raised. Is there a central processing unit? I think that is indeed an extremely interesting question. If you ask a geneticist what the properties of mutation in a central processing unit would be, they are mostly likely to cite properties that prevent its easy detection. Examples of processes that can be central processing units have been raised, like the Cyclin-Cdk machinery that drives the eukaryotic cell division cycle. There is a question from Nicole as to whether when physicists classify, they apply artificial division to continuums that exist in biology. There is a question from Ned Wingreen that no one answered about whether the heterogeneity of responses viewed in eukaryotic cells and tissue culture dishes tells us anything at all about the world inside organisms. So, with many questions left open, I would declare our session closed and thank the rapporteurs, the prepared remarkers and all of you for discussion. Thank you.

Session 3

Inter-cellular Interactions and Patterns

Chair: *William Bialek*, Princeton, USA
Rapporteurs: *Gürol Süel*, UCSD, USA and *Eric Siggia*, Rockefeller, USA
Scientific secretaries: *Lendert Gelens*, KULeuven, Belgium and *Han Remaut*, VUB, Belgium

W. Bialek This is the third of our sessions, entitled Inter-cellular Interactions and Patterns. It fits somewhere in between yesterday's cell behavior and control and morphogenesis that will be discussed tomorrow. I think you will find that between our rapporteur's talks and "prepared remarks" that the span of the biological systems discussed here will be very broad, so this is not a session where we have taken a slice through the biological world in any traditional way. And I hope that that will be a stimulus for us all to think about the conceptual physics problems that have similarly cut across many different systems. Much of what you'll be hearing today is about the flow of information between different cells, different organisms, and I hope we will be inspired to think about whether there are general things that stand behind the phenomenology. We start with a rapporteur talk by Gürol Süel which is followed by a prepared comment by Ned Wingreen.

Rapporteur Talk by G. M. Süel: Intercellular Interactions in Microbial Biofilms

Communication among cells is typically associated with higher organisms, but even simple bacteria exhibit fascinating cell-to-cell signaling mechanisms that can give rise to collective behavior as a function of time and space, resulting in higher order physiologies and morphologies.

1. Introduction

The fitness of any living cell or organism is directly tied to its environment. One of the most important advances that occurred during evolution is the ability of cells to alter the environment through their actions, thereby influencing the fitness and behavior of other cells. Over time, the process became more sophisticated and specialized giving rise to what we refer to today as cell-to-cell signaling. It is important to note that signaling among cells is not exclusive to multicellular organisms, but even evolutionary more ancient unicellular organisms such as bacteria had already developed communication mechanisms.[1]

Bacteria are unicellular organisms, but this description should not be seen to imply that bacteria are solitary creatures. Studies have shown that bacteria are social organisms that live in crowded communities.[2] One of the themes to emerge from this conference is that we need to consider the context of biological organism we are investigating. In the case of bacteria, we have to consider that most bacteria on our planet live in densely packed communities (also known as biofilms or mats). Therefore, the context of the bacterial cell is not just the external environment, but also the existence of other cells around it that can modify the extracellular conditions through their actions. This context proves an important perspective to guide our interpretations and consequent understanding of biological processes.

Many fundamental biological processes, such as DNA replication and gene expression were first characterized and understood by studying bacteria. In fact, one could even go so far as to say that most of our fundamental biological insights come from the study of microorganisms. Yet there are many more open questions of perhaps equal importance that have yet be uncovered, because studies of bacteria in the past have been typically performed not in the context of communities. Specifically, many previous studies utilized batch culture measurements and population averages of cells growing in low density shaking liquid cultures. Therefore, depending on the question we are asking, it seems rather likely that we could learn new insights by looking at bacteria under different contexts, especially in the context of the community.

Recent advances in the field of microbiology arose from a shift of focus to study bacteria in densely packed communities, commonly referred to as mats or biofilms. Here I will use the term biofilm. Biofilms are communities that contain cells that excrete extra-cellular matrix components that provide a scaffold that adheres to cells and binds them together.[3] When cells are in these densely packed communities they appear to engage in behaviors that would not be evident in the context of studies of the past that utilized well mixed low-density liquid cultures to study bacterial behavior.[4] It has been speculated that such bacterial communities enable cooperative metabolic interactions and an enhanced collective defense against environmental stress conditions.[5] These interesting behaviors that bacteria may execute in communities would imply some sort of cell-to-cell signaling. I will describe below such signaling among cells and illustrate how it enables coordination of bacterial behavior in space and time to improve collective fitness.

There are many ways to categorize cell-to-cell signaling. To describe signaling from the perspective of physics, one may want to understand the properties of relevant signals, or the physical aspects of signal propagation between cells. Another way to categorize signaling among cells is to consider whether the outcome of the signaling generates linear or nonlinear cellular behaviors. Keeping these perspectives in mind, we will discuss here two cell-to-cell signaling processes in bacteria, namely quorum sensing and electrochemical signaling. While we will discuss the differences between these two distinct signaling mechanisms, we also realize similarities between these signaling processes that can both enable emergent behavior. In other words, the transition from an independent, individualistic, solitary lifestyle, to a collective behavior that can only be accessed in the context of a community of cells.

In Figure 1, you can see that a bacterial community is visible to the naked eye and exhibits features that are quite intriguing. Specifically, the image shown is that of a biofilm grown on an agar surface, which was formed by the soil bacteria *Bacillus subtilis*. If you do not know anything about *B. subtilis*, do a quick exercise: Raise your index finger, and now, just touch your eyebrow and then look at your finger. If you could zoom in, you would see *Bacillus subtilis* cells on your skin.

Fig. 1. An image of a Bacillus subtilis biofilm.

While *B. subtilis* is categorized as a soil bacterium, they can be found in and on tremendously diverse places, inert or alive such as on the surface of furniture, or your gut and skin.[6] *Bacillus* are one of those bacteria that can basically be found everywhere humans live. If we look at the scale bar on the figure above that indicates one millimeter, we can see that bacteria grow into massive structured communities. We want to understand how bacteria, that are approximately 2 micrometers in length, give rise to such large and organized communities and how the development and operation of a biofilm community depends directly on cell-to-cell signaling. Let us begin with the bacterial cell-to-cell signaling mechanism known as quorum sensing.

2. Quorum Sensing

An important question is how unicellular bacteria transition to a community lifestyle that exhibits characteristics of a multicellular organisms, such as spatial organization of cell types and coordinated metabolic states among cells. This transition from planktonic to community lifestyle appears to be regulated, at least in part, by a cell-to-cell signaling process known as quorum sensing.[7] I will briefly describe the key features of quorum sensing and then discuss that quorum sensing is a diffusion limited process that generates a threshold response.

2.1. *The basic steps of quorum sensing*

Quorum sensing (QS) is the best characterized bacterial communication system to date. The study of quorum sensing has been pioneered by scientists such as Bonnie Bassler, at Princeton, who unfortunately could not come to Brussels. However, we do have here Professor Ned Wingreen, who is also an expert on quorum sensing! The bacterial QS response typically involves two-component circuits comprised by a receptor and a transcriptional response regulator.[8] This means that extracellular chemical signals that interact with the receptor can modulate the activity of a downstream transcriptional regulator. This signal transduction between receptors and transcriptional regulators typically involves post-translational modifications such as phosphorylation, which can alter the conformation, or dynamics of proteins and thus their activity.

The basic principle of quorum sensing is that bacteria excrete small chemical molecules as seen in the left panel of Figure 2. These molecules exhibit a very diverse range of chemical structures that include small peptides. Such quorum sensing molecules are then detected by designated membrane protein receptors (middle panel of Figure 2). These receptors are usually quite specific and can discriminate among a diverse class of QS molecules.[9] This specificity allows bacteria to respond to specific signals, even when many different QS molecules can be present in the environment.

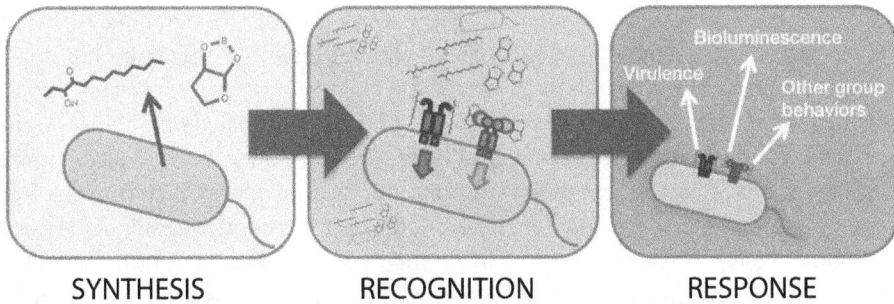

SYNTHESIS   RECOGNITION   RESPONSE

Fig. 2.   Schematic representation of quorum sensing.

2.2. *Quorum sensing response*

As indicated above, the response to quorum sensing signals typically results in a change in gene expression. Such gene expression changes determine the behavior of bacteria (right panel of Figure 2). For example, bacteria can switch to producing antibiotics or bioluminescence.[10,11] They can also express virulence factors, which is of great interest to the biomedical community.[8] Quorum sensing has also been shown to regulate expression of extracellular matrix components that are required for the initial formation of a bacterial biofilm community. In fact, Stephen Chu published a paper, "Molecular Architecture and Assembly Principles of *Vibrio cholera* Biofilms,"[12] which is a beautiful demonstration of the initial aspects of the formation of a biofilm. This paper describes how cells that have attached to a surface begin to secrete proteins that enable bacteria to adhere to one another resulting in the onset of a community structure. Identification of the molecules that are responsible for the physical binding of cells to one another is of interest to researchers that are seeking to control biofilm formation and perhaps even resolve such communities. It turns out that the expression of such extracellular matrix components that are responsible for cell-to-cell adhesion are driven by quorum sensing.[13] Therefore, QS has been suggested to play an important role in the transition to the biofilm lifestyle of bacteria.

2.3. *Quorum sensing and surface attachment*

Biofilms typically form at physical interfaces, such as air-liquid, liquid-solid or air-solid. In all these cases, the biofilm appears to adhere to the surface of the interface. Consequently, an important process that is commonly regarded as the first step in biofilm formation is the attachment of cells to such surfaces. Here again, quorum sensing appears to play a role. First, attachment to surfaces is facilitated by extracellular matrix components, whose expression is in turn regulated by QS. This means that the expression of ECM molecules would be more likely in the presence of a high concentration of QS molecules, which in turn would correspond to higher cell densities. In addition, it has also been proposed that surfaces can act as reflective boundaries to locally increase QS.[14] If there is no flux of QS signaling molecules into the surface, the concentration of the signaling molecule would be highest at the surface. Bacteria that make it to the surface would then experience a high QS signal which would stimulate expression of ECM components and thus biofilm formation. However, this idea remains to be tested more directly. It is also possible that there is simply a cell-to-cell heterogeneity of ECM expression that arises from fluctuations in gene expression,[15] such that some cells are expressing higher levels of ECM components, and when these cells "bump" into a surface, they are more likely to stick to it. Either way, the role of QS in biofilm initiation is worth pursuing, as it can provide a better understanding of biofilm initiation.

2.4. *Spatial aspects of quorum sensing*

Even biofilms formed by a single species, contain cells that reside in distinct physiological or morphological states. For example, some cells may be expressing ECM components, while others are expressing flagella motor components for motility.[16] Yet, other cells can differentiate into morphologically distinct spores, that represent a dormant and extremely resilient state of bacteria to survive extreme temperatures or draught. Interestingly, for *B. subtilis* biofilms growing on agar surfaces, it has been shown that these distinct cells types are not arbitrarily distributed in space.[17] Rather, ECM expressing cells are more likely to be observed near the agar surface, while spores reside near the top of the biofilm closest to the air interface. Quorum sensing may regulate the differentiation of cells into distinct states, and thus it is important to consider the role of QS in the spatial organization of distinct cell types. Let us remember that quorum sensing is based on the passive diffusion of signaling molecules among cells. Consequently, to understand how quorum sensing can affect the spatial organization of cells in a biofilm, we have to consider factors such as the diffusion and degradation of the signaling molecules. For example, if the signal is rapidly degraded, then the distance over which a signal released from a given cell can be effective will be limited. The degradation of the signal will thus also affect its spatial concentration gradient. Another factor is the diffusion of QS molecules within biofilms. Biofilms are very densely packed with cells that are encapsulated by extracellular matrix components. Over 200 different molecules can comprise the extracellular matrix, with the majority being polysaccharides, amyloid fibers and DNA.[18] These molecules can interact (bind) with QS signals such as peptides, and impact their diffusion. Conditions inside the biofilm can thus not be regarded to be analogous to liquid media conditions. Diffusion of short peptides that typically serve as QS molecules is thus likely to be limited in biofilms when compared to liquid cultures. Therefore, the physical properties of the biofilm can limit the effective range of QS. This limitation on QS does not preclude its role in the spatial organization of biofilms, but rather raises critical questions that have to be considered and it also provokes the question whether there could be other signaling mechanisms that play a role in the spatial organization of biofilms.

2.5. *Threshold response and synchronization of cellular behavior*

A key feature of quorum sensing is that cells exhibit a threshold dependent response.[19] In other words, when the extracellular concentration of quorum sensing molecules reaches a certain threshold, cells respond by changing their gene expression. Therefore, quorum sensing typically promotes the synchronization of cellular behavior within a population. In other words, when the global signal reaches a certain concentration, a synchronized population-level response is triggered. Such synchronization appears to play a critical role in pathogenic behavior of bacteria, where bacteria will only trigger expression of pathogenic genes when they have reached a sufficiently high cell density.[20] Presumably, this synchronization enables bacteria to execute a more effective attack on the host by first reaching a higher density to then induce a more potent effect on host cells. The higher cell density may also allow some bacteria to escape the host immune response. Interestingly, cells within the biofilm community exhibit a high degree of heterogeneity, where cells in different regions of the biofilm engage in different behaviors at different times. For example, cells in the periphery of the biofilm can have a different metabolic activity compared to bacteria within the biofilm interior.[21] These differences can also change over time as the biofilm develops and cells communicate with each other and also engage in complementary metabolic cooperation.[22] Furthermore, some cells within the biofilm can express motility genes, while others express ECM components.[17] This indicates that the overall biofilm population is not synchronized in its behavior. This heterogeneity does not rule out a role for QS in mature biofilms, but raises the question of whether other mechanisms are required to regulate the spatial and temporal organization of cell types within biofilms.

Next, I will discuss that bacteria in a biofilm utilize a different form of long-range communication that is based on ion channel-mediated signaling. This electrochemical signaling mechanism enables oscillations in biofilm growth dynamics. Furthermore, I will discuss how this electrochemical signaling is propelled by an active cell-to-cell relay mechanism that is reaction limited.

3. Ion Channel Mediated Electrical Signaling: Long-range Signaling and Overcoming the Limits of Diffusion

When we think of electrical signaling in biology, we typically think of the brain. The molecules that underpin brain function are ion channels that regulate the flux of charged ions across the cell membrane. It is these ion channels that allow neurons in the brain to generate action potentials and communicate with each other. There are two main ion channels involved in the generation of action potentials, namely sodium and potassium channels, and we will focus here on potassium channels. In fact, potassium ion channels are also critical for the operation of muscles and many other tissues such as the heart and liver.[23,24] The functional significance of potassium ion channels is also emphasized by its evolutionary exploitation, where predators, like spiders, scorpions and snakes, have evolved small molecules that they secrete into a prey to inhibit these ion channels. [25]

The first structure of any ion channel was that of the potassium ion channel. Obtaining the x-ray crystallography structure of this channel was by no means a straightforward task. Rod MacKinnon and his group made amazing strides in being able to obtain the first structure of a potassium ion channel. This paper, published in *Science*, is one of the most amazing papers published. [26] I highly recommend this paper if you want to read a really elegant and striking paper. The structure of the potassium ion channel is right up there with the structure of DNA, in that both structures revealed how the protein performs its biological function. The ion channel is basically mimicking the hydration shell around a potassium ion, and that is the secret to its selectivity and conductance.

What most people may not be aware of is that the first structure of a potassium ion channel was that of a bacterial ion channel. Scientists have since obtained many other bacterial ion channel structures. [27] These bacterial ion channel structures have even been used to design drugs that target neurological diseases. One of the ironies was that even though many bacterial ion channel structures were obtained, it was unknown what potassium ion channels do in bacteria.

In fact, here is a statement by Ching Kung, a leader in the field of microbiology, who wrote in his review on the tenth anniversary of the potassium ion channel structure [27]: *"It is ironic that the puzzling basis of ion specificity is finally solved in concrete terms with a channel from a 'lowly' bacterium and its true function in the life of the bacterium is unknown."* It is noteworthy that the word *"lowly"* is in quotation marks. While bacteria are unicellular organisms, they are not to be confused with solitary organisms. In fact, most bacteria on our planet reside in communities, such as biofilms. [2] The social existence of bacteria provoked the question whether bacterial ion channels may have a native function in the context of the community lifestyle, such as a biofilm.

3.1. *Quantitative measurement of biofilm dynamics*

To investigate the function of bacterial ion channels in communities, we developed a microfluidic based approach to quantitatively study bacterial biofilms. [22] This advance was born out of necessity to address the challenge to obtain quantitative measurements in these densely packed communities. With an appreciation of the importance of bacterial communities, most biofilm studies over the last two decades have focused on biofilms grown on agar plates. While this approach is highly feasible from an experimental perspective, biofilms grown on agar plates have limited optical accessibility and thus these studies have only provided gross morphological information on biofilms. Recent studies have begun to utilize new experimental platforms to achieve single-cell resolution imaging. However, due to technical challenges, these approaches were applied to only biofilms that contain thousands of cells. [28] The innovation in our device is that we have an optically accessible large growth chamber, which allows us to grow a bacterial community that contains millions of cells. In addition, by utilizing a shallow chamber height of approximately 9 microns, we are able to grow large biofilm, while still being able to resolve individual cells within the biofilm and perform quantitative measurements.

We then identified a way to measure the membrane potential of bacteria in biofilms, by using a fluorescent dye that acts as a Nernstian potential indicator (the higher the membrane potential, the brighter the fluorescence signal in cells). Specifically, we used Thioflavin-T, which is a fluorescent reporter that indicates changes in the membrane potential.[29] As a biofilm grows we can see oscillations in the fluorescence of the dye. This result shows that a bacterial community of approximately two million cells is collectively changing its membrane potential. The collective oscillations in the biofilm are not uniform in space and time. Specifically, we can see that the fluorescent signal first increases in the center of the biofilm, and then this increase in the signal propagates from the interior to the periphery of the biofilm. These spatio-temporal changes within the biofilm community give rise to wave-like propagation through the community.[30]

3.2. *Mechanism for electrical signaling in biofilms*

The above described observations raise two important questions: How does it work and what is it doing? One of the points that led us to the answer was the realization that bacterial biofilms are incredibly densely packed. This dense packing results in a fundamental problem of nutrient access for the cells that are buried deep within the bacterial community. In particular, we identified that the interior cells are being starved for glutamate, which is a critical nitrogen source for all bacteria and many other organisms. Studies have shown that the majority of proteins synthesized in bacteria derive their nitrogen from glutamate.[31] Furthermore, glutamate is uniquely positioned between the TCA cycle and biomass synthesis, and thus serves as a link between catabolism and anabolism. In fact, bacteria have a designated stress response to glutamate limitation to ensure protein synthesis and biomass production. What we uncovered is that glutamate starvation in the interior of the community results in the opening of an ion channel that is called YugO in *Bacillus subtilis*. YugO is a potassium ion channel with a gating domain called TrkA (Figure 3).

The TrkA domain is sensitive to the metabolic state of the cell.[32,33] When bacteria are starved for glutamate, these ion channels open. *B. subtilis* cells contain approximately 350 mM potassium,[34] while the extracellular concentration of potassium is 5 mM. Consequently, opening of the YugO ion channel results in an outward flux of potassium that is driven by the chemical gradient across the cell membrane. This process is described in detail in the literature and this efflux of positively charged potassium ions increases the negative membrane potential, causing hyperpolarization. The potassium efflux occurs until the equilibrium described by the Nernst equation is reached. Specifically, given the extracellular concentra-

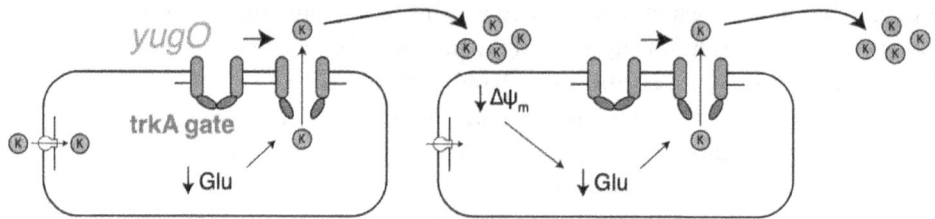

Fig. 3.   Schematic diagram of the mechanism for electric signaling in biofilms.

tion of potassium of 4 mM and the estimated intracellular concentration of potassium of 450 mM, the Nernst equation indicates that efflux of potassium ions will reach equilibrium once the membrane potential of the bacterial cell changes from its resting state value of approximately −90 mV to −110 mV. At this point, the concentration-driven efflux of potassium cations is opposed by the inside negative electrical potential of the cell. For the detailed mathematical description of this process please see (Prindle *et al.* 2015).[30]

The next question raised is once a change in membrane potential is triggered, how is this signal propagated through the biofilm? The process is best described as pulse coupling, where a signal propagates among discrete units (cells) as a function of space and time. Specifically, efflux of potassium from one cell will alter the membrane potential of the neighboring cell. This process has been described in neurons and is known as ephaptic transmission. Essentially, the local increase in extracellular potassium that result from cellular potassium efflux, result in influx of potassium ions in the immediately neighboring cell. This influx occurs because the cell is negatively charged and ion channels permit ion flux bi-directionally. According to the literature, it takes approximately 10,000 monovalent ions to generate a very large potential difference of 100 mV across a bacterial membrane.[35] Here is how we arrive at this number: Let us assume that the membrane voltage difference is due to monovalent cation transport only and that the cell is a parallel plate capacitor. Taking membrane width $d = 4$ nm, relative permittivity of the bilayer $\mathcal{E}r = 2$ (Appendix 2 in Ref. 36), vacuum permittivity $\mathcal{E}0$, and voltage change $V = 100$ mV, one can calculate areal charge density σ. Using this value, one can calculate total charge q needed to be moved for 100 mV change as areal charge density times surface area (for *E. coli* = 5 mm^2), divided by electron charge. If we plug in the numbers, we find that number to be $q = 10,000$ monovalent ions. This value is consistent with membrane capacitance Cm of 1 mF/cm^2, where Cm = A*Csp, and Csp is specific membrane capacitance = 1 μF/cm^2.[37] Therefore, a 10 mV change in membrane potential would require 1,000 cations to be moved across the membrane. The change in the membrane potential can increase (hyperpolarize), or decrease (depolarize) the resting membrane potential of the cell depending on whether cations are moving in or out of the cell. In fact, it is known that influx of potassium ions depolarizes cells.[38] Such depolarization has been experimentally observed for many cell types that are exposed to high extracellular potassium ion concentrations. A sudden increase in extracellular potassium ions will result in its cellular influx, presumably through importers. This outcome appears to be generic as it has been reported in many diverse biological systems, ranging from neurons to bacteria.[30] Depolarization is also known to reduce the proton motive force, which is required for bacteria to take up glutamate.[39] Specifically, glutamate is a charged amino acid that cannot simply diffuse across the membrane of the cell. *Bacillus subtilis* has evolved a specific transporter (GltP) that co-transports glutamate together with two protons into the cell.[40] Therefore, glutamate uptake directly depends on the proton motor force, which is a function of the membrane potential. Consequently, depolarization reduces glutamate uptake, which in turn means this cell is going to experience glutamate limitation (as reported by stress response promoter activity[22]). Once again, glutamate limitation results in the opening of the YugO potassium ion channels, and efflux of potassium ions. In this way, the signal propagates by being re-amplified by each cell, a process that can be likened to a domino effect or a chain reaction, where the signal is relayed from cell to cell.

These findings show that in addition to the structural similarity of ion channels between bacteria and mammalian cells, they also have functional similarity. Bacteria seem to be using ion channels for the purpose of cell-to-cell signaling, similar to neurons in the brain. This was unexpected. Interestingly, there is a process in mammalian brains known as Cortical Spreading Depression (CSD), which also results in depolarizing wave of extracellular potassium that propagates through the brain and interferes with the metabolic activity of distant neurons. Intriguingly, CSD is triggered by glutamate starvation, which also triggers electrical signaling through depolarizing extracellular potassium waves in bacterial biofilms. One might speculate that CSD could have some relation to electrical signaling in biofilms as the same metabolic pathways, ion channels and principle of long-range transmission are involved.

The active cell-to-cell signaling relay enables the biofilm to transmit long-range signals by overcoming the limits of processes that exclusive rely on diffusion of signals. Action potentials in neurons are the best characterized and understood reaction-limited signal transmission system that also overcome limitations of diffusion. Therefore, we developed a model of bacterial electrophysiology to describe electrical signaling in these communities. To model signaling in the biofilm, we reached back to the gold standard of electrophysiology: the model originally developed by Hodgkin and Huxley in the 1950's.[41] This mathematical framework accounts for the observed membrane potential dynamics that propagate within the biofilm with constant amplitude.[30] It is noteworthy that other biological systems have also overcome limitations of diffusion. For example, elegant work by James Ferrell's group at Stanford described mitotic "trigger waves" as the underlying mechanism to allow for a frog oocyte to achieve long range coordination that would not be possible through a process based only on diffusion.[42] Such trigger waves and action potential propagation have similarities in that they are both based on excitable dynamics with positive feedback loops that enable long-range signaling without a decay in the amplitude of the signal.[43] While diffusion is clearly a critical part of many biological processes, it appears that spatially extended systems also utilize active relays such as action potentials and trigger waves to achieve more effective long range communication and coordination.

3.3. *Timing of action potentials in bacteria versus neurons*

You might ask: is the action potential very similar to what happens in neurons? The answer is no. In neurons, action potentials require not only potassium but also sodium. Essentially, two ions are needed that move in opposite directions across the membrane. However, bacteria can generate a more primitive action potential without a counter ion.

So what gives? The answer is the characteristic time of the response. If you need to generate an action potential on the timescale of milliseconds because you are a fly trying to navigate in real time during flight, you need to react very quickly. For this speed you are willing to pay the price of having a counter ion to quickly reset your membrane potential and generate very fast action potentials. The bacterium, on the other hand, is trying to solve a problem that does not require millisecond resolution. It is trying to solve a problem of metabolic starvation. Bacteria are able to resolve this problem at a slower time scale and so the potassium ion channel by itself is sufficient. The characteristic response time of the action potential in biofilms has been determined to be approximately 20 min.[44] This means that the duration of an action potential in biofilms is about five orders of magnitude slower than the typical duration of an action potential in neurons. Noteworthy is that potassium ion channels evolved before sodium ion channels,[45] and so bacterial action potentials may constitute ancient, primitive and much slower action potentials that arose hundreds of millions of years before neurons evolved.

3.4. *Bacteria coordinate their membrane potential in space and time over different scales*

One of the conclusions that arises from this research is that bacteria can coordinate their membrane potential in space and time. Bacteria can thus collectively modulate their bioenergetic state. The depth of this process goes back to the origins of life and all living cells utilize membrane bioenergetics. It may thus not be surprising that changes in membrane potential can regulate many biological functions. In fact, we uncovered a whole range of functions and behaviors that take place across different time and length scales, which I briefly describe below.

3.4.1. *Within the biofilm*

What is the biological purpose of ion channel-mediated electrical signaling at the level of the biofilm? One function it performs appears to solve the problem of starvation. As mentioned before, the interior of the biofilm becomes starved for nutrients, not just because it is buried in the interior, but because the peripheral cells are consuming nutrients and thus actively starving the interior cells of food. This starvation of interior cells is thus not just a problem of diffusion but also consumption.

Interior cells respond to starvation by sending out electrochemical signals. When these signals arrive at the biofilm edge, they reduce the growth of cells at the periphery. Consequently, nutrient consumption by peripheral cells is reduced and allowing more nutrients to trickle into the interior, alleviating the stress. As the stress is reduced, the signal decays and the peripheral cells start growing again. When the peripheral cells grow, they starve the interior cells again. This gives rise to a spatially extended negative feedback loop, which generates collective oscillations of biofilm growth and electrochemical signaling that are experimentally observed. For a detailed description of the quantitative aspects of this process, please see Liu *et al.* and Prindle *et al.*[22,30,46]

3.4.2. *Beyond the edge of the biofilm*

An interesting question is whether these waves of extracellular potassium ions that propagate through the biofilm cease when they arrive at the edge of the biofilm? The speculation was that these waves could propagate beyond the edge and extend some distance away from the edge of the biofilm. Biofilms form in aqueous environments and can thus be surrounded by planktonic or motile cells that are not part of the community. As the electrical waves periodically hit the edge of the biofilm, bacteria that are swimming within some distance of approximately 100 cell length, can respond to the electrochemical signals by swimming towards the edge of biofilm.

Here is a brief description of the underlying mechanism: (1) The biofilm is releasing potassium waves that extend past the biofilm. (2) This causes a change in the membrane potential of motile cells at some distance away from the biofilm, which is similar to the membrane potential response generated by cells within the biofilm. (3) Since motility is driven directly by the proton motor force, which turns the flagella, these electrical waves washing over the motile cells modulate their tumbling frequency and thereby recruit bacteria from a distance to the biofilm through this electrical signaling. This process is also accounted for by what is known as an agent-based model, where each cell is considered as an independent entity and the motility of cells in response to a dynamic potassium gradient is simulated mathematically.[47]

Since this form of signaling is different from quorum sensing in that it does not require designated receptors, one can thus reason that this process would be generic and apply to other bacterial species as well. In fact, we were able to show that even gram-negative bacteria (*Bacillus subtilis* is gram-positive bacteria) such as *Pseudomonas* can be attracted and then incorporated into a gram-positive biofilm. When these motile cells are attracted to the biofilm edge, they get stuck because of the sticky extracellular matrix. Over time, as the biofilm grows, these motile cells become incorporated. Action potentials generated by biofilms thus provide a potential mechanism for the formation of mixed species biofilms, which we know exist in nature.[3]

Why would a biofilm under stress recruit other bacteria to its community? Similarly, one may ask why a motile cell would want to seek out a biofilm community with densely packed cells and limited resources? I liken this process to stress induced mutagenesis, where cells under stress will increase the probability of generating mutations within their genetic code. Why would a cell want to introduce mutations into its genome during a period of stress? Mutations will generate diversity for natural selection to act upon cells with higher fitness. Therefore, the generation of diversity under stress seems to be an important part of evolution by natural selection to overcome challenges. Along the same line of thought, it is conceivable that by recruiting different cells to its community, the biofilm generates diversity in its make-up that can then be subject to natural selection. Such evolutionary questions are not easy to test, but we may approach this problem by setting up a more defined laboratory evolution experiment. We can control the diversity of species in the biofilm and then determine whether the heterogeneity affects biofilm fitness (as defined by simple metrics such as biofilm growth) under defined environmental conditions. Such approaches are worth pursuing in future studies to advancing our understanding of mixed species biofilms.

3.4.3. *Coupling between two biofilms*

Given that electrochemical signals can extend beyond the edge of the biofilm, this provokes the question of what happens if there is another biofilm community nearby? To answer this question, we grew two biofilms separated by approximately 0.75 mm in the same growth chamber and observed their membrane potential using the same fluorescent dye as described before. What we observed is that the biofilm with the slightly larger size begins oscillating first, but then once the other biofilm also initiates oscillations, the two biofilm oscillations become immediately synchronized. In other words, the two biofilms oscillate in phase. What we were observing is the sharing of the stress signal between the two biofilms, which synchronizes their oscillatory stress dynamics. So, the stress of one biofilm imposes stress on the other through electrical signaling. However, there is another process that couples the two biofilms, namely competition for nutrients. This competition arises because the two biofilms are growing in the same chamber and are together experiencing the limited nutrient environment.

A mathematical model of coupled phase oscillators predicted that if competition is increased, it would push the two biofilms to oscillate precisely out of phase rather than in phase. That is because the competition would be stronger, dominating the coupling between the two biofilms. While the electrochemical signaling is driving the two biofilms to synchronize their phase, competition is pushing them out of phase.

Experiments validated this prediction under conditions where nutrient concentrations were reduced. In particular, the biofilms start to oscillate out of phase; anti-phase to be precise. This arises when biofilms are further starved for glutamate, which is the key amino acid that triggers action potentials. What is the biological significance of this? We hypothesized that the two biofilm colonies were engaging in a strategy known as timesharing. Meaning, instead of splitting the nutrients, which would be the case if they were oscillating in phase, anti-phase oscillations allow each biofilm to take turns consuming whatever nutrients are available at that time. We constructed different mutants where we manipulated the competition strength or the communication strength and in all cases we see that the phase relationship directly affects the growth rate of biofilms. This gives rise to the counterintuitive result that biofilm pairs growing in lower nutrient concentrations grow better. Let us consider a scenario to illustrate how a pair of biofilms engaged in timesharing under low nutrient conditions grow better than two biofilms under higher nutrient concentrations. The important point to note here is that each biofilm grows and thus consume nutrients periodically. This means that they are able to utilize the available nutrients only during their growth phase. Let us then assume one unit of nutrient concentration. If the growth of two biofilms is synchronized, their nutrient consumption is also synchronized. During their synchronized growth phase, the two biofilms will essentially split the available nutrients and thus have access to half the resources (0.5). When the two biofilms are in their non-growing phase, the available nutrient is basically "wasted" as it flows out of the chamber and does not contribute to biofilm growth. Now let us consider two biofilms that are engaged in timesharing and thus their growth and nutrient consumption are out of phase. We provide a 25% reduced nutrient concentration (0.75) for these antiphase oscillating biofilms. Given the antiphase oscillatory growth, the two biofilms do not have overlapping nutrient consumption and instead take turns utilizing available nutrients. Therefore, each biofilm will have access to all the nutrients available during its growth phase (0.75). In this way, timesharing biofilms growing at a 25% reduced nutrient concentration (0.75) will experience greater nutrient access, compared to two biofilms that are resource splitting at full nutrient levels and thus only have access to 0.5 nutrients. Consistent with this, it has been shown that timesharing biofilm pairs exhibit a quantifiably higher average growth rate than two resource splitting biofilms growing under more nutrient rich conditions.[48] It is fascinating to discover such surprising outcomes that in hindsight can be easily understood.

4. Other Types of Electrochemical Communication in Biofilms

Ion channel mediated electrical signaling is not the only type of electrochemical communication in biofilms. There are different types of electrical communication between bacteria.[49] Bacteria can, for example, make nanowires that support direct electron transport. When bacteria are densely packed, they can also use cytochromes that allow electrons to hop from one cell to the other. Victoria Orphan and Diane Newman are working on these types of processes.[50-52]

5. Future Challenges: The "Dark Matter" of Bacterial Cell-to-Cell Signaling

5.1. *Bacterial and biofilm electrophysiology*

It appears that bacteria, while being known for their unicellularity, operate as a multicellular organism in densely packed biofilm communities. This higher-order coordination is made by communication among bacteria. These findings emphasize the importance of considering the context of community-level existence when studying these "simple" life forms.

In terms of future challenges, I think it is important to point out that we still understand extremely little about bacterial electrophysiology. In their book, *Cell Biology by the Numbers*,[35] Rob Phillips and Ron Milo point out that after water, the most common component of a cell is inorganic ions. The cell hoards potassium, magnesium and calcium ions. We have a rather murky understanding of what these ions do in the cells. A simple hypothesis may be that cells import or export ions to maintain the intracellular environment within a narrow range that could be described as biologically optimal. However, it is also possible that cells modulate ion flux to change the activity of various proteins and cellular functions. For example, cells may increase the expression of ion transporters or channel to control protein activity in a post-translational manner. Such processes emphasize the importance of determining how many ions of a given species are bound to molecules, versus how many are in free "solution". Pursuing these questions may reveal what biological processes are regulated by changes in ion flux. It is worth noting that these questions extend beyond bacteria, since ions are the most common component after water molecules for any living cell, even human cells.

Even though the scientific community accumulated many insights into the electrophysiology of neurons, comparatively little is known about the electrophysiology and the role of membrane potential in many other cell types and species. The role of membrane potential has been studied in muscle, cardiomyocyte cells and neutrophil cells,[53,54] but there is a whole range of processes and functions that remain to be uncovered simply because we have not explored them carefully, or because we lacked the technology to do so. While the membrane potential can enable for example uptake of nutrients against chemical gradients in diverse cell types,[55,56] other functional roles for the membrane potential may yet to be elucidated.

5.2. *Coupling through shared resources*

I would also like to point out other mechanisms of cell-to-cell signaling that are not very well understood. For example, communication among cells can also arise from processes that do not involve specific signaling molecules, but rather environmental or extracellular resources that are being shared. Competition for resources can couple cells and represent a different form of communication among cells. Perhaps we have to be careful to refer to such coupling as signaling, because such processes are not producing a specific signal, it is just that the action of one cell that affects the behavior of another cell. This is not through some direct signal but just by the fact that one cell, for example, might change the pH around another cell, thereby causing a response.

5.3. *Heterogeneity*

One of the intriguing aspects of cellular behavior is variability at the single cell level. Responses of cells to quorum sensing signals as well as electrochemical signals can be variable, giving rise to cell-to-cell heterogeneity in populations of cells. This heterogeneity is evident when we look at individual cells inside a biofilm.[44] We do not fully understand the processes that give rise to such heterogeneity and we also do not understand whether heterogeneity has biological consequences or serves biological consequences. Why is this type of heterogeneity observed for both electrical signaling and quorum sensing? This may indicate that there could be common mechanisms that are responsible for heterogeneity. Such general principles would be a great breakthrough in our understanding of cellular behavior. I think it was a beautiful question that Ned Wingreen asked about how much of the heterogeneity is something that is real in terms of the biology and about how much might be just an artifact of the type of conditions or experimental measurements scientists are performing in the lab. This is an important challenge for the future of the field that can provide deeper insights.

5.4. *Local weak interactions*

Another challenge for the future is in regards to weak interactions between cells that are not easy to measure experimentally. In particular, interactions at very short length and time scales can be difficult to measure experimentally, but could still have a critical impact on the biology at very long length or time scales. These types of interactions might elude detection as weak biochemical interactions have always been tricky to measure experimentally. But I would like to emphasize that it is worthwhile to pursue weak interactions in biology. We have learned in physics that, for example, interactions at the sub-atomic scale can influence fundamental properties of the cosmos at length scales that span light years and time scales that span billions of years.

5.5. *Pattern formation and signaling that is not based on biomolecules*

Cells can also interact or even communicate through mechanical forces. There is building evidence that such mechanical forces play a role in the coordination and organization of bacterial communities.[57,58] Studies have shown that such mechanical forces can lead to the formation of 3D morphological features in bacterial biofilms. Since the conference is about space and time, I am going to end by saying that biofilms do amazing processes and generate fascinating patterns in space and time. If we look at one of these biofilms (such as the one in Figure 1) and if I did not tell you that this was a biofilm you might think that you are looking at mammalian development. In fact, there is buckling of tissues because of mechanical forces that are very reminiscent of what happens during development in higher organisms including human development.

6. Conclusion

I hope that I was able to provide a sense of what bacteria are capable of in terms of communication and organization in space and time. It is clear that we have much to learn about bacteria and we even need to change our paradigm and start thinking and approaching bacteria not as solitary creatures, but social creatures that reside in densely packed bacterial communities where they can access surprising emergent behaviors. Signaling among cells through quorum sensing or electrochemical signaling provide an experimentally accessible framework that is ready for application and development of theories that can allow us to understand the range of behaviors accessible to one of the most prevalent life forms in our planet (and perhaps even on other planets).

References

1. C. D. Nadell, J. B. Xavier, and K. R. Foster, The sociobiology of biofilms, *FEMS Microbiol. Rev.* **33**(1): 206–24 (2009).
2. H.-C. Flemming, Bacteria and archaea on Earth and their abundance in biofilms, *Nat. Rev. Microbiol.* **17**: 247–260 (2019).
3. H. Flemming, J. Wingender, U. Szewzyk, P. Steinberg, S. A. Rice, and S. Kjelleberg, Biofilms?: an emergent form of bacterial life, *Nat. Publ. Gr.* **14**(9): 563–575 (2016).
4. P. Stoodley, K. Sauer, D. G. Davies, and J. W. Costerton, Biofilms as complex differentiated communities, *Annu. Rev. Microbiol.* **56**: 187–209 (2002).
5. K. Jefferson, What drives bacteria to produce a biofilm?, *FEMS Microbiol. Lett.* **236**(2): 163–173 (2004).
6. A. M. Earl, R. Losick, and R. Kolter, Ecology and genomics of Bacillus subtilis, 269–275 (2008).
7. D. G. Davies, The Involvement of Cell-to-Cell Signals in the Development of a Bacterial Biofilm, *Science* **280**(5361): 295–298 (1998).
8. C. M. Waters and B. L. Bassler, Quorum Sensing: Cell-to-Cell Communication in Bacteria, *Annu. Rev. Cell Dev. Biol.* **21**(1): 319–346 (2005).

9. A. Eldar, Social conflict drives the evolutionary divergence of quorum sensing, *Proc. Natl. Acad. Sci.* **108**(33): 13635–13640 (2011).

10. B. A. Duerkop *et al.*, Quorum-sensing control of antibiotic synthesis in Burkholderia thailandensis, *J. Bacteriol.* **191**(12): 3909–3918 (2009).

11. W. C. Fuqua, S. C. Winans, and E. P. Greenberg, Quorum sensing in bacteria: the LuxR-LuxI family of cell density-responsive transcriptional regulators, *J. Bacteriol.* **176**(2): 269–275 (1994).

12. V. Berk *et al.*, Molecular architecture and assembly principles of vibrio cholerae biofilms, *Science* **337**(236): 236–239 (2012).

13. Y. Sakuragi and R. Kolter, Quorum-sensing regulation of the biofilm matrix genes (pel) of Pseudomonas aeruginosa, *J. Bacteriol.* **189**(14): 5383–6 (2007).

14. A. Trovato, F. Seno, M. Zanardo, S. Alberghini, A. Tondello, and A. Squartini, Quorum vs. diffusion sensing: A quantitative analysis of the relevance of absorbing or reflecting boundaries, *FEMS Microbiol. Lett.* **352**(2): 198–203 (2014).

15. M. B. Elowitz, A. J. Levine, E. D. Siggia, and P. S. Swain, Stochastic gene expression in a single cell, *Science* **297**(5584): 1183–1186 (2002).

16. P. S. Stewart and M. J. Franklin, Physiological heterogeneity in biofilms, *Nat. Rev. Microbiol.* **6**(3): 199–210 (2008).

17. H. Vlamakis, C. Aguilar, R. Losick, and R. Kolter, Control of cell fate by the formation of an architecturally complex bacterial community, *Genes Dev.* **22**(7): 945–953 (2008).

18. L. Hobley, C. Harkins, C. E. MacPhee, and N. R. Stanley-Wall, Giving structure to the biofilm matrix: an overview of individual strategies and emerging common themes, *FEMS Microbiol. Rev.* 1–21 (2015).

19. Q. Seet and L. H. Zhang, Anti-activator QslA defines the quorum sensing threshold and response in Pseudomonas aeruginosa, *Mol. Microbiol.* **80**(4): 951–965 (2011).

20. T. R. de Kievit and B. H. Iglewski, Bacterial quorum sensing in pathogenic relationships, *Infect. Immun.* **68**(9): 4839–4849 (2002).

21. J. Cole, L. Kohler, J. Hedhli, and Z. Luthey-Schulten, Spatially-resolved metabolic cooperativity within dense bacterial colonies, *BMC Syst. Biol.* **9**(1): 1–17 (2015).

22. J. Liu *et al.*, Metabolic codependence gives rise to collective oscillations within microbial communities, *Nature* **523**(7562): 550–554 (2015).

23. M. Tristani-Firouzi, J. Chen, J. S. Mitcheson, and M. C. Sanguinetti, Molecular biology of K^+ channels and their role in cardiac arrhythmias, *Am. J. Med.* **110**: 50–58 (201AD).

24. A. Ramírez, A. Y. Vázquez-Sánchez, N. Carrión-Robalino, and J. Camacho, Ion channels and oxidative stress as a potential link for the diagnosis or treatment of liver diseases, *Oxid. Med. Cell. Longev.* **2016**: 1–17 (2016).

25. S. Bajaj and J. Han, Venom-derived peptide modulators of cation-selective channels: Friend, foe or frenemy, *Front. Pharmacol.* **10**: 1–12 (2019).

26. D. A. Doyle *et al.*, The structure of the potassium channel: molecular basis of K^+ conduction and selectivity, *Science* **280**(5360): 69–77 (1998).

27. B. Martinac, Y. Saimi, and C. Kung, Ion channels in mircrobes, *Physiol. Rev.* 1449–1490 (2008).

28. R. Hartmann *et al.*, Emergence of three-dimensional order and structure in growing biofilms, *Nat. Phys.* **15**(3): 251–256 (2019).

29. M. Haidekker and E. Theodorakis, Environment-sensitive behavior of fluorescent molecular rotors, *J. Biol. Eng.* **4**: 11 (2010).

30. A. Prindle *et al.*, Ion channels enable electrical communication within bacterial communities, *Nature* **527**(7576): 59–63 (2015).

31. M. C. Walker and W. A. van der Donk, The many roles of glutamate in metabolism,

J. Ind. Microbiol. Biotechnol. **43**(2–3): 419–430 (2016).

32. Y. Cao *et al.*, Gating of the TrkH ion channel by its associated RCK protein TrkA, *Nature* **496**(7445): 317–322 (2013).

33. T. P. Roosild, S. Miller, I. R. Booth, and S. Choe, A mechanism of regulating trans-membrane potassium flux through a ligand-mediated conformational switch, *Cell* **109**(6): 781–791 (2002).

34. A. M. Whatmore, J. A. Chudek, and R. H. Reed, The effects of osmotic upshock on the intracellular solute pools of Bacillus subtilis, *J. Gen. Microbiol.* **136**(12): 2527–2535 (1990).

35. R. Milo and R. Phillips, *Cell Biology by the Numbers*, 1st ed. (Garland Science, 2015).

36. W. D. Stein, *Channels, Carriers, and Pumps?: An Introduction to Membrane Transport* (San Diego: Academic Press).

37. O. S. Andersen, Cellular electrolyte metabolism, in *Encyclopedia of Metalloproteins*, eds. R. H. Kretsinger, V. N. Uversky, and E. A. Permiakov (New York: Springer, 2013), pp. 580–587.

38. H. F. Lodish, A. Berk, S. L. Zipursky, P. Matsudaira, D. Baltimore, and J. E. Darnell, Section 21.2 the action potential and conduction of electric impulses, in *Molecular Cell Biology*, 4th ed. (New York: W. H. Freeman, 2000).

39. T. Krulwich, G. Sachs, and E. Padan, Molecular aspects of bacterial pH sensing and homeostasis, *Nat. Rev. Microbiol.* **9**(5): 330–43 (2011).

40. B. Tolner, T. Ubbink-Kok, B. Poolman, and W. N. Konings, Characterization of the proton/glutamate symport protein of Bacillus subtilis and its functional expression in Escherichia coli, *J. Bacteriol.* **177**(10): 2863–9 (1995).

41. A. L. Hodgkin and A. F. Huxley, A quantitative description of membrane current and its application to conduction and excitation in nerve," *J. Physiol.* **117**: 500–544 (1952).

42. J. B. Chang and J. E. Ferrell, Mitotic trigger waves and the spatial coordination of the Xenopus cell cycle, *Nature* **500**(7464): 603–607 (2013).

43. L. Gelens, G. Anderson, and J. E. Ferrell, Spatial trigger waves: positive feedback gets you a long way, *Mol. Biol. Cell* **25**(22): 3486–93 (2014).

44. J. W. Larkin *et al.*, Signal percolation within a bacterial community, *Cell Syst.* **7**(2): 137–145.e3 (2018).

45. P. A. V. Anderson and R. M. Greenberg, Phylogeny of ion channels: Clues to structure and function, *Comp. Biochem. Physiol. – B* **129**(1): 17–28 (2001).

46. R. Martinez-Corral, J. Liu, A. Prindle, G. M. Süel, and J. Garcia-Ojalvo, Metabolic basis of brain-like electrical signalling in bacterial communities, *Philos. Trans. R. Soc. B Biol. Sci.* **374**(1774): 20180382 (2019).

47. J. Humphries *et al.*, Species-independent attraction to biofilms through electrical signaling, *Cell* **168**(1–2): 200–209.e12 (2017).

48. J. Liu *et al.*, Coupling between distant biofilms and emergence of nutrient time-sharing, *Science* **356**(6338): 638–642 (2017).

49. D. D. Lee, A. Prindle, J. Liu, and G. M. Süel, SnapShot: Electrochemical communication in biofilms, *Cell* **170**(1): 214–214.e1 (2017).

50. S. E. McGlynn, G. L. Chadwick, C. P. Kempes, and V. J. Orphan, Single cell activity reveals direct electron transfer in methanotrophic consortia, *Nature* **526**(7574): 531–535 (2015).

51. L. E. P. Dietrich, T. K. Teal, A. Price-Whelan, and D. K. Newman, Redox-active antibiotics control gene expression and community behavior in divergent bacteria, *Science* **321**(5893): 1203–1206 (2008).

52. K. C. Costa, N. R. Glasser, S. J. Conway, and D. K. Newman, Pyocyanin degradation

by a tautomerizing demethylase inhibits Pseudomonas aeruginosa biofilms, *Science* **355**(6321): 170–173 (2017).

53. K. Chen, D. Zuo, Z. Liu, and H. Chen, Kir2.1 channels set two levels of resting membrane potential with inward rectification, *Pflugers Arch. Eur. J. Physiol.* **470**(4): 599–611 (2018).

54. N. Demaurex, J. Schrenzel, M. E. Jaconi, D. P. Lew, and K.-H. Krause, Proton channels, plasma membrane potential, and respiratory burst in human neutrophils, *Eur. J. Haematol.* **51**(5): 309–312 (1993).

55. L. Chen, B. Tuo, and H. Dong, Regulation of intestinal glucose absorption by ion channels and transporters, *Nutrients* **8**(1): 1–11 (2016).

56. M. J. Chrispeels, N. M. Crawford, and J. I. Schroeder, Proteins for transport of water and mineral nutrients across the membranes of plant cells, *Plant Cell* **11**(4): 661–76 (1999).

57. M. Asally *et al.*, Localized cell death focuses mechanical forces during 3D patterning in a biofilm, *Proc. Natl. Acad. Sci. USA* **109**(46): 18891–6 (2012).

58. J. Yan *et al.*, Mechanical instability and interfacial energy drive biofilm morphogenesis, *eLife* **8**: 1–28 (2019).

Prepared comments

N. Wingreen: Future directions in quorum sensing

It is now appreciated that bacteria communicate with each other via small diffusible molecules in a process called quorum sensing. Traditionally, quorum sensing has been studied among planktonic cells in a well-mixed culture. However, in nature, bacteria are typically found in biofilms, namely surface-associated bacterial communities embedded in an extracellular matrix. In recent work from the Bassler lab and colleagues at Princeton, improved imaging has allowed single-cell resolution of living growing biofilms, revealing striking order of biofilm cells, with a 2D-to-3D transition as a consequence of directional cell division and anisotropic pressure caused by cell-to-surface adhesion.[a] The next step is to explore the origins and consequences of behavioral heterogeneity within growing bacterial biofilms. The aim is to answer previously intractable questions regarding the distinct roles played by individual cells inside growing biofilms: Which inputs to cells — quorum-sensing signals, mechanical stresses, nutrient availability — regulate matrix production? In turn, how do variations in gene expression and matrix production among cells in the biofilm control local morphogenic processes and, ultimately, global biofilm morphology? An important tool for these studies will be optogenetic activation of individual cells within growing biofilms.

Another new direction in quorum sensing, also developed in the Bassler lab, concerns interkingdom communication. In the pathogen *Vibrio cholerae*, multiple quorum-sensing circuits control pathogenesis and biofilm formation. A recent study[b] identified and characterized a new quorum-sensing signal-receptor pair. The signal, DPO, is made from threonine and alanine. Through a signal transduction pathway DPO represses genes required for biofilm formation and toxin production. The production of DPO relies on threonine that is released by the action of microbiome bacteria that digest a host protein, providing an example of interkingdom cooperation between host and microbiome to repress virulence by a pathogen.

Discussion

W. Bialek OK. Unprepared Comments? Daniel.

D. Fisher General questions on both of these. How much are some of these things epiphenomena that happen to be seen, but are not really what they may be evolved for. Particularly about Gürol Süel's comment toward the end about

---

[a] J. Yan, A. G. Sharo, H. Stone, N. S. Wingreen, B. L. Bassler, *Proceedings of the National Academy of Science* **113**(36): E5337–43 (2016).

[b] K. Papenfort, J. E. Silpe, K. R. Schramma, J. P. Cong, M. R. Seyedsayamdost, B. L. Bassler, A Vibrio cholerae autoinducer-receptor pair that controls biofilm formation, *Nature Chemical Biology* **13**(5): 551–557 (2017).

the *B. subtilis* eating some other colony and the electrical signal that may be killing them. Superficially seen, there may be more plausible scenarios of what's going on than the communication aspects of it. And also, quorum sensing, I can imagine for *V. cholera* it plays an important role. But then there are also ideas that in some contexts, what you want to know is — do you have neighbours? You want to know, is it worthwhile releasing something, or digesting something in the extracellular environment if that is going to get washed away or you are not going to get any benefit from it, then it is not worthwhile, so you put a bit out there and detect whether it comes back. And so in some sense, is it self sensing rather than quorum sensing? So a comment really on these issues on how one might separate things as they occur and can be seen in the lab, but determining whether they are functionally relevant or community relevant for the organisms.

G. Süel I'll start. Yes, it is a beautiful question, and we don't know the answer perfectly. But what I can tell you, is for example in the case of what I showed everybody on *B. subtilis*. One thing that is striking to us is that the expression of this particular potassium ion channel is directly tied to quorum sensing and is regulated such that the cells that are in the biofilm are the ones that express it. So, if they are in a different type of lifestyle, motile cells and so forth, they do not express this ion channel, they express it only in the context of the biofilm. Which is interesting and might suggest that it serves a purpose in the context of this bacterial community. But again, I cannot give you a 100 percent answer.

D. Fisher But that could be the killing part of it, more than the long-distance signalling part of it.

G. Süel Oh yes, and I hope that I was able to allude to the fact that there is a whole range of things that happen as a consequence of cells being able to coordinate their membrane potential. That can be for the good or for the bad, so to speak. Just like with many processes, they can be used for doing something beneficial, or maybe more for the predatorial aspect. And to be honest, we do not know which one came first, if one came first, and by no means is what I showed you is the end of the story. There seem to be many more things. And it is not surprising, the membrane potential is such a fundamental feature of any cell, all cells are polarised, so if you tap into it, it clearly makes a difference.

A. Walczak I would like to ask a question about what are the rules that these signalling systems have to obey and this is motivated by trying to think about more complicated systems. Because except for the last one Ned described, they are still pretty simple in terms of how they work. So obviously, if you are going to signal by diffusion, the information cannot propagate faster than the molecule diffuses, you can't break the speed of light, there are some basic rules you have to obey. Where are we now in this list of fundamental rules that are not just the rules of the road, where you can break the speed limit, like the ribosomes that Terry talked about yesterday, where you can't make molecules if you don't have ribosomes is another one. Do we have a list of rules right now that could be helpful for understanding more complex systems such as signalling by cytokines in the immune system for example.

H. Levine I guess I don't understand Daniels' skepticism. Maybe it is just natural.

D. Fisher I'm always skeptical.

H. Levine I know, that's what I said. But, I also want to make a historical remark, which is that the idea that bacteria behave in communities, with large degrees of cooperation, has been an idea that has been around since for example James Shapiro's work of several decades ago. It was very controversial at the time because there was no molecular or cellular basis for understanding of what were just observational correlates. It looked like the bacteria were coordinating their behaviour, but there was no idea how that worked or what the signals were. And I think the field was sort of broken open when quorum sensing was discovered, because here was something, that at least the bacteria cared about their density that then they could measure. And so as the work from Gürol's talk showed, now we see that bacteria care for other things in their neighbours. They care about their metabolic state of their neighbours, presumably they care about other factors of their neighbours by mechanisms that remain to be elucidated. So I for one think that this does play on a very much stronger footing than what are sort of observational results of several decades ago. And really, the difference between prokaryotic cells and eukaryotic cells will be a measure of degree rather than a dichotomy at the end of the day.

W. Bialek With the history. You describe the field as being broken open with the discovery of molecular mechanisms, but then another thing that resonates and has been rattling around in the discussion, is that we often have our hands on particularly mechanistic pieces, but are less clear about function.

H. Levine Sure, and again, that was a problem with the earlier work. They were all in vitro systems and you got these examples showing that the bacteria could communicate with each other under laboratory conditions that tried to push them to do that. But how often that was relevant in an environmentally relevant biofilm was I think uncertain. And again, in the quorum sensing world at least in some context it was clearly physiological. So I think that also was part of why that really pushed the field forward. We can now argue about whether a particular realisation of this wave or this colony — colony interaction is a physiologically relevant thing, but we can't argue any more whether there aren't examples of physiologically relevant communication.

S. Chu A couple of comments. First, in the study that Gürol Süel mentioned, where we made the cell target four different proteins to the biofilm. This was a superresolution study with movies 15 nm resolution of life biofilms. We showed that specific proteins were the key for the first initial biofilm sticking, and other proteins were then expressed. Because we were doing superresolution, we were discovering many things. The first obvious ones go back to the starvation — nutrient thing, which was that the biofilm left very large channels, connected channels, not percolated channels, so that nutrients can diffuse in. But as the biofilm got bigger, a very natural question went back to communication. The bacterial electrical communication is great work, and that can go over long distances, but a lot of the other protein stuff that is being ejected had to go by diffusion, and we always suspected that there could be this budding of vesicles that would take things longer distances. The final thing I want to comment on is that these biofilms are actually partially a defence mechanism for the bacteria. We were studying *V. cholera* and *Pseudomonas*. When the body attacks these bacteria, they hunker down, it is like a little fort. And we were assuming that when it is a little fort, you had to break this fort because the antibodies could no longer attack it. It is a defence mechanism so that those bacteria could survive, whether it is on your teeth, or on a knee implant, very very serious and chronic infections are caused by biofilms because they hunker down and they don't disappear. And so we were trying to see of a way to lay siege on the fort, and starve it to death. We were looking for mechanisms to stop those channels to actually put in the nutrients, because then you could kill the biofilm, and it would be great because biofilms are the tartar on your teeth, they are the stuff that fouls the bows of ships and they cause very many problems.

S. Eaton I was struck about what was said about different species of bacteria actually using the same potassium signal, which leads you to wonder of how they can compete with each other in some way. But then I started wondering about, what is the fate of those bacteria that get sucked up into these other species' biofilm. They are probably not cooperating, I don't know, and so what is determining the period of these oscillations, and could different species have different periods, and what if you put two colonies together that had a different inherent period of oscillation. Would they compete, would they ...? How does that work?

G. Süel It is a great question. In terms of the fate once the cells are incorporated, we are actively working on that. You can think of it from two perspectives, you can think of it from the perspective of the community, why am I interested in recruiting members, maybe even strange members, and then from the perspective of the motile cells, you can say why would I want to go there. Those are very beautiful questions and I think one thing that potentially might come out is that the answers are going to be complicated and not just one fits all, as you might expect. The period of the oscillations that you asked, is determined by the slow step, which is this negative feedback loop that depends on the consumption, sort of how quickly nutrients can get back. And it also depends on the size of the community, as the size grows the period actually changes. And in terms of different species having different frequencies, Massimo and I have been pondering this question and is something we are very interested in and pursuing.

U. Alon There is a fragility to these cell communication circuits, where a mutant that misreads the signal could have a growth advantage. For example, a mutant that isn't repressed by the signal can start growing. I just want to say more generally, and following Alexandra's question, there have to be principles also in tissues of our body that do feedback loops by cytokines, where you have a feedback loop and you could have a cell that is mutant in the receptor for the cytokine, misreading the signal and having a growth advantage. There have to be principles to defend against that fragility. One principle that emerges in human cells is what is called biphasic responses, where the feedback signal is toxic at both low levels and high levels. So in a case where you lose the receptor, or where it is locked "on", the cells kill themselves. The mutants, they think, there is too much or too little signal. And that is actually linked to diseases like diabetes, where glucose kills beta cells and there is cytotoxicity. So there is a tradeoff between protection against mutants, and dynamic instability that can lead to diseases. And I wonder in your cells, how often do you find mutant *B. subtilis* that just grow without inhibition.

B. Shraiman I was trying to take stock of these different signalling mechanisms, and it looks as if, as a communication channel, this electrical signalling channel is a channel without a password. Since the channel is so ubiquitous, presumably many other species will be listening in on the signal. Quorum sensing presumably has the capacity to be password protected, different molecules will be detected by different species. But Gürol also already mentioned the existence of coupling through shared resources. That is an indirect sort of signalling, and it is both universal, in the sense that whoever is interested in responding to oxygen concentration or carbon dioxide concentration will sense it, and yet obviously, is specific because the sensitivities will be different between different species. That business gets us also into another kind of microbial communities and interactions that will surely come up tomorrow, when we talk about ecology.

K. Wüthrich What determines the size of the biofilms? I mean, they grow to the outside, and the guys outside have nutrient, they have everything, and the guys in the middle cannot determine the size. So how is the size determined?

G. Süel That is a beautiful question. I don't know the answer, let me start with that. But we know that in nature, people have observed these types of biofilms on oceans that can literally span hundreds of meters. These are some of the weird planktonic communities that swim around that are some kind of a biofilm. They are also limited sometimes by physical means and so forth, but I don't think anybody has really been able to deeply understand if there is a size limitation, and what determines the size. Those are beautiful questions, but I think they remain unaddressed as of now.

Prepared comment

M. Vergassola: Behavior and memory

Understanding the basis of human and animal behavior is a scientific frontier in a diverse range of disciplines, and has major technological applications, viz. for biomimetic control of automated processes.

A major element in the above endeavor is the identification of the memory involved in the behavioral responses. One extreme is given by cues that are very short and trigger an immediate stereotyped reaction by the animal. Classical work[a] highlighted the importance of reflexes as the basis for instinctive behavior. At the other extreme stand behaviors that involve learning from a combination of factors that possibly go a long span back in the past. D. Kahneman[b] provides an excellent popular introduction to the importance of time scales for human behavior and thinking.

Identifying the extent of memory, the time span of the input signal that controls the output behavioral responses, the representation of that memory, are basic fundamental questions that underlie the understanding of behavior. While humans obviously excite our anthropocentric interest, there is also a great deal to be learnt from animals like insects, and in fact, even from unicellular bacteria.

Indeed, the term "bacterial social intelligence" was coined[c] to recognize the capacity of bacteria to extract informative cues about their environment, so as to rapidly adapt and even anticipate its future changes. A single bacterium has of course no brain, and limited representation and storage capacity due to its unicellular nature. For instance, the memory of a single bacterium during chemotaxis, i.e. while chasing chemical cues to direct its motion, is just a few seconds, and its embodiment has been worked out in its molecular details.[d] Still, large and structured colonies have much wider and powerful possibilities for distributing tasks, learning from experience to

---

[a] I. P. Pavlov. Annals classics in *Annals of Neurosciences* **17**, 136–141 (2010).
[b] D. Kahneman. *Thinking, Fast and Slow* (Farrar, Strauss and Giroux, 2011).
[c] E. Ben Jacob, I. Becker, Y. Shapira, H. Levine. *Trends in Microb.* **12**, 366–72 (2004).
[d] H. C. Berg. *E. coli in Motion* (Springer, New York, 2003).

make informed decisions and anticipate the future. The key resides in the communication among individuals, which endows them with much wider scales in space and time. In particular, degrees of freedom/fields additional to their density and velocity, store representations of the colonies' history that allow for better inferences and decisions. Recent research is discovering a number of those means of communication, e.g. quorum sensing,[e] long-distance electrical signaling,[f] exchange of metabolites,[g] and unveiling their role in the social intelligence of bacterial colonies. Even in the classical setting of a colony expanding in a Petri dish initially full of nutrients, couplings among bacteria lead to modifications of their physiological state, and are essential to quantitatively understand the outward expansion of the colony.[h]

As for insects, their collective behavior is an endless source of fascination, yet their brain allows them astonishing feats even at the individual level. While many of those prowesses have been reported and described, their basic neural mechanisms remain largely mysterious. For instance, male moths locating females from hundreds of meters by following sporadic and intermittent cues of pheromones is a feat that humans hardly match with automated machines. However, while their trajectories have been reported and described,[i] we do not know what is the underlying algorithm that controls their flight in response to the history of pheromone cues that they receive. Even more, reactive and learning schemes have been proposed in the literature, yet we do not really know what is the memory of past detections that male moths are using and its neural representation. The state-of-the-art for the insect model organism par excellence, *Drosophila melanogaster*, is similar. The modulation of a surprisingly large number of walking parameters in response to odors has been described in recent years.[j] Experimental recordings of the flies' locomotion (even in the absence of odors) has evidenced behavioral sequences with complex temporal dynamics, multiple time scales and a structure that hints at a hierarchy of internal states.[k]

Physics has a major role to play as behavioral cues are often rooted in the physical world, and elucidating the nature of the input stimuli and their algorithmic processing requires systematic methods that are so far missing. The long-term goal is to inject physical approaches into the long-

[e]S. T. Rutherford, B. L. Bassler. Cold Spring Harb. *Perspect. Med.* **2**:a012427 (2012).

[f]D.-Y. Lee, A. Prindle, J. Liu, G. M. Suel, *Cell* **170**, 214–214 (2017).

[g]Z. Long, B. Quaife, H. Salman, Z. N. Oltvai, *Sci. Rep.* **7**:12855 (2017).

[h]J. Cremer, T. Honda, Y. Tang, J. Wong-Ng, M. Vergassola, T. Hwa, Growth and expansion of chemotactic bacterial populations in nutrient-replete environments (2018).

[i]C. T. David, J. S. Kennedy, A. R. Ludlow, *Nature* **303**, 804–806 (1983).

[j]S.-H. Jung, C. Hueston, V. Bhandawat, *ELife* **4**:e11092 (2015).

[k]G. J. Berman, W. Bialek, J. W. Shaevitz, *PNAS* **113**, 11943–48 (2016).

standing behaviorist[1] conceptual underpinning of experimental analyses for a quantitative theory of behavior.

Discussion

J. Howard I've got a point about the overall metabolism of bacteria in this colony. What fraction of their metabolism is going into coordinating with other bacteria and what fraction is going into whatever they are doing, I guess, growth of the whole colony. I wonder if that is something that can be measured, or estimated, how much food is coming in and what is it all going into in terms of going into growth, or how it is all partitioned out.

G. Süel I'm not aware of a perfect answer to your question, but what is interesting is that these communities spend a lot of energy making this extracellular matrix. Meaning, a fraction of the cells, and that may be some kind of a hint, become factories that literally spew out polypeptide, sugars, polysaccharides, and that is stuff that costs energy to make. We were talking yesterday about how energy might be cheap, meaning ATP is not so critical, but biomass is critical. These communities are defined by having this extracellular matrix, clearly they are using a certain amount of energy to make it, but in terms of how much goes in and having these specific numbers, I'm afraid that is something for the future.

T. Hwa Just to add to that. When you think about starvation, you have to be careful about the growth that it is limited by. In Güroll's case that was nitrogen, so carbon was in surplus, so that is not a problem of making matrix or so forth. So it is more than cells not growing.

A. Perelson The other model system that people have studied over decades is the single cell amoeba moving to form fruiting bodies in *Dictyostelium*, where there is signalling through cyclic AMP and there is chemotaxis, and it is very important in the development of a multicellular organism. This is another beautiful example of cell signalling.

O. Leyser I wanted to mentioned that in terms of theory of behaviour, there is obviously a huge body of theory of behaviour in behavioural ecology. A lot of the benefits they have is that it is much easier to see what optimum is, because they are much closer to this reproductive success phenomenon that is the driver for evolution. So I think a lot of what is going on here is the kind of struggle we have overall, to understand systems behaviour in the context of what it is actually trying to deliver, what its function is, in the absence of a really clear understanding of the selective landscape in which the organisms that we are talking about are working. I think trying to kind of square that circle is going to be an important thing for the field.

---

[1]B. F. Skinner. *Science and Human Behavior* (Simon and Schuster, 1953).

W. Bialek What I would like to do is to hear one more prepared remark, and that will leave us with the remainder of this part of this session.

Prepared comment

Herbert Levine: The challenge of cancer

Cancer is the dark side of multicellular behavior. That is, individual cells must subjugate their Darwinian drive to proliferate and invade new territories for the good of the overall organism. These capabilities have been inhibited since the developmental stage; they are gone but not forgotten. During tumor progression, mutations can loosen these inhibitions and cells can recover these forgotten capabilities. In the context of metastasis, these include the ability to move through tissue and the ability to take up residence in new locations and re-initiate growth.

Our group has studied the genetic/epigenetic networks that control the aforementioned phenotypic transitions. By constructing tractable models of the core circuits underlying these transitions (see Figure 1a for an example of the circuit controlling EMT, the motility transition), we have learned several important lessons concerning cancer cells. First, cells not only can trans-differentiate, going from a specialized epithelial phenotype to an equally specialized mesenchymal one, but also reset to a less specialized hybrid state (see Figure 1b). This state was predicted to exhibit collective motility rather than individual cell motion, a prediction that has been verified in a number of experiments in recent years. It is now generally accepted in the cancer community that cells can exist in a spectrum of such states.

Furthermore, our approach shows that there is a natural correlation between this dedifferentiation along the epithelial-mesenchymal axis and the ability to initiate new growth in foreign soil. To quote from a recent review by Weinberg *et al.*, "there is growing evidence that a cell that has only undergone partial EMT is best positioned to acquire stem cell properties". Thus, epithelial plasticity extends over multiple biophysical characteristics and the hybrid cell phenotype may be the one most likely to initiate new tumors.

Finally it is worth noting that this developing story dovetails nicely with recent evidence that metastasis can often take place via small clusters of cells. After all, being in a small cluster requires adhesion in addition to motility and hence can be expected to favor cells with hybrid phenotypic properties. Our group is currently working on tissue-scale models which directly couple cell motility and cell-cell interaction to proteomics data, and which would then be a tool for studying cluster formation and translocation.

In the end, cancer for eukaryotic cells is somewhat reminiscent of bacterial colony formation. In the latter, cells both compete and collaborate,

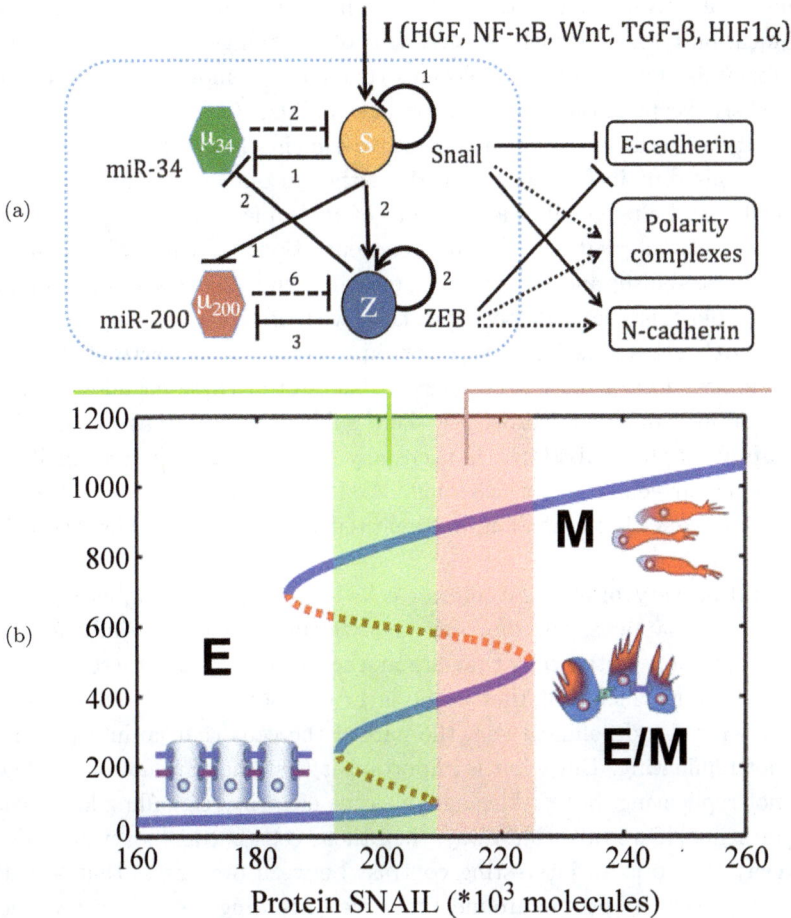

Fig. 1. (a) A circuit model of the epitheliel-mesenchymal transition (EMT) as it relates to cells in the primary tumor. (b) This class of models can give rise to phenotypic multi-stability especially when including the possibility of a stable hybrid E/M branch with mixed biophysical properties.

attaining a fair degree of multicellular coordination but never totally abrogating an individual cell's desire to proliferate and spread. Hence one can hope that studying cancer progression and learning its vulnerabilities ultimately prove easier than coming to grips with the immense complexity of fully regulated developmental biology.

Discussion

W. Bialek We have time for a few short comments, and I would like to see hands we haven't really seen so much. You can try to raise a different hand Daniel.

N. King I have been trying to think about of how this topic of intercellular communication intersects with physics. One of the things that keeps coming back to me is that there are molecules that have to move from one cell to the other, except in the case of electrical signalling. One of the things I struggle with is the fact that we do these experiments in the lab with very simple systems, but in fact, in the world, either in our guts or out in the ocean, cells are being bombarded by lots of molecules, and often the molecules that act, can act at very small concentrations. To me a big challenge is signal-to-noise, and seems like something that physicists can comment on that biologists haven't been wrestling with as much.

J. Lippincott-Schwartz This is a question related to the growth of the bacterial colonies. Gürol, have you looked if bacteria in these colonies are actually dividing. In other words, are there subpopulations that are dividing, or are all of them dividing. Is there any sort of spatial heterogeneity in the particular cells that are dividing. And does that change over time as the colony gets bigger, making it connected in some way to how the colony is remodeling.

G. Süel I'll be very brief. The interior cells stay viable as long as this communication happens, but they are not actively replicating. And then there is a transition to the cells that are at the periphery, and there is a layer, it is not very thick, but that is the layer that allows the biofilm to continue growing. And behind them, the part of the cells that go interior, they are not replicating. But what is important is that they are not dead. They are not replicating, but viable as long as the electrical signalling keeps communicating. If we take that away, then those cells become starved to death.

A. Murray There is an interesting contrast between organisms that are studied extensively by people around the table, including me, which are microbes which are easily genetically manipulable, about whose natural history we know effectively nothing; and then organisms where we really understand function and ecological properties, and behaviour and connections to survival. People like Terry are trying to close the gap, but I think admitting that that gap exists and talking to people that sit on the other side of it might profit this community substantially.

W. Bialek So we are actually going to have to think about the signals that actually occur in nature, that the organisms have evolved to deal with. In particular, the responses of many neurons to the signals that you encounter in nature are very different, qualitatively different, from the kind of responses that you see of what had been the standard physiologist's toolbox. So I think that trying to make the ideas of matching mechanism and function to the natural context, trying to make those ideas precise in ways that I think the physics community can resonate with, is something that at least in that context drove new ideas for experiments to produce qualitatively surprising things. Independent of whether any of the ideas that the physicists had were right, they lead to the discovery of new phenomena. And so, let us pause for coffee, and we will reconvene in 20 minutes or so.

Rapporteur Talk by Eric Siggia: Inter-cellular Interactions and Patterns: Vertebrate Development and Embryonic Stem Cells

Development from egg to embryo to adult is a fascinating instance of biological self-organization for which genetics has supplied us with a parts list. It remains to find the principles organizing the assembly of those parts. In the last decade, embryonic stem cells (ESC) have provided the material from which to build the mammalian embryo. This review, for a quantitative audience, explains why colonies of ESC are an ideal system with which to peal back the multiple layers of regulation that make embryonic development such a robust process.

1. Introduction

It is foolish to summarize a subject as vast as vertebrate development, yet a more focused discussion would sacrifice the bits of generality I will try to convey. If physicists are fond of 'self organization' and 'symmetry breaking' then biology offers no more dramatic example than embryology. It puts to shame any of the contrived systems invented for systems biology; real physiology remains more interesting. The reader looking for the universal theory uniting just some of the topics in our session: biofilms, flocking behavior, and development should look elsewhere. Attempting to treat them together leads to a degree of superficiality that illuminates nothing. Slogans that biology, is robust, modular, evolvable, etc., are too vague to be useful.

These remarks are aimed towards the student of biology from the mathematical and physical sciences, who wishes for a few provisional guideposts as to what problems seem most approachable at the current instant. In almost all cases, autonomous first principles theory is a fool's errand. It would appear to outsiders that biological data is infinite (e.g., there are upwards of 20,000 papers in Pubmed that mention each of the six or so intercellular signaling pathways that pattern the early vertebrate embryo), yet it has been the experience of most in the field, that theoretical ideas require new data. So this review aims to provide the skeleton of concepts that could motivate the next round of experiments, and highlight the systems most likely to provide answers.

In searching for principles, why study vertebrates and not arthropods; all the signaling pathways are present in arthropods, without the huge degeneracy of components. Genetics is easier, and evolution moves more quickly and has created fascinating variety, (see remarks of Nipam Patel). But there is a natural interest in our selves, common interests mean more shared reagents, techniques cell lines, and it is not a sin to be medically relevant. But the real advance that makes vertebrate development interesting for the quantitative class is pluripotent stem cells specifically in what follows human embryonic stem cells, hESC. These cells quite literally give rise to all cells of the adult. Basic cell culture taught us about intracellular signaling and organelles (see the report of Lippincott-Schwartz) exploiting what are basically cancer cells, HeLa[1] being the most notorious example. Such systems are a very dubious starting point for problems of cell communication and embryology, even if one can engineer them with some of the right constituents. Biological components do many things in-vitro that do not happen in-vivo. The same caveat applies to stem cells and at crucial points an embryological comparison is needed, but in the appropriate context stem cells do the appropriate thing, as shown by the canonical grafting experiments.

2. Gastrulation

A favorite system for experimental vertebrate embryology from the early 20th century is the frog *Xenopus*. The eggs are 1.2 mm in diameter, they are easily fertilized on demand in the lab, and become swimming tadpoles in two days. No special regents needed, just pond water. The reader is invited to view one of the gastrulation movies on Xenbase or YouTube. The egg begins with top and bottom (animal, vegetal) hemispheres distinguished, sperm entry defines the future dorsal side. Signals from the vegetal side, induce a band of mesoderm cells around the equator from the multipotent animal cap (hemisphere) cells. At gastrulation, this band closes like a purse string, by converging towards the dorsal side. The converging cells dive under the epidermis, and form a stiff bundle, the notochord, that elongates and literally builds the anterior-posterior axis. The vegetal hemisphere is pulled inside and the cavity formed from the outside inward by the so-called convergence-extension movements becomes the future gut (the online movies essential here). The master of *Xenopus* gastrulation is Ray Keller at University of Virginia and his papers provide the best description we have for the forces driving these morphogenic movements e.g., Ref. 2.

While the embryo is dramatically changing shape, it also is laying down, very literally, a coordinate system defined by the HOX genes along the anterior-posterior axis. That morphogenesis and fate assignment happen simultaneously is quite essential, since the cues for position come precisely from the cell movements. The HOX genes are located in contiguous cluster in the genome and are expressed sequentially in time in the converging mesoderm band by very complex regulation tied to their genomic organization (see papers of D. Duboule Lausanne). The HOX expression is locked down when a cell goes through the point of convergence on the future dorsal side, the Spemann organizer, (see Wikipedia). Thus a temporal signal is converted into a spatial coordinate as the embryo builds its body axes, Figure 1.[3,4] The organizer should not be thought of as defined structure like the gut, but rather a reaction center through which cells transit and change state. Although the organizer can be surgically transplanted to induce a second body axis, in the chick it can also regenerate following excision.[5] Exactly how the juxtaposition of tissues surrounding the organizer recreates the organizer is not understood, though recently an ectopic organizer was created in the chick by transplanting a patch of cells derived from hESC (preprint Martyn, Kannno, Ruzo, Siggia, Brivanlou).

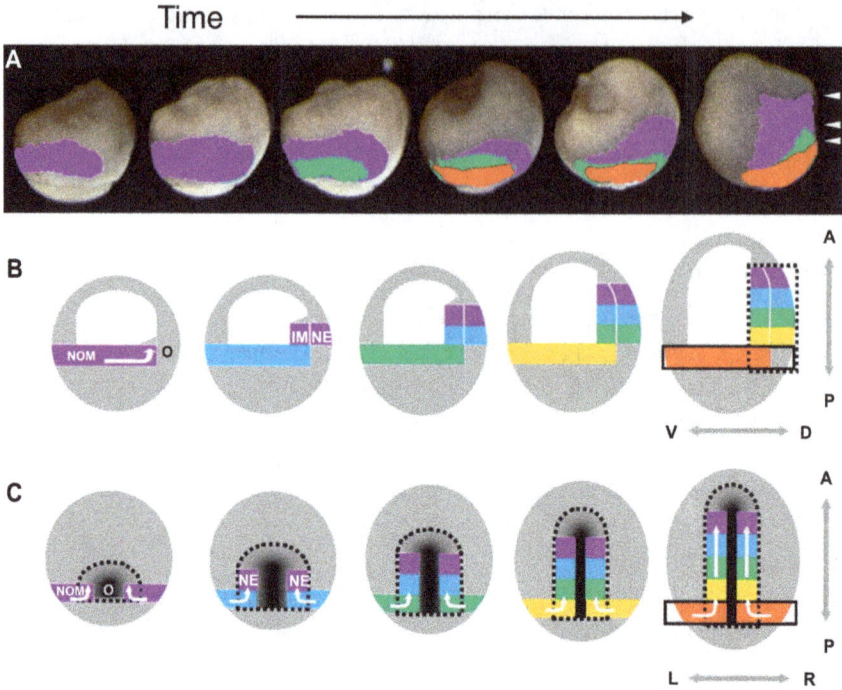

Fig. 1.   Temporal progression of HOX gene expression in the equatorial mesoderm is locked down on the anterior-posterior (A,P) axis. Sagittal sections are shown on the top two rows and a dorsal view on the bottom (V,D ventral, dorsal; L,R left,right). From Ref. 4, Figure 6.

The dorsal-ventral axis is established by the signaling pathways that recur through out development BMP, Nodal, WNT, and FGF/MAPK. They are very dynamic prior to gastrulation, Figure 2,[6] and more so afterwards, and it would be perilous to approximate the embryo as one-dimensional in such circumstances.[7] The data in Figure 2 is derived by sectioning *Xenopus* embryos and staining the slices with antibodies for the transcription factors that move to the nucleus in response to the signals. Thus one records the net effect of the secreted morphogens and their inhibitors in the embryo. One might have hoped for more modern data from light sheet microscopy on the transparent zebra fish embryo, but as of this writing nothing comparable in scope to the 2002 Schohl paper is available. Modern technology consumes its creators.

3. Positional Information and the Community Effect

The cells in embryos have to accomplish two feats. They need to express discrete fates in the right places in response to continuous signals. This pattern formation process is naturally broken into 'positional information' a term coined by Lewis Wolpert (see his Developmental Biology textbook), and 'community effect', introduced by John Gurdon.[8]

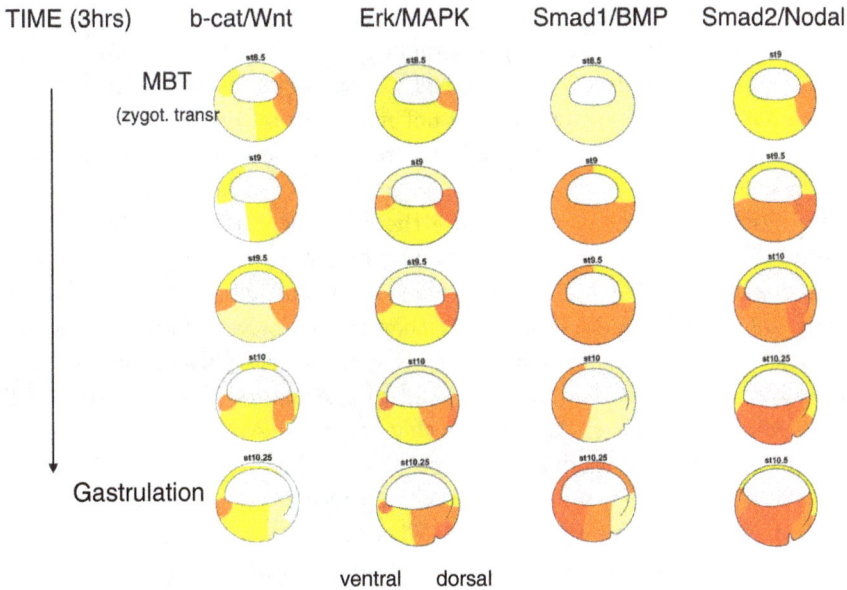

Fig. 2. Sagittal views of the activity of the canonical signaling pathways just prior to gastrulation in Xenopus, from Ref. 6, Figs. 9, 10. MBT or mid blastula transition denotes the beginning of general zygotic transition followed 3hr later by gastrulation. The color scale for intensity places red highest and yellow lowest.

One way for position to regulate fate is via a secreted signal, a so-called morphogen, whose level initiates some transcriptional cascade resulting in a defined fate. If the morphogen is activating, some intracell inhibitions have to operate down stream of the primary signal to exclude the low morphogen fates from regions of high morphogen. By far the best data we have on this paradigm is in *Drosophila* from the Gregor lab at Princeton.

The situation in vertebrates is more complex. Classical experiments in *Xenopus* from Smith[9] (for Activan/Nodal) and Brivanlou[10] (for BMP), used multipotent cells obtained by dissociating the *Xenopus* animal cap prior to gastruation. Graded levels of ligands were applied, the cells re-associated and gene expression compared against similarly timed intact embryos. A 10–20x range of concentrations elicited the full range of fates in fairly discrete bands. Hence Activin/Nodal and BMP were declared morphogens. However in contrast to Drosophila, it is difficult to imagine these morphogens as static around the time of gastrulation, and none have been directly visualized at WT levels. Furthermore the classic experiments from Smith and Brivanlou assayed expression at a convenient endpoint, and already in the mid 1990s papers from J. B. Gurdon showed the dynamics of morphogen interpretation was more complex than assumed.[11] (A general aside: most genetic screens normalize to an endpoint that is well removed from the time at which the gene operates; this obscures the dynamic role we believe those genes should have.) Thus one may ask whether its just morphogen levels that define fates.

An alternative view of morphogen signaling, with almost no in-vivo data, posits that cells respond to morphogens adaptively, in analogy to E. coli chemotaxis. That is the absolute level of morphogen does not matter at all provided it is static. The transcriptional network down stream of the receptor has a fixed point independent of morphogen level (some simple examples in Ref. 12). Then by continuity, if the adaptive system does not simply ignore the stimulus, the transcriptional output is determined by a smoothed time derivative of the input where the time scale is set by the feedbacks. While negative feedbacks at multiple levels are the norm for signaling pathways, this does not imply they are adaptive. However it is easy to imagine that position relative to an unsteady source of morphogen could be inferred from the received signal. For those inclined to information theory, there is a literature on communication via a diffusive channel, but clearly the information is in the rate of change of the signal so an adaptive receiver is called for.[13] Note from the embryo's point of view both the source and receiver can be tuned by evolution to work together to define the position. There is no reason to consider the information theoretic limits on reception for a presumed source of diffusible morphogen since properties of the source may also be tuned. The classical experiments on Activin/Nodal and BMP as morphogens are completely compatible with reception by an adaptive system. An adaptive transcriptional response was demonstrated for a myogenic cell line by microfluidic control of the signal in Ref. 14, and in hESC in a preprint from the Warmflash lab.

The community effect is more mysterious since multiple mechanisms contribute. Perhaps the best understood vertebrate example is the transition from the 8-cell mouse embryo where cells are nearly equivalent to the preimplantation embryo with three distinct lineages,[15] Figure 3. One should perhaps digress here and define some terms from the pre-molecular era of embryology.[16] A cell is said to be:

- Competent if it is able to respond to a signal,
- Specified or committed if it will assume its normal fate in the absence of further signals,
- Determined if its fate is unchanged even if challenged with new signals,
- Differentiated if it visibly changes its morphology or identity.

Cells in each of the three lineages in the mouse blastoderm are determined, in the above nomenclature. They will only graft into the layer from which they came, which is generally how these properties were assayed in the pre-molecular era.

At the 8 cell stage the embryo 'compacts' and the cells acquire a basal (in) and apical (out) polarity.[17] By a combination of oriented cell divisions, mechanics,[18] and probably mutual inhibition at the transcriptional level, the trophoblast separates from the inner cell mass (Ref. 19 and recent papers from the J. Rossant lab). A second stage of transcriptional bistability mediates the splitting of the inner cell mass. A combination of cell sorting (analogous to phase separation driven by surface

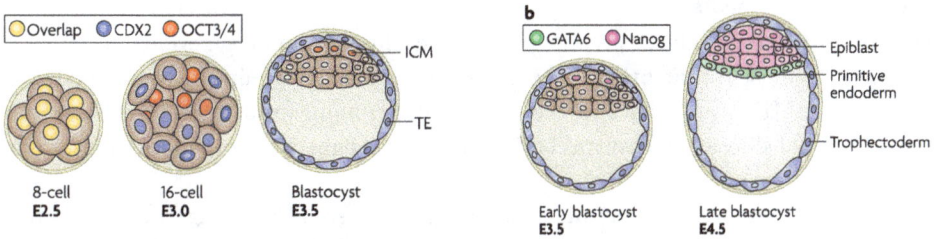

Fig. 3. Schematic of mouse embryo from 8 cells to 128 cells preimplantation showing the progressive emergence of the epiblast (which gives rise to the body proper) and the extraembronic lineages, primitive endoderm and trophectoderm, along with some of the distinguishing markers [15] Figure 1.

tension differences between the cells) and potentially chemotaxis driven by FGF4, separates the epiblast from the primitive endoderm, [20] (and earlier papers from the Hadjantonakis lab). Finally there are isolated examples of cell death driven by cell competition, a still mysterious process at the molecular level whereby minority cells are eliminated. Thus all imaginable mechanisms contribute to lineage separation in the mouse blastula.

Hypothetically a reaction-diffusion system with a nonlinear self-activation of one species and its inhibition by a second activated species with a larger diffusion constant could convert a mixture of cells to two pure populations. [21] These are also the ingredients for a Turing system, and with suitable nonlinear saturation it will give rise to two discrete phases. Evidence for cooperative fate determination in a small hESC system was provided in Ref. 22, without elucidating all the molecular players.

4. Signaling Pathways are Reused

In spite of what one might read in a textbook, signaling pathways do not work in isolation in the vertebrate embryo. There is a cascade from BMP to Wnt to Nodal in the mouse that initiates primitive streak formation, [15] and the same chain of induction in hESC (Ref. 23 and to appear), with similar consequences. The neural crest delaminates from the neural plate before it closes and under the control of BMP, Wnt, FGFs cells stream out and reconstitute mesoderm derivatives (bone, muscle, cartilage) and ectoderm derivatives (peripheral nerves, melanocytes). They play a major role in the morphogenesis of the vertebrate face.

The dorsal-ventral axis of the neural tube is defined by Sonic hedgehog (Shh) from the notochord and floor plate (ventral) and BMP4 from the roof plate (dorsal). Somites form in a head to tail sequence mediated by a retinoic acid gradient anteriorly and a Wnt, FGF gradient posteriorly. [24] (The A. Aulehla lab has developed somite-forming explants as an interesting model for spatial patterning.) In-vivo, the somites first condense as epithelial balls by a mesenchyml to epithelial transition (MET), which can also occur ectopically. [25] They subsequently undergo an

EMT on their medial-ventral side and wrap around the spinal cord to form the vertebrae, cartilage and a second population shifts by half a period and creates the skeletal muscle that bridges the vertebrae. All these gymnastics are under the control of BMP, Wnt, Shh and their inhibitors coming from three directions: the dorsal-ventral sides of the central body axis (neural tube and notochord) and lateral mesoderm.[26]

Wieschaus remarked that much of morphogenesis is like origami, the folding of epithelial sheets, but morphogenesis also entails a back and forth between the mesenchyml and the epithelial state. The transition from the presomitic mesoderm, to the somites, and back to the mobile precursors of bone and cartilege is a good example. Is this in part a mechanism to enforce discrete fates on a continuum of cells? Certainly the HOX genes must be expressed in registry with the discrete somites.[27]

The point of this jumble of jargon is to delineate a broad question in the spirit of the Solvay conferences. Biologists do not ask why certain pathways are deployed in certain contexts and in certain combinations, it is too easy to rationalize it all as evolutionary artifact. The literature abounds in just-so stories, none as entertaining as Kipling. What more can be done? There is almost no biophysical and dynamical characterization of the canonical signaling pathways in an embryonic context. Are they simple ON/OFF switches, because it is assumed that disconnected cells on a dish properly report pathway response? But this ignores the fact that much of development involves epithelial layers that may be apically-basally polarized. A polarized epithelium could control the reception of activators and inhibitors,[23] but almost nothing is known in-vivo. To a first approximation, the embryo is still conceived as empty space where any signal can go wherever it is needed. The practical or engineering reason to address the 'why' question is that it may yield a useful phenomenological description of the interrelated processes of morphogenesis and fate determination. These can be fit to data and become predictive. Even half correct theory, that really addressed global questions of pattern formation with molecular details, would greatly accelerate progress in embryology and regenerative medicine.

5. Stem Cell Biology

This subject is practically infinite, and the next three short sections serve just to delineate some concepts and open questions for a quantitative audience and provide a few key references. The subject incites a gold rush fervor with a concomitant inattention to detail, since commercial applications beckon, but in my view the best work remains well grounded in developmental biology.[28,29]

5.1. *Organoids*

One of the most spectacular examples of organoids, and indeed the first, are the mini-guts of Sato and Clevers.[30] The human gut has a surface area of several hundred square meters, formed by a meshwork of protrusions, villi, that continuously turn over. There are specialized stem cells that occupy the base of the crypts, the slender cavities that are mixed among the villi. They self renew, and their descendants include all the more specialized cells that populate the villi. The Clevers lab found molecular markers for these stem cells and to prove their regenerative capacity they cultured isolated cells in a 3D matrix. To their surprise they made mini-guts with crypts and villi! Furthermore stem cells isolated from the mini-guts would repeat the generation process indefinitely. This system is simple enough that molecular pathways can be dissected, e.g., Ref. 31, and the usual players are at work, Wnt, EGF, Notch in the crypt, opposed by a BMP gradient from the villi. The last triumph of this system is medical.[32] Here mini-guts made from a patient biopsy were used to screen approved drugs against a rare mutation in the cystic fibrosis gene. The patient improved within hours after receiving the screening candidate.

Working from both mouse and human ESC, the J.M. Wells and J.R. Spence labs have created embryonic gut and stomach. The H. Snoeck lab has created embryonic lungs, and A. Grapin-Botton grows pancreas from stem cells. All these systems beg for quantitative modeling.

More dramatic to the public at least than these endoderm derivatives, are the optic-cups from the Sasai lab[33] from hESC, following their work in mouse. One begins from a ball of cells, its invaginations from the surface and after several additional weeks, six types of neural retinal cells form in appropriate configuration with plausible connections.

There is not yet a full convergence of groups studying mammalian embryonic development and those recapitulating parts of the process with ESC. But organoid systems for the quantitatively minded, are the best compromise between reality and tractability to study the relation of morphogenesis and differentiation. They realize the mantra 'if you built it you understand it'.

5.2. *Adult stem cell niches*

Systems that renew routinely such as blood, the immune system, skin as well as those that renew upon injury, such as skeletal muscle, or the liver all have dedicated populations of so-called adult stem cells that can recreate the necessary tissue. Typically these cells reside in compartments distinguished by structure and accompanied by specialized signaling, always involving the canonical pathways we know from development plus perhaps some specialized growth factors. Hence the question, can these niches be understood, or better predicted from what we know about development? Some prominent biologists in the field would say no.

Of particular interest in this regard are stem cell niches that can be reconstituted in-vitro, the mini-guts mentioned above being a prominent example. A second case is the satellite cells that regenerate the myotubes of skeletal muscle. They normally are dispersed among the myotubes and not in any obvious specialized structures.

The entire system of stem cells and myotubes was recreated from ESC in Ref. 34, and the functionality of the satellite cells verified by grafting.

For the systems that can only be studied in-vivo, haematopoiesis, is perhaps the most challenging since the relevant stem cells constitute of order 0.01% of the bone marrow (S. Morrison 9/28/2017 Rockefeller lecture) and reside in a structurally complex environment.[35] The medical implications of preparing haematopoietic stem cells would be immense if the homing problem could also be solved, i.e., how to get them in the right place. One could potentially cure all blood/immune cancers. Another niche studied by the E. Fuchs lab at Rockefeller is for hair cells, and the signals are Wnt, Shh, and BMP. Finally this volume has a commentary from BD Simons on branching morphogenesis in the kidney where the stem cell niche resides at the tips of the growing endothelial networks.

5.3. *Micropattern culture of hESC*

A group of physics postdocs and students working jointly with me and Ali Brivanlou at Rockefeller are exploiting hESC to recapitulate the earliest steps of embryonic patterning. Stem cells differentiated on a slide with canonical morphogens assume a variety of fates in a spatially disorganized fashion. Early endoderm protocols tolerated a lot of death but still generated useful numbers of cells for subsequent assembly steps (papers from Wells and Spence labs noted above). Our primary discovery was that mere spatial confinement in 2D micropatterned colonies induced the cells to self pattern in a reproducible way.[36] Thus cells communicated with each other in preference to the primary morphogen that was manifestly uniform in the solution. The following paragraphs summarize some results from these systems, most in the process of publication, with an emphasis on technique. The potential of these systems in reviewed in Ref. 37.

The micropatterns are 0.5–1 mm in diameter and display four fates in a radially symmetric pattern that from outside to center correspond to: extraembryonic, endoderm, mesoderm, and endoderm. Their order matches that derived by projecting the cup shaped mouse epiblast onto a disk (P. Tam in Ref. 38). The mes-endoderm cells plausibly arise by gastrulation for which both the morphogenic movements and molecular markers correspond to what we expect from the (mouse) embryo, though nothing is known molecularly about human gastrulation and only a little from non-human primates. The $\sim$2000 cells in each micropattern define their fate by distance from the colony boundary, as shown by comparing disks of different radius. As the size shrinks, the inner fates disappear and the outer territories retain their dimensions. The same secondary morphogens and secreted inhibitors operate on the micropatterns as in the embryo.

A second paper,[23] clarified in molecular terms how cells sensed the colony edge and measured distance from it. The pluripotent colonies are apical-basal polarized epithelia, and they restrict their BMP and Activin/Nodal receptors to their baso-lateral side, thus rendering them inaccessible to apically supplied morphogens, except on the colony boundaries where the receptors become apically accessible. Growing cells on filters is a very clean way of distinguishing apical from basal responses. The second mechanism restricting signaling to the colony edge are secreted inhibitors that come into play when the BMP morphogen is applied, move laterally in the colony and leak out the edges. Pattern formation was examined in an exhaustive zoo of shapes, and all could be predicted from the data collected on disks plus the assumption of 2D diffusion with zero boundary conditions.

Cell lines with both activator and inhibitors under DOX control can readily be generated, as well as homozygous knock out lines for genes that are essential for pattern formation. Using filter grown colonies with sparsely seeded DOX inducible cells, it is possible to watch the local spreading of both activators and inhibitors and how they interact with the same components applied selectively to the apical or basal sides of the colony. In a very natural context it is possible to dissect the influence of cell polarity on signaling.

There are live reporters for the BMP, Activin/Nodal, and WNT pathways. Thus signaling history can be related to cell fate. Patterning can be triggered with a secondary morphogen such as WNT, the same germ layer arrangement obtained with BMP stimulation, less the outermost extra embryonic ring, as one would infer from data in a mouse.

Three-dimensional differentiation from balls of ESC, so-called embryoid bodies (EB), is a common starting point for organoid development. The same technology has been used to explore the emergence of germ layers, but the results are not nearly as standardized as micropattern culture, imaging is more complex, and there has been far less molecular dissection of the signaling.[39] But by far the biggest problem with these systems as a model for gastrulation related events, in mammals is that an epithelial cell population, the epiblast, initiates the process and gives rise to the entire adult body. (Incorrect inferences from EB as to how the mouse's inner cell mass cavitates to form the epiblast were only corrected in Ref. 40.) Human ESC are technically an easier starting point for gastrulation since they naturally propagate in a state very analogous to the post implantation epiblast, while mouse ESC resemble the preimplantation inner cell mass. The mESC can be converted to epiblast cells, but the resulting state is not entirely stable and seems more variable than the normal hESC (details are technical).

Our own technique for work in 3D, seeds single cells in a specially tailored matrix and allows them to grow into an epithelial shell with basal out and apical in while remaining pluripotent (Simunovic *et al.*). A very gentle BMP stimulus results in spontaneous polarization of the epithelial cysts into a primitive streak region, showing all the markers of gastrulation (that define the future posterior), and a complementary region with anterior epiblast (future ectoderm) markers. This is a true symmetry breaking and does not require an asymmetric source of BMP as in prior work with mouse.[41] Another variant on this method of 3D culture even results in morphological symmetry breaking prior to gastrulation.[42]

An important step in taming mESC to explore gastrulation and beyond in the mouse has been taken in a forthcoming paper by Morgani and Hadjantonakis. They devised a protocol to recreate the pre-gastrulation mouse epiblast on micropatterns. They then added the BMP and Wnt morphogens that would normally come from the extraembryonic tissues and observed radial patterning. The same antibody combinations could be applied to the mouse embryo at successive time points, and the correspondence of patterns and fates mapped. The details are too voluminous to recount here but are very encouraging. In the absence of any data from human embryos undergoing gastrulation it is essential to benchmark the micropattern technique against some embryonic system.

The technology used in the mouse micropattern paper is conventional antibody stains for triples of markers. This scores the proteins and allows co-stains for signaling effectors. Space-time specific expression data resolved down to a few cells for the mid-late streak mouse embryo can be found in Ref. 43. Single cell RNA-seq is appealing technology but in development it needs to retain its time-space label. Not all genes deserve equal weight, the Hadjantonakis study focused on those with an interesting phenotype.

6. Phenomenology

The physical reader should realize this term is a pejorative in biological contexts. It denotes a return to 19th century biology and the absence of the methods that made 20th century biology great: genetics, biochemistry, structural biology, molecular biology, etc. However the modeler who embraces these advances literally is doomed, at least in the area of development. A glance at the molecular constituents for any of the signaling pathways (e.g., Wnt homepage maintained by the Nusse lab) reveals 5–10 core constituents decorated by another 10–50 modifiers. The molecular complexity frustrates transferring actual numbers between systems, and the most common description of reactions with the Michaelis-Menten system requires many parameters. The solution sometimes adapted is to randomly sample parameters and select those behaviors obtained most frequently. This shows that random equations can do many things, but more fundamentally is contrary to the incrementalism that we believe is inherent in Darwinian evolution, unless you think that Diana sprung fully formed from the head of Zeus as depicted on ancient Greek vases.

Examples of successful phenomenology in the context of development are rare and I myopically mention some examples of mine and from my immediate collaborators. The foundation of the approach goes back to a book written by a student of Waddington,[16] and are based on translating the embryological concepts of competence, commitment, and determination to the language of dynamical systems. The necessary mathematics is embodied by the subject of Morse-Smale dynamical systems (`http://www.scholarpedia.org/article/Morse-Smale_systems`). Colloquially these are systems of equations whose limiting behavior both forward and backward in time are a finite set of fixed points and periodic orbits. They are rich enough to cover anything we can hope to measure and describe in developmental biology, even if we put aside the periodic orbits. Gradient flows with some technical assumptions are Morse-Smale.

The mapping between classical embryology and mathematics equates an equivalence group of cells[16] to the direct product of the model used for one cell. Commitment is, with various nuances, to flow into a fixed point. The power of this brand of phenomenology, and also its point of failure is whether the parameters can be fit to encompass all available data. Typically gene knockouts, and overexpression data is available, but more informative is always dynamic interventions made while the system is poised among multiple outcomes. If a multivariable system has two stable states, then the simplest phenomenological model would replace it with the relaxational dynamics induced by a quartic potential in one dimension. Various morphogens would tilt the potential and favor one state over the other, ultimately by annihilating one fixed point with the saddle in a reverse saddle node bifurcation. It is clear how to add noise to the system (partial penetrance in genetic language), whose biological source could be environmental, epigenetics (molecular tags on DNA and chromatin that vary between animals), or true molecular noise. The problem becomes interesting if there are multiple experimental handles on the relative stability of the two states. The challenge is to represent them all in terms of the coefficients in the potential. The first guess would be a linear combination exactly parallel to what is done for computational neural nets, where a linear weighted sum of inputs is put through a nonlinear function.

While such a representation seems very antithetical to a Michaelis-Menten network it is not so far removed from development. The interesting mutations in development do not create fundamentally new structures, but rather permute known ones. Genetics is based on quantifying recognizable characters. A fried embryo is not informative, but the old observation of genetic assimilation, that environmental results often phenocopy genetic ones is profound. Thus we suppose that evolution has added multiple layers of regulation, many still unknown, to insure the stability of the two states in the above example. Phenomenology accepts those states and concentrates on the simpler problem of parameterizing the dynamics mediated by the morphogens (during the competence period but prior to commitment) that control the competition among them.

Phenomenology becomes more interesting when three states are in play. It is informative to enumerate a hierarchy of models by enumerating critical points and their connections in various spatial dimensions, and parameterizing the vector fields to within topological equivalence. A nontrivial example for vulva development in *C. elegans* is given in Ref. 44, and a forthcoming paper by the same authors in eLife. An application to intermediate range signaling by Notch-Delta was presented in Ref. 45.

Phenomenology should also be the preferred description for moving boundary problems e.g., Ref. 46. When two locally stable states are separated by an interface and some component of the system can move between cells giving rise to defacto diffusion, then its very appealing to model it as the relaxation of a bistable free energy with a spatial gradient following Kolmogorov. A more prosaic use of phenomenology parameterized the cellular response to a morphogen and coupled it with a reaction-diffusion system for the secreted inhibitor to the nominally uniform morphogen. [23]

Another avenue for phenomenological reasoning codifies the continuity of outcomes with variable morphogen levels by a phase diagram. Clearly the entire signaling history impacts the pattern of embryonic fates, and for present purposes we plot terminal outcomes as a function of signal levels imposed by genetic means. (We ignore the specifics of those genetic manipulations here, so as to succinctly illustrate the ideas.)

The representation in Figure 4, which is actually computed from a model of vulva patterning, exemplifies that it would be surprising given a complex model to observe boundaries strictly parallel to the coordinate axes. Thus it is logically impossible to assert that N controls just the green fate and EGF just the red, as one might infer from a casual reading of biological papers, but rather there is simply more green on top and more red on the lower right. Once phase boundaries are freed from alignment with the coordinate axes, they generically meet in triple points. These are points where conventional genetic analysis becomes complex and thus interesting. It is evident that the most informative data for fitting the dynamical model that underlies the phase diagram, are precisely those genetic backgrounds that yield mixed fates (partial penetrance in the jargon). Thus merely codifying the obvious yields insights.

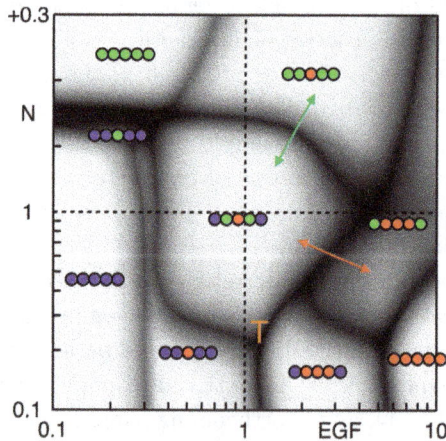

Fig. 4. Phase diagram for the states of five cells each of which can assume three states (RGB) under the control of two morphogens N(otch) and EGF. Pure states are bounded by grey boundaries showing zones of mixed fates. The green arrow shows a generic transition where only one cell (reflection symmetry is imposed) changes state, the red arrow shows a correlated change and the T shows one example of a triple point where three boundaries meet (Corson & Siggia, to appear).

The developmental geneticist uses a *sensitized* background to define the activity of a new mutant that has no effect in the wild type background. In mathematical terms the sensitized state is one near to the inset of a saddle point, that is the ridge ending in the pass that separates two basins of attraction.[44] It is typically difficult to infer by verbal reasoning alone the activity of the silent mutation from the identity of the terminal states. But such data can be very useful in model fitting. More generally viewed from the perspective of modeling, the most informative experiments apply time dependent perturbations to the system while the decisions among states are being made. This modality is the complete converse to the typical genetic screen where time is eliminated and only a late terminal state is recorded. Although genetics furnishes us with a parts list for development, the description of those parts is rather removed from the context in which they function. Imposing a Morse-Smale description on development may ultimately prove to be incomplete, but in the interim, certainly suggests many informative dynamic experiments.

Can theory do more than principled data fitting in cell and developmental biology? An old article by Jacob[47] reminds us that evolution works by tinkering, rearranging existing parts. Darwin, in his oft-quoted passage on the evolution of the vertebrate eye, observes that complex structures can be created rapidly by gradient search. Models for various slices of development were derived by gradient optimization in a series of papers by P. Francois and me. Some rather nonobvious dynamical models emerged with their specific parameters and no appeal to parameter space volume. A template for how to generalize these ad-hoc simulations to general theory is provided by models in machine learning.[48] Within a defined learning environment, some rules are shown that can be learned from only a polynomial number of examples, while others require an exponential number. One would expect the evolutionary tinkerer to discover only the former class.

The biologist's aversion to phenomenology has a specific connotation in development. If genes are the atoms of biology, can a phenomenological model ever constitute fundamental understanding? However if the genetic description is infinitely complex, do we really learn anything from an equally complex model? There is a parallel debate about the uses and abuses of phenomenology in neuroscience (Ref. 49 and the Oct. 27 2017 issue of *Science*).

7. Perspectives

To categorize an embryo as an instance of non-equilibrium symmetry breaking, reduce embryology to banal physical categories that hide the interesting phenomena. In the words of C. H. Waddington in his 1966 *Principles of Development and Differentiation*

> *"To anyone with his normal quota of curiosity, developing embryos are perhaps the most intriguing objects that nature has to offer. If you look at one quite simply ... and without preconceptions ... what you see is a simple lump of jelly that ... begins changing in shape and texture, developing new*

parts, sticking out processes, folding up in some regions and spreading out in others, until it eventually turns into a recognizable small plant or worm or insect...

Nothing else that one can see puts on a performance which is both so apparently simple and spontaneous and yet, when you think about it, so mysterious."

While no one believes that new chemistry or physics is required to treat biology, some thought is necessary to arrive at an informative level of description, just as in neuroscience.[49] It is often the case that gene centric models for development drown in the number of unknown parameters, so often a simpler more phenomenological description suffices to explain the data, and supplies some intuition. An instance noted in the discussion comment by Prof. Alon is a model for the approximate scaling of the Xenopus tadpole when the embryo is halved at the few cells stage which contains order 30 parameters.[7] A later paper[51] from the lab of a prominent biologist in Xenopus development found the elaborate model wanting, and noted a simple Turing reaction-diffusion model was more informative, recalling a prior remark to the same effect.[52]

Turing patterns are invoked in a wide variety of contexts. I prefer to use the term in the strict sense of a reaction-diffusion system that leads to spontaneous spatial pattern with a wavelength determined by diffusion constants and reaction rates. The regulatory system that controls the three axes of the vertebrate limb, proximal distal, AP (thumb is anterior) and DV (palm is ventral) has a long history in embryology,[53] and the amphibian limb is currently a key model system for the molecular understanding of regeneration (E. Tanaka lab). Turing physics has been an appealing explanation of vertebrate digits since their periodicity can be decoupled from their identity.[54] However a very instructive rebuttal to molecular data for a Turing origin of the vertebrate digits[55] was given in Ref. 54. Alternative models of periodic patterns may involve mechanics.[56]

The reason for concluding this opinion piece with examples with flawed models of developmental systems published in visible forums, is to impress upon the reader the diversity of facts that can impinge upon a model and desirability to partner with a lab conversant with those facts. Failed models are a sign of progress, since data and theory are engaged.

Acknowledgment

This work was supported by NSF grant PHY 1502151.

References

1. R. Skloot, *The Immortal Life of Henrietta Lacks.* (Broadway Books, 2011).
2. R. Keller, D. Shook, and P. Skoglund, The forces that shape embryos: physical aspects of convergent extension by cell intercalation, *Phys. Biol.* **5**(1): 15007 (2008).

3. C. D. Stern, J. Charité, J. Deschamps, D. Duboule, A. J. Durston, M. Kmita, J.-F. Nicolas, I. Palmeirim, J. C. Smith, and L. Wolpert, Head-tail patterning of the vertebrate embryo: one, two or many unresolved problems?, *Int. J. Dev. Biol.* **50**(1): 3–15 (2006).

4. S. A. Wacker, H. J. Jansen, C. L. McNulty, E. Houtzager, and A. J. Durston, Timed interactions between the Hox expressing non-organiser mesoderm and the Spemann organiser generate positional information during vertebrate gastrulation, *Dev. Biol.* **268**(1): 207–219 (2004).

5. K. Joubin and C. D. Stern, Molecular interactions continuously define the organizer during the cell movements of gastrulation, *Cell* **98**(5): 559–571 (1999).

6. A. Schohl and F. Fagotto, Beta-catenin, MAPK and Smad signaling during early Xenopus development, *Development* **129**(1): 37–52 (2002).

7. D. Ben-Zvi, B.-Z. Shilo, A. Fainsod, and N. Barkai, Scaling of the BMP activation gradient in Xenopus embryos, *Nature* **453**(7199): 1205–1211 (2008).

8. J. B. Gurdon, A community effect in animal development, *Nature* **336**(6201): 772–774 (1988).

9. J. B. Green, H. V. New, and J. C. Smith, Responses of embryonic Xenopus cells to activin and FGF are separated by multiple dose thresholds and correspond to distinct axes of the mesoderm, *Cell* **71**(5): 731–739 (1992).

10. P. A. Wilson, G. Lagna, A. Suzuki, and A. H. Hemmati-Brivanlou, Concentration-dependent patterning of the Xenopus ectoderm by BMP4 and its signal transducer Smad1, *Development* **124**(1): 3177–3184 (1997).

11. J. B. Gurdon, A. Mitchell, and D. Mahony, Direct and continuous assessment by cells of their position in a morphogen gradient, *Nature* **376**(6540): 520–521 (1995).

12. P. Francois and E. D. Siggia, A case study of evolutionary computation of biochemical adaptation, *Phys. Biol.* **5**(2): 026009 (2008).

13. P. J. Thomas, D. J. Spencer, and T. Sejnowski, The diffusion mediated biochemical signal, (2004).

14. B. Sorre, A. Warmflash, A. H. Hemmati-Brivanlou, and E. D. Siggia, Encoding of temporal signals by the TGF-β pathway and implications for embryonic patterning, *Dev. Cell* **30**(3): 334–342 (2014).

15. S. J. Arnold and E. J. Robertson, Making a commitment: cell lineage allocation and axis patterning in the early mouse embryo, *Nat. Rev. Mol. Cell Biol.* **10**(2): 91–103 (2009).

16. J. M. W. Slack, From egg to embryo: Regional specification in early development, (1991).

17. R. O. Stephenson, J. Rossant, and P. P. L. Tam, Intercellular interactions, position, and polarity in establishing blastocyst cell lineages and embryonic axes, *Cold Spring Harbor Perspectives in Biology* **4**(11) (2012).

18. J.-L. Maître, H. Turlier, R. Illukkumbura, B. Eismann, R. Niwayama, F. Nédélec, and T. Hiiragi, Asymmetric division of contractile domains couples cell positioning and fate specification, *Nature* **536**(7616): 344–348 (2016).

19. T. Rayon, S. Menchero, I. Rollán, I. Ors, A. Helness, M. Crespo, A. Nieto, V. Azuara, J. Rossant, and M. Manzanares, Distinct mechanisms regulate Cdx2 expression in the blastocyst and in trophoblast stem cells, *Sci. Rep.* **6**(1): 27139 (2016).

20. M. Kang, V. Garg, and A.-K. Hadjantonakis, Lineage establishment and progression within the inner cell mass of the mouse blastocyst requires FGFR1 and FGFR2, *Dev. Cell* **41**(5): 496–510.e5 (2017).

21. H. Bolouri and E. H. Davidson, The gene regulatory network basis of the 'community effect,' and analysis of a sea urchin embryo example, *Dev. Biol.* (2009).

22. A. Nemashkalo, A. Ruzo, I. Heemskerk, and A. Warmflash, Morphogen and community effects determine cell fates in response to BMP4 signaling in human embryonic stem cells, *Development* **144**(17): 3042–3053 (2017).

23. F. Etoc, J. Metzger, A. Ruzo, C. Kirst, A. Yoney, M. Z. Ozair, A. H. Hemmati-Brivanlou, and E. D. Siggia, A balance between secreted inhibitors and edge sensing controls gastruloid self-organization, *Dev. Cell* **39**(3): 302–315 (2016).

24. A. Aulehla and O. Pourquie, Signaling gradients during paraxial mesoderm development, *Cold Spring Harbor Perspectives in Biology* **2**(2): a000869–a000869 (2010).

25. A. S. Dias, I. de Almeida, J. M. Belmonte, J. A. Glazier, and C. D. Stern, Somites without a clock, *Science* **343**(6172): 791–795 (2014).

26. A. E. Brent and C. J. Tabin, Developmental regulation of somite derivatives: muscle, cartilage and tendon, *Curr. Opin. Genet. Dev.* **12**(5): 548–557 (2002).

27. T. Iimura and O. Pourquie, Hox genes in time and space during vertebrate body formation, *Dev. Growth Differ* **49**(4): 265–275 (2007).

28. Y. Sasai, Cytosystems dynamics in self-organization of tissue architecture, *Nature* **493**(7432): 318–326 (2013).

29. Y. Sasai, Next-generation regenerative medicine: organogenesis from stem cells in 3D culture, *Cell Stem Cell* **12**(5): 520–530 (2013).

30. T. Sato and H. C. Clevers, Growing self-organizing mini-guts from a single intestinal stem cell: mechanism and applications, *Science* **340**(6137): 1190–1194 (2013).

31. H. F. Farin, I. Jordens, M. H. Mosa, O. Basak, J. Korving, D. V. F. Tauriello, K. de Punder, S. Angers, P. J. Peters, M. M. Maurice, and H. C. Clevers, Visualization of a short-range Wnt gradient in the intestinal stem-cell niche, *Nature* **530**(7590): 340–343 (2016).

32. A. Saini, Cystic fibrosis patients benefit from mini guts, *Cell Stem Cell*, (2016).

33. T. Nakano, S. Ando, N. Takata, M. Kawada, K. Muguruma, K. Sekiguchi, K. Saito, S. Yonemura, M. Eiraku, and Y. Sasai, Self-formation of optic cups and storable stratified neural retina from human ESCs, *Cell Stem Cell* **10**(6): 771–785 (2012).

34. J. Chal, M. Oginuma, Z. Al Tanoury, B. Gobert, O. Sumara, A. Hick, F. Bousson, Y. Zidouni, C. Mursch, P. Moncuquet, O. Tassy, S. Vincent, A. Miyanari, A. Bera, J.-M. Garnier, G. Guevara, M. Hestin, L. Kennedy, S. Hayashi, B. Drayton, T. Cherrier, B. Gayraud-Morel, E. Gussoni, F. Relaix, S. Tajbakhsh, and O. Pourquie, Differentiation of pluripotent stem cells to muscle fiber to model Duchenne muscular dystrophy, *Nat. Biotechnol* **33**(9): 962–969 (2015).

35. G. M. Crane, E. Jeffery, and S. J. Morrison, Adult haematopoietic stem cell niches, *Nat. Rev. Immunol.* **17**(9): 573–590 (2017).

36. A. Warmflash, B. Sorre, F. Etoc, E. D. Siggia, and A. H. Hemmati-Brivanlou, A method to recapitulate early embryonic spatial patterning in human embryonic stem cells, *Nat. Methods* **11**(8): 847–854 (2014).

37. I. Heemskerk and A. Warmflash, Pluripotent stem cells as a model for embryonic patterning: From signaling dynamics to spatial organization in a dish, *Dev. Dyn.* **245**(10): 976–990 (2016).

38. C. D. Stern, *Gastrulation: from cells to embryo* (Cold Spring Harbor, N.Y.: Cold Spring Harbor Laboratory Press, 2004).

39. S. C. van den Brink, P. Baillie-Johnson, T. Balayo, A.-K. Hadjantonakis, S. Nowotschin, D. A. Turner, and A. Martinez Arias, Symmetry breaking, germ layer specification and axial organisation in aggregates of mouse embryonic stem cells, *Development* **141**(22): 4231–4242 (2014).

40. I. Bedzhov and M. Zernicka-Goetz, Self-organizing properties of mouse pluripotent cells initiate morphogenesis upon implantation, *Cell* (2014).

41. S. E. Harrison, B. Sozen, N. Christodoulou, C. Kyprianou, and M. Zernicka-Goetz, Assembly of embryonic and extra-embryonic stem cells to mimic embryogenesis in vitro, *Science* (2017).

42. Y. Shao, K. Taniguchi, R. F. Townshend, T. Miki, D. L. Gumucio, and J. Fu, A pluripotent stem cell-based model for post-implantation human amniotic sac development, *Nat. Commun.* **8**(1): 208 (2017).

43. G. Peng, S. Suo, J. Chen, W. Chen, C. Liu, F. Yu, R. Wang, S. Chen, N. Sun, G. Cui, L. Song, P. P. L. Tam, J.-D. J. Han, and N. Jing, Spatial transcriptome for the molecular annotation of lineage fates and cell identity in mid-gastrula mouse embryo, *Dev. Cell* **36**(6): 681–697 (2016).

44. F. Corson and E. D. Siggia, Geometry, epistasis, and developmental patterning, *Proc. National Academy of Sciences* **109**(15): 5568–5575 (2012).

45. F. Corson, L. Couturier, H. Rouault, K. Mazouni, and F. Schweisguth, Self-organized Notch dynamics generate stereotyped sensory organ patterns in Drosophila, *Science* (2017).

46. J. B. Chang and J. E. Ferrell, Mitotic trigger waves and the spatial coordination of the Xenopus cell cycle, *Nature* **500**(7464): 603–607 (2013).

47. F. Jacob, Evolution and tinkering, *Science* **196**(4295): 1161–1166 (1977).

48. M. J. Kearns and U. V. Vazirani, *An Introduction to Computational Learning Theory* (MIT Press, 1994).

49. Y. Frégnac, Big data and the industrialization of neuroscience: A safe roadmap for understanding the brain?, *Science* **358**(6362): 470–477 (2017).

50. S. Piccolo, *Yoshiki Sasai: stem cell Sensei.* **141**(19): 3613–3614 (Oxford University Press for The Company of Biologists Limited, 2014).

51. H. Inomata, T. Shibata, T. Haraguchi, and Y. Sasai, Scaling of dorsal-ventral patterning by embryo size-dependent degradation of Spemann's organizer signals, *Cell* **153**(6): 1296–1311 (2013).

52. P. Francois, A. Vonica, A. H. Hemmati-Brivanlou, and E. D. Siggia, Scaling of BMP gradients in Xenopus embryos, *Nature* **461**(7260): E1–discussion E2 (2009).

53. C. J. Tabin and L. Wolpert, Rethinking the proximodistal axis of the vertebrate limb in the molecular era, *Genes Dev.* **21**(12): 1433–1442 (2007).

54. T. W. Hiscock, P. Tschopp, and C. J. Tabin, On the formation of digits and joints during limb development, *Dev. Cell* **41**(5): 459–465 (2017).

55. J. Raspopovic, L. Marcon, L. Russo, and J. Sharpe, Modeling digits. Digit patterning is controlled by a Bmp-Sox9-Wnt Turing network modulated by morphogen gradients, *Science* **345**(6196): 566–570 (2014).

56. A. E. Shyer, A. R. Rodrigues, G. G. Schroeder, E. Kassianidou, S. Kumar, and R. M. Harland, Emergent cellular self-organization and mechanosensation initiate follicle pattern in the avian skin, *Science* **357**(6353): 811–815 (2017).

Prepared comment

Sharad Ramanthan: Control of size and time during development

"The most obvious differences between different animals are the differences in size, but for some reason zoologists have paid singularly little attention to them," J. B. S. Haldane, *On Being the Right Size* (1927).

If one had to describe the diversity of mammals, the one variable that is most prominent is the size: from a field mouse to a blue whale: I quote here from Martin Raff.[a] Just as important as the question of size is the question of time: the times for which animals develop in an egg or a womb, even among mammals varies from a few weeks to many months. How are the variations in the time of development generated through evolution and regulated in individual species? In fact, it is unclear whether the questions about size and time of development are closely related or independent. While, perhaps, zoologists had ignored the question about size early in the 1800's, it is unclear whether molecular biology has made substantial progress in answering the questions about size or time. This is not for lack of trying these are really difficult questions to answer.

Classical embryology experiments involving embryo fusions of mouse embryos before gastrulation suggest that there might be an intricate relationship between these questions: fused embryos undergo gastrulation earlier than expected, as if they knew that their size after fusion was larger. On the other hand, recent experiments made possible through stem cell biology to generate different neuronal progenitors suggest that there might be an aspect of temporal control that is cell autonomous. Human neural progenitors transplanted into a developing mouse cortex develop into cortical neurons that populate and send projections to the correct layer. However, the human neural progenitors develop on human timescales and do not speed up in a mouse.[b] Even more recent experiments that involve co-cultures of mouse and human embryonic stem cells suggest that even the epiblast cells have some cell intrinsic mechanism of controlling time as they go through development.

The question is how one makes progress on these problems about time. I believe that ideas from physics and chemistry that have served us well in understanding other dynamical systems might serve us here. In chemistry one has an idea of a reaction co-ordinate (which initially arose from quantum chemistry), that allows for a mechanistic measure of the progress of a chemical reaction. Think back for example to the SN2 reactions from undergraduate organic chemistry. In physics, the idea of an order parameter reflecting the broken symmetry of the physical systems has allowed

---

[a] I. Conlon, M. Raff, Size control in animal development, *Cell* **235**, 96(2) (1999).
[b] S. Temple, The development of neural stem cells, *Nature* **122**, 414 (2001).

to follow the dynamics of phase transitions in high dimensional complex systems. If one had such order parameters for development, whereby at the single cell and population level one could look at the progression of symmetry breaking, fate specification and fate choice, one could potentially make progress. Following physical systems, one could ask: what controls the dynamics of these order parameters, what controls the time scales, and hence what controls time.

Recent experimental methods allow the acquisition of the levels of thousands of genes in single cells as well as epigenetic modifications throughout the genome over the course of development. From these data, the challenge then is to identify the broken symmetries, the order parameters and the demonstration that by measuring the few key variables, one can follow the progression of a lineage decision. The computational methods to address such a challenge are significant. Statistical analysis of noisy high dimensional data poses unique challenges. First, as the dimensionality increases, maintaining the density of data points requires exponentially larger amounts of data, which is unfeasible. Second, not all the dimensions are equally informative, meaning that computational analysis can be dominated by variability in dimensions that are not biologically meaningful, thus leading to incorrect conclusions. If however, we can overcome these challenges, identify the key genes whose levels serve as order parameters for a particular lineage decision, measure the levels of the corresponding proteins in real time to read the mind of single cells in the embryo as it is making lineage decisions, see how the state of the order parameter is passed on to the daughters as it divides and then try to determine what control the dynamics of these order parameters, we might be able to make progress. When it comes to size and time, given how hard these questions are, many paths appear quixotic. Genetic screens have led to heterochrony but it has been very hard to speed up development. The thought of identification of order parameters and understanding how their dynamics are controlled is a path we are following. It is as quixotic as any other, but we need to find the answers!

Discussion

W. Bialek Unprepared remarks?

H. Levine I wanted to ask a question, which is: what signals are relevant for what type of information. As I think you said in your talk, there are a variety of different signals that couple in principle to different pathways, but somehow together they were conveying whatever the information that was needed, at least in this particular embryological state. Michael Elowitz in his talk yesterday showed five different signaling pathways that were classified by what molecule was involved in it: Notch vs. Wnt vs. whatever. There are other more physical based classifications, e.g. the Notch pathway operates in this mode whereby the cells actually have to touch each other, whereas the signals you were talking about operate in a diffusive mode typically. Diffusion on the other hand can be different, it can be diffusion that passes through cells. Also, there can be exosome-based transport, which nobody mentioned, where signals are released in pre-packaged time-release capsules and exactly why that is a useful mode for different types of information is not known. I think the idea of nanotubes that connect cells together was mentioned. So there is in addition to the molecular classification of these different signals with their own particular forms and the way they couple, there is also this different physical transport classification of these different signals. And I think we just do not know in a general sense which of those is relevant, which of those is capable of passing what type of information. And I think that is a very general question that cuts across several of the talks that we have heard over the last day or so.

E. Siggia In the case of these 2D micro-patterns, we can be very quantitative about all the very good questions you have raised. To summarize very crudely, there are elements of that positional measuring system, which on the patterns is very comparable to the embryo, which involves cell biology, where you put the receptors, reaction-diffusion of activators, secondary activators, and also various inhibitors. All these things have different ranges of signaling and there is just a large plethora of numbers, all of which conspire it seems to give you the 200 microns for these patterns. There are many knobs, all of which seem to get tuned by evolution to give 200 microns. But it is a very robust system, which is hard to sort out because there are many redundancies.

S. Eaton I agree that size control is one of the most mysterious things left out there. I think in terms of understanding how morphogens establish patterns, this has made amazing progress. And the kind of answers we are going to get are becoming clear, but still how they are controlling size is a big mystery. You mentioned the development of a new cell type, these basal progenitors in the human cortex that somehow seem to explain a change in the size they want to reach. Somehow that is an explanation that just says "Every cell by itself has a program. They are not listening to their neighbors. They are just doing a few more cell divisions." However, we also know that cells in a tissue must be measuring the volume. There must be some way they communicate to measure a volume of a tissue. At least for *Drosophila* we think this is true. If you make cells that are twice as big, they make a tissue of the same size with fewer cells. There are things that are going on to measure the volume and the cells are cooperating to do so, and we have no idea what these things are. I think that would be something to think about further and it is one of the big remaining mysteries.

T. Lecuit I would like to make two small comments. The first one is: what do you think the impact is of tissue geometry in signaling? You have one representation of a mouse embryo which is a cup, and then you had a model version which is a disk. And of course in primates and chicken it is a disk. Does it have an impact in the way that we understand patterning, is it just a simplification, or is it something important to bear in mind? The second comment is: in thinking about growth control, one interesting case is the case of symmetry that is observed in many animals. We know deviation from symmetry, such as left-right asymmetry. And how do you make something symmetric? Then the question of feedback and precision comes to mind, whether at biochemical or circuit level or biomechanical. I think it is a fascinating process. If you think about a face, the millions of operations that build it up over time, and any error or change could be amplified massively. I view it as a truly fascinating problem that the field of cell competition, which was mentioned to you and originates from work by Pat Simpson and Morata in the seventies. It is still a very active field of research in which we know some molecules, but hardly the logic. And that takes place within organs, where if you perturb a growth pattern in compartments, then the animal would have a normal organ, like a wing in flies. And if you perturb the growth pattern on the right, then the animal will be symmetric nonetheless. So I think there are many feedbacks within and between organs to make sure that symmetry is ensured, and I view it is a remarkable and open field of research.

E. Wieschaus Just to basically rephrase and readdress some of the points that Suzanne and Thomas have raised: the work on size regulation in *Drosophila* has argued basically that the process is endpoint controlled, rather than process controlled. So you do not control how many cells divide or any individual signaling process. You grow until you reach a particular size, you measure that size, and then you stop. You can change the relative rates of cells in the primordia, you can manipulate it in all different kinds of ways, but you always stop growth at a point when you are at the appropriate size. The way the fly field has come to think about this is that what you are actually measuring is the slope of a gradient (or the slope of a positional field) that extends across the epithelium. In early stages when the primordium is small, the slope (or the change from one cell to the next) is steep, and the steepness of that slope (that is the juxtaposition of a cell with a neighbor that is very different from it) will induce growth. As the field gets larger the slope decreases to a point when the juxtaposition between individual cells is below the threshold that will drive growth. So it is a very simple view for how you always end up with the same size because what is genetically determined is your sensitivity or threshold to differences to your neighbor.

S. Grill One further comment to elaborate on these thoughts that we have just heard. Probably a good general question to ask is whether in these processes that give rise to structure and a pattern, how much is even possible to separate growth mechanics deformation from the regulatory processes.

S. Eaton In response to what Eric said, a slope was an original idea that was proposed. But of course when people then measured what these gradients look like, they are not linear gradients, but instead they are really exponential gradients, which makes that model a bit more difficult. And I think there have been subsequently also other models proposed by Frank and Marcos Gonzalez-Gaitan about rates of change of morphogen signaling over time, and other models. So there are quite a few things that might work, but I do not think we yet know what does work.

E. Wieschaus I just think that the crucial idea is the endpoint regulation of growth control, rather than the process of controlling individual cell divisions. Whether we measure the right morphogen gradient when we look at a growing imaginal disk, or where is the actual property that is being measured, has been hard to determine. But I think we have to figure out is what is that feature that is being measured at the end.

M. Elowitz I just want to pick up on the theme that Herbert Levine brought up in reference to Eric Siggia's talk about signaling pathways. I want to mention that what is amazing about these signaling pathways, but sometimes gets lost, is that we only have a small handful of pathways that work in so many different contexts. That is not an obvious thing. I think we can all imagine a world in which the brain has its own signaling pathway, and the liver has its own, and the skin has its own, etc. So there is something very special about this set of pathways that allows them to do an incredibly versatile range of different activities. I think Eric did a great job at articulating the conventional view that in development we think about this. We think about the pathways as different channels to convey information and one context might use BMP, while another context might use Wnt. But I think it is really important to look for a more principled explanation that would really tell us what one pathway can do well that another pathway cannot do as well and why we need to have a Wnt-like pathway and a BMP-like pathway and a Hedgehog-like pathway, and it is not good enough to have seven different pathways that work like BMP, but are orthogonal to each other. I think this is a really fundamental question and it can be figured out.

O. Leyser I would like to zoom out even a little bit further from that and talk about the epistemological points that you raised that I think were sort of related to the things I was trying to say yesterday. Our desperate desire to linearize and discretize has led to this very dominant "DNA makes RNA makes protein" idea, which is then reinforced by the fact that those are the things that we find easy to manipulate over the years. We think about this very gene-centric and then pathway-centric world, whereas I think what is really important in the field now is to move beyond that and to think about this much more nuanced set of interactions that is very context dependent. I would make it maybe a linguistic comparison that we need to move from a semantics-based epistemological framework to a pragmatics-based one, where we think much more broadly about the interactions between these things and much less about the blobs and arrows. Because after all "DNA makes RNA makes protein", but actually it is proteins that make DNA and you never get DNA in life out of the cytoplasm in the context of all the information that is in that place, so that linearity is misleading.

J. Howard I just wanted to make a quick remark to also back up what Michael Elowitz said about these different pathways. I think we need more bio-chemical information about the details of these pathways. There must be a reason why a particular pathway is chosen for a particular thing, because it has some biochemical properties, and I think that is a very important open question. Why these pathways for these particular things? It needs biochemical information, but it also needs a physical way of thinking what it is good for, because often biochemists do not ask that question.

J. Lippincott-Schwartz Related to that point, I am curious as to whether peo-ple who have been thinking about this have also been thinking about the metabolism. How do these systems as they are differentiating get closer or further away from a nutrient source? These organoids are growing in a medium that contains glucose and other metabolites. As these organoids and structures are growing they are getting further away from the nutrient sources. Just as in the bacterial biofilms, a similar thing might be going on in these organoid systems where cells are getting further away from the nutrient source. They are sensing some type of nutrient gradient and this is something that should be focussed on.

N. King In response to the idea that there is something special about the develop-mental regulatory pathways or any of the molecular machinery, I will give the devil's advocate view. All animals have these developmental regulatory pathways. They have evolved very early. My expectation is that there is nothing special about them. They were there and that in evolution they were simply tinkered with. As another example I give the Hox genes, which evolved from an ancestral homeobox, which is a very boring transcription factor. There is nothing special about it at all, it sticks to DNA. And yet, it happened to have been regulating something interesting through duplica-tion and divergence. So, I suspect there are not any rules that we might get out looking at these pathways. What is interesting is that they can be rede-ployed in so many different ways. Think about two component systems in bacteria, all the things these systems regulate in bacteria! There is nothing special about these molecules. Bacteria just took them and repurposed.

R. Neher I wanted to make a similar point to Nicole King. There has been a bit of dichotomy — that you emphasized — between the different signaling pathways and what they might do well and what they might not do well. And then there is the context, the geometry, the size, etc., which also seems to be important. This is quite distinct from the biochemical features of these different pathways. If you look at how we communicate, on my cell phone I probably have five or six ways of communicating with different people. It is really the context of these people, depending on who I communicate with, that will determine the medium I use. It is not because one of these channels is better than any other, but it is just how it came about and what kind of habit we got into. I think there is a lot of evolutionary serendipity in this. I would not try to put particular emphasis on why this is good and why this is bad, or why it might be good for this purpose or the other. I think there are a lot of accidents that happened along the way and redundancies that get reinforced.

Prepared comment

M. Cristina Marchetti: Mechanical coordination of cell migration and patterning

Collective motion is the hallmark of active systems — collections of self-driven entities that individually dissipate energy and collectively organize in coherently moving structures at large scales. Examples of this emergent behavior range from bird flocking to the coordination of cell motion in developmental processes, such as morphogenesis and wound healing. Work over the last two decades has demonstrated that this type of large scale organization does not necessarily require complex interactions or biochemical signaling, but can arise from simple local rules. Motivated by this progress, my group has been using ideas from active matter physics to show that simple mechanical interactions can give rise to collective patterns where large groups of cells organize in coherent structures or move together in the same direction. Our work quantifies the behavior of tissues in terms of single-cell properties and provides insights on the role of mechanical mechanisms, complementary to well explored biochemical ones, for regulating collective migration and pattern formation in dense tissue.

Three types of models have been used to describe dense cell collections: particle-based models, cellular Potts models and Vertex models. In particle-based models, cells are described as particles of fixed (often circular) shape that repel each other when overlapping and adhere to their neighbors through short-range interactions (Figure 1a). In cellular Potts models and Vertex models, in contrast, cells can adjust their shape to optimize the interplay of cortex contractility and cell-cell adhesion. In Vertex models

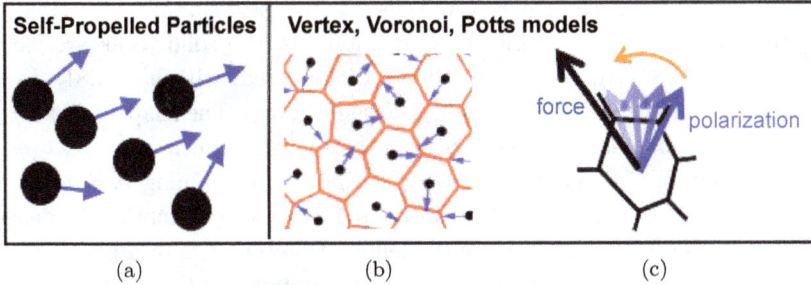

Self-Propelled Particles	Vertex, Voronoi, Potts models

force   polarization

(a)   (b)   (c)

Fig. 1.   Cells as self-propelled particles (a) and as polygons of irregular shape in the Vertex or Voronoi model (b). In both systems the cell polarization is shown as a blue vector. Frame (c) displays the tendency of cells to align their polarization with the force due to neighboring cells.

of confluent tissue monolayers, cells are modeled as irregular polygons that cover the plane (Figure 1b) and interact through mechanical forces controlled by an energy that describes the mechanical cost for cells to change their shape. For a single cell or area A_i and perimeter P_i the shape energy is given by $\epsilon_i = k_A(A_i - A_0)^2 + k_P(P_i - P_0)^2$, where A_0 and P_0 are target area and perimeters controlled by cortex contractility and cell-cell adhesion, and k_A, k_P are associated stiffnesses. The shape tissue energy is then $E_{\text{shape}} = \sum_i \epsilon_i$, where the sum is over all cells. Importantly, the interactions encapsulated by the shape energy are not pairwise additive, nor restricted to nearest neighbors, but rather determined by the cell network topology.

Minimization of E_{shape} has been used, for instance, to successfully quantify the relation between cell shape and mechanical forces in the *Drosophila* embryo.

In all three models, cell motility can be introduced as a self-propulsion force applied to each cell. The direction of the propulsive force describes the direction of polarization of individual motile cells, where the front-rear symmetry may be broken by asymmetries in the cytoskeleton. In the simplest model, this cell motility or polarization is assumed to have fixed length and direction randomized by noise. Our group has recently shown that a model of dense tissues dubbed Self-Propelled Voronoi (SPV) model, where polygonal cells move like self-propelled particles by exerting traction forces on a frictional substrate, but interact mechanically through the shape energy, exhibits a rigidity transition between solid-like and liquid-like states.[a] In the solid, cells are jammed and intercalation is energetically impossible, while in the liquid, cells frequently change neighbors. The transition is controlled by the target shape parameter $s_0 = P_0/\sqrt{A_0}$ that captures the interplay of cortex contractility and cell-cell adhesion, and by cell motility. Importantly, this work identified the mean cellular shape $q = \langle P_i/\sqrt{A_i}\rangle$, where

---

[a]D. Bi, X. Yang, M. C. Marchetti and M. L. Manning, *Phys. Rev. X* **6**, 021011 (2016).

⟨...⟩ denotes the average over all cells, as an experimentally accessible structural order parameter for the transition. In the solid, cells are isotropic and $q < 3.81$. In the liquid, they are elongated, with $q > 3.81$. In other words, cell shape is a signature of tissue rigidity. The shape-driven rigidity transition has been observed in bronchial epithelial cells.

Building on the SPV model, we are now examining two mechanisms that can account for mechanical coordination of cell motion in dense tissues.[b] The first captures the tendency of motile cells to align their polarization with the local mechanical force exerted by other cells, as sketched in Figure 1c. With this aligning interaction, the SPV exhibits, in addition to stationary liquid and solid phases, liquid and solid *flocking* states, where cells coordinate their motion on large scales. The SPV flocking transition is continuous, as in models of bird flocking where agents interact with their topological neighbors. One may then speculate that epithelial cells interact topologically whereas mesenchymal or non-confluent cells may, like particles, interact metrically, for instance through the surrounding medium. The flocking liquid shows nematic order of cell orientation and anisotropic streaming flows of size quantifiable in terms of cellular displacements, as observed in certain epithelia. The SPV flocking transition and associated large-scale cell streaming can provide a purely mechanical strategy for enhancing cell migratory capability in cancer, without requiring genetic alterations. The second mechanism embodies coordination of cell motion via the sensing of local tissue rigidity. We have implemented this in a continuum model where cells coordinate their motion through the coupling of their polarization to variations in cellular shape, which in turn corresponds to gradients in tissue rigidity. We have coined the name *morphotaxis* to describe the tendency of cells to polarize and collectively migrate towards stiffer or softer regions of the tissue. This mechanism is related to the so-called *plithotaxis*, a term coined by X. Trepat to describe the observed tendency of cells to migrate along directions of minimal shear stress. The model embodies a purely mechanical mechanism for cell patterning: positional sensing not through chemical signals, but through mechanical cues that are transmitted via gradient in local tissue rigidity as embodied by cell shape.

Discussion

F. Jülicher To connect the points of Christina Marchetti with the discussion we had before on signaling pathways, I would like to highlight that we can distinguish signals that are scalar in nature from signals that have a

---

[b]F. Giavazzi *et al.* arXiv:1706.01113; M. Czajkowski *et al.* arXiv:1710.09405.

vectorial/tensorial nature. An important example are the planar cell polarity systems, which are signaling pathways that convey vectors to each cell and that can organize anisotropies in a tissue. Vectorial/tensorial signals naturally also couple to stress and cell shape, which provides a very natural link between geometry, mechanics and different types of signaling pathways. There is a lot of information about these signaling pathways and their vectorial/tensorial nature in cells from biology, which will be very important.

A. Auhlela I would like to come back to a slightly earlier point about the potential role of dynamic signaling dynamics and relative timing in encoding space. Like Eric Wieschaus said, as developmental biologists we like to think of the result and what are the constraints of the system. One thing that I wanted to bring up is that patterning is in many cases temperature independent, so creatures can grow at different temperatures, at different rates, but still the patterning comes out pretty much normal. I think about very nice work of Andy Oates who studied the zebrafish embryo segmentation at different temperatures. The oscillations that actually underlie this segment formation are actually very much changed in their period, yet the segments come out with the same spatial dimension. I think this relative encoding in the dynamics that Michael Elowitz mentioned yesterday, could be a very elegant way to think about the constraints of the signaling machinery to bring about such temperature independence. This is by no means a new idea, in fact Brian Goodwin proposed this fifty years ago. He said you could use the relative encoding between pulsatile oscillatory signals to encode space. I think that now with all we have heard and discussed, technology and also viewpoint, it could be a very good time to think about the role of dynamics in more depth.

C. Marchetti I think Frank Jülicher brought up a very good point. One of the things we have tried to highlight is exactly the distinction between anisotropy between individual cells (which is more a scalar quantity) and the alignment of elongated cell shapes (which is more a tensorial quantity). It seems that both mechanisms can actually drive coordination of cell motion, as well as coordination of local patterns.

T. Gregor We have heard about multiple different ways how cells communicate: signaling channels, diffusion, biochemical pathways. All of our description of these is very phenomenological or observatory. I have not heard of ways in which we could find a number to characterize these channels, similar to what Rob Philips has shown yesterday when he showed 6–10 different relationships and there was one model that could characterize all of them. There must be a way to do this with signaling channels and I wonder whether anyone is starting to go in this direction, finding channel capacity, or information or energy flow. We should be able to quantify and hence compare the different systems. If you have several numbers that allow us to quantify these signaling channels, we might also understand the intricacies of how they interact with each other.

U. Alon I wanted to mention that signaling pathways should work across many geometries, and many situations (temperatures, etc.), which means they need to be extremely robust. I wanted to mention work of Naama Barkai, seminal work that was not in Eric Siggia's presentation, where you look at a pathway and the way it is structured. E.g. for BMP you have the morphogen and you have the inhibitors. And the inhibitors do something amazing, they both inhibit signaling, but they enhance its range. So it is a bifunctional molecule. When you look at that property, you get signaling that is extremely robust to geometry, to parameters, and that might hint why BMP pathway is built the way it is. Along these lines, there is work on special properties of the Notch pathway and the Hedgehog pathway that Michael Elowitz is doing, where you can look at molecular details that seem arbitrary and unexplainable, but when you put them together you get cancellations that give you remarkable robustness. Therefore this suggests why these pathways could work in different geometries and different situations. In my opinion that is a research program that eventually will help us make sense of these details in terms of a larger theory of robust patterning.

R. Brangwynne One of the things we have heard about is mechanical patterning. There is a lot of signaling and of course there is much work over the years on mechano-chemical signal transduction and the intimate relationship between these. One thinks about focal adhesion for example and signaling pathways, in a conventional sense they are sensitive to mechanical stress. It occurs to me that in thinking about the cell as a information storage and processing unit, I think a framework for trying to understand how information is stored and processed in mechanical signals as well as more conventional signal transduction pathways would be really helpful. From an information theoretical and computer science perspective, the cell is ultimately doing computation with these different signals and it seems to me that some sort of framework for understanding information and how information is processed and stored in living matter both through mechanical and chemical signals is needed. I do not hear this very much and I think something along these lines should be discussed.

J. Lippincott-Schwartz One thing that could fit that question relates to protein crowding in the cytoplasm. Recently we published a paper with Dave Weitz showing that changing the osmolarity of a stem cell could change its fate. For instance, it is known (from Dennis Discher) that when you put stem cells on a stiff or soft substrate they turn into different lineages (bone vs. muscle vs. fat). We made the observation that cells on these different substrate stifnesses change their volume and their crowded state. Changes in osmolarity thus impact the crowded state of the cytoplasm.

Prepared comment

Irene Giardina: Collective behavior in animal groups

Biological aggregates often display collective behavior, where individuals move or function in a coordinated way. Such collective phenomena occur across different scales, from the microscopic one of cell colonies and bacterial clusters, up to the macroscopic scale of animal groups. A number of intriguing questions arise, which are at the core of current research in living active matter: How does collective behavior occur? How general are its properties across scales and species? What (if any) is its biological function? More broadly, can we develop a theoretical framework to explain such phenomena? In condensed matter systems scaling laws and the renormalization group approach tell us that, if we are interested in the large scale behavior, the details of the interactions between the individual constituents do not matter. Only a few features are relevant, and simple models are able to describe a whole class of systems. Our hope is that something similar also holds for living interacting aggregations. This assumption, however, cannot be taken for granted. This is particularly true for animal groups, where individuals are complex organisms and interactions occur through complicated cognitive processes. The roadmap towards a statistical physics approach therefore requires an empirical validation of its premises.

Recent experiments on flocks of birds and swarms of insects provide important results in this respect. Flocks and swarms represent qualitatively different instances of self-organized collective behavior. In the first case, the system is globally ordered resulting in the collective motion of the group. Swarms, on the contrary, do not display directional ordering and remain spatially confined. Both systems, however, exhibit large scale collective patterns. The connected correlation function $C(r)$ of the velocity fluctuations — measuring the degree of movement coordination between individuals at distance r — is indeed long-range in both cases.[a,b] The correlation length ξ scales with the group's size indicating scale free behavior (see Figure 1).

Correlation functions can be computed also in time, quantifying the influence of an individual in the group on another individual at a different location and time. Surprisingly, such correlation functions obey scaling laws. Scaling, and the dynamic scaling hypothesis in particular, is a very powerful concept. It tells us that when the correlation length is large, the set of parameters $\mathbf{P}$ (biological and environmental) controlling the group's dynamics determine the behavior of correlations only through the correlation length. In formulas:

---

[a] A. Cavagna *et al.*, *Proc. Natl. Acad. Sci. USA* **107**, 11865 (2010).
[b] A. Attanasi *et al.*, *Phys. Rev. Lett.* **113**, 238102 (2014).

$$C(k, t; \mathbf{P}) = C(k, 0; \mathbf{P}) \, F\left(k\,\xi(\mathbf{P}), t/\tau_k\right) \tag{1}$$

$$\tau_k = k^{-z} g(k\,\xi(\mathbf{P})) \tag{2}$$

where k is the wavenumber, τ_k is the relaxation time of $C(k, t)$, F and g are two scaling functions, and z is the dynamic critical exponent. If the above equations hold, when we consider systems with different values of k and ξ but the same product $k\xi$, the reduced correlations $\hat{C}(r, t) = C(r, t)/C(r, 0)$ must obey the same decay behavior as a function of the scaling variable $k^z t$. This is what happens for wild swarms of midges, as displayed in Figure 1, where data from three different species, with different sizes and external conditions are collapsed into a single curve.[c]

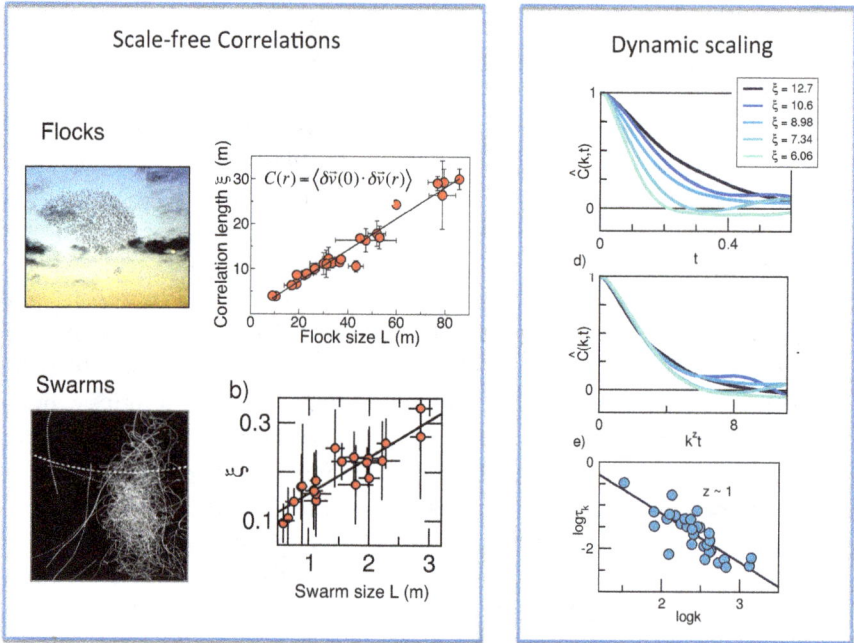

Fig. 1.   Left panel: Scale free correlations in flocks (top) and swarms (bottom). The correlation length is plotted against the group's size. Each point corresponds to a different flock (swarm) and it is an average over several instants of time.[1,2] Images from the Cobbs group, ISC-CNR, www.cobbs.it. Right panel: Dynamic scaling in swarms of midges. Correlations are plotted for swarms with different ξ, fixing $k\xi = 1$ (top figure). Correlations rescale when plotted as a function of the scaling variable $k^z t$ (middle figure) with $z \sim 1$. The relaxation time τ_k scales as $k^{-z} = \xi^z$ (bottom figure).

---

[c] Cavagna *et al.*, *Nat. Phys.* **13**, 914 (2017).

The validity of scaling laws enormously simplifies the theoretical task. It also allows defining 'classes of behavior' corresponding to different values of the dynamic exponent z. Systems in the same class can de described with the same effective model. These results support the statistical physics perspective adopted in active matter, where simplified models have been investigated to characterize collective behavior. At the same time, it puts some constraints on the validity of such models in that the experimental value of z must be reproduced. In the case of swarms, for example, the value of z found in natural groups is different from the one of standard models, suggesting that new ingredients must be included in the theoretical description. Similar conclusions can be drawn for flocks when comparing experimental data on static scaling and on the dispersion relation with the predictions of flocking models. In both cases, it appears that the missing ingredient is related to a behavioral inertia and second order terms in the dynamics, which would be ineffective in the thermodynamic limit but do matter for finite sizes. Far from being an irrelevant detail, in flocks inertia produces quick linear propagation laws allowing finite groups to exploit scale free correlations to maintain global coherence in time.[d] Its role in swarms is still to be fully investigated. Once we understand what are the effective parameters regulating the system's dynamics on the relevant time-scales we can investigate where the system is located in parameters' space, what is special about its behavior from a mechanistic perspective, and whether this is related to some kind of biological efficiency.

Discussion

A. Chakraborty I have a question that pertains to the last comment made by Jennifer Lippincott-Schwartz and also a comment that Frank Jülicher made. Is there a single example or more where it is clearly understood how cells sense mechanical stress and change their differentiation pattern beyond generalities?

S. Eaton I wonder whether it would be productive to think about the mechanisms that allow the flocking behavior to scale with the size of the system. Another thing that scales with the system size is morphogen gradients. You can make a big animal or a small animal. The pattern elements are proportionally distributed, suggesting the morphogen gradients must somehow scale with the size of the system. A fascinating example of that is planaria that can vary hugely in size and yet they have this Wnt gradient that spans from the head to the tail and it just scales. So it is interesting to think about the

---

[d]A. Attanasi *et al.*, *Nat. Phys.* **10**, 691 (2014).

rules that allow your system to scale and then which type of interactions between cells would we have to postulate to make things scale.

S. Chu Short answer to the question of Arup Chakraborty. In synaptic plasticity, the formation of synapses, what happens is that as the synapse begins to form there is actually mechanical tension built and then sensed to make sure that you are close and you have that very small gap. So, at least at that level, there is real mechanical motion that sends back a signal.

A. Hyman I think Irene Giardina brings up a really good point, which is, with the flocking of birds and insects you can take a coarse grained approach where you are effectively ignoring how the bird flaps its wings or how it undergoes sensing. The properties of the flock emerge from simple interactions. At what level should one be considering development, because Eric Siggia stated that you could only understand development to the level of genes, which I am still surprised to hear, because I still think one can only understand development at the level of cells interacting with each other, and you coarse grain the whole process so you do not worry too much about the individual genes. You just coarse grain how the cells interact with each other and you want to define input and output.

S. Mayor I just want to make a comment about the notion of mechanical sensing during tissue morphogenesis or even at the level of a cell. Wherever we have looked, we find that the mechanical signal is transduced in some sense into a chemical signal, very much like an electrical signal at the chemical synapse. I do not know how this would influence some of the mechanisms or some of the theories that one might want to build around how mechanics could influence patterning and behavior of cells and tissues.

W. Bialek The clock is working against us I am afraid. Let me make a couple of summary remarks and then we will go to be photographed. The first observation is that I think in the course of the morning we have actually seen eight orders of magnitude in length scale, which is worth keeping in mind. In the first half, two ideas which came up were issues about placing signaling systems in their natural context, and the search for rules that would govern these signals. I think in this half two related ideas have come out. One is the problem of putting very different kinds of signaling systems on the same basis. Whether you are talking about chemical signals, mechanical signals, the ability of organisms to observe each other even when they are not in contact as in the case of flocking, and there are obvious candidate frameworks for doing that. We know that information is supposed to be measured in bits, but whether that is actually a useful way of doing this is another question. In response to the notions of context and the search for rules, there is the question of how much is accidental and how much is to be understood as tuned to the particular context. That is also something that came up in other ways yesterday. It would be nice to at least ask the question precisely in each case, not assuming that the answer is the same in each case. I wanted to end by noting one thing, and I want to say this in a way that is value neutral. The different systems that we have heard about in particular incarnations that we have seen, there is an enormous variation in the extent to which they are being characterized qualitatively or quantitatively. Maybe some of us have the prejudice that there is a linear thing where you start with something qualitative and you are aiming for something quantitative. I think that is interesting, particularly at a gathering of physicists interested in biological problems, that there is this enormous variation in how quantitative our descriptions are, and I am inclined to think that has an impact on the depth of theoretical understanding we can reach, but there are different views. So in the same way as one wants to answer questions about accidents vs. principles with an open mind, we should think about this issue of what does it mean to characterize communication among ourselves and organisms either quantitatively or qualitatively, and how that relates to the type of understanding we are looking for. So let me thank all the people who spoke. My apologies to those of you who did not get to say what you wanted. Our thanks to all of the speakers.

Session 4

Morphogenesis

Chair: *Thomas Lecuit*, College de France, France
Rapporteurs: *Eric Wieschaus*, Princeton, USA and *L. Mahadevan*, Harvard, USA
Scientific secretaries: *Sophie De Buyl*, VUB, Belgium and *Bortolo Mognetti*, ULB, Belgium

T. Lecuit Good morning everybody. So the session is entitled morphogenesis. And maybe this is the opportune time to have a session on morphogenesis reflecting on hundred years of D'Arcy Thompson. I will just sketch up three important steps in the study of morphogenesis, which is probably one of the oldest questions that scientists have been interested in looking into in biological systems because you could look at morphogenesis with naked eyes years ago.

The first step, I would say, was taken by D'Arcy Thompson who made the first explicit and somewhat detailed attempt to describe morphogenetic processes in mathematical and physical terms to explain the laws of organization mechanically with, for instance, a focus on surface tension, as well as — in his "transformation theory" — hypothesized mathematical laws in the transformations of shape.

The second important step was the discovery of the laws of heredity whereby you can explain how the information, which was difficult to understand in physical terms at the beginning of the 20th century — it was actually explained by the evolved chemistry of DNA replication and transmission — and which is required for the emergence of shapes in the physical world.

The third important step I would say, on the theory side, was the work of Alan Turing "On the chemical basis of morphogenesis", which introduced the chemical reaction-diffusion system as a possible mechanism of pattern formation in biology. The paper was not an attempt to vindicate the role of mechanics, actually he did say that mechanics was important but was too complicated to be addressed.

And maybe a fourth step is the discovery of genes which underlie pattern formation in organisms. Genes define spatial patterns in a dynamic way as we have seen in our third session. A very important concept articulated clearly by Lewis Wolpert in 1968 is the concept of "positional information" which attempts to explain in general terms how cells receive spatial coordinates that then identify specific cell behaviors, differentiation and dynamics.

So today I would say we are in the best conditions to bring together physics, and I mean by that, mechanics in particular and the evolved chemistry, that we call biochemistry, to understand how genes encode shapes. We want to understand how genetics and the evolved chemistry provide regulatory control over mechanics to produce geometry and dynamics. And to cover this broad field we have two rapporteurs, the first is Mahadevan from Harvard University who will begin. Thank you.

Rapporteur Talk by L. Mahadevan: Multicellular Morphogenesis

How do genes encode functional geometry in living matter? To answer this question, we need to combine our understanding of the biochemical and the biophysical processes at molecular, cellular and tissue levels in terms of quantitative predictive theories. This review attempts to summarize the current state of our understanding of multicellular morphogenesis from a physical perspective.

1. Introduction

Morphogenesis, the origin of form in (physical and) biological systems, is a much written about subject, given the relative ease of casual observations that it affords, and the consequent plethora of myths that it has engendered, most famously in Kipling's *Just So Stories*, and its mimics.[1] And just like other familiar phenomena that are easy to observe, but hard to understand, it is far more complex than meets the eye, because it intimately links multiple spatial and temporal scales, both in developmental and evolutionary settings.

A hundred years ago, the biologist, mathematician and philologist D. W. Thompson used his magnum opus, "On growth and form,"[2] to emphasize the importance of geometry and (classical) physics in addressing the title, stating that shape is a consequence of "matter that is moulded, moved or conformed according to the laws of physics," emphasizing the role of physics in morphogenesis. Fifty years ago, the developmental biologist C. Waddington convened a set of meetings[3] that led to an edited set of proceedings titled *Towards a Theoretical Biology*, where he wrote "No conceptualization of a living system is adequate unless it includes at least four importantly different time scales, those of metabolism, *development*, heredity and evolution," emphasizing the role of dynamics. And in between, inspired by both Waddington and Thompson, the mathematician A. Turing authored the prescient paper "The chemical basis for morphogenesis," in which he discussed "a possible mechanism by which genes of a zygote may determine the anatomical structure of the resulting organism," and proceeded to provide a minimal mechanism for biochemical patterns.[4] Turing recognized that the chemical perspective was important given the chemical nature of gene action, but was well aware that a complete answer would require linking this to a mechanical and geometric view. So where are we today?

Over the last half century or so, the modern revolution in biology has given us the tools to measure and manipulate events at multiple scales spanning molecules, cells and tissues. Simultaneously, a framework centered around the notions of gene regulation, signaling and transcription has been brought to bear on the processes that characterize tissue organization at the multicellular level. However, the formation of organs is not only well orchestrated biochemically, it is also a consequence of transport, deformation and flow generated by growing tissues. Thus one must combine the power afforded by molecular and cellular approaches with the conceptual and theoretical advances of physics in understanding how matter, information and energy are patterned in space and time in nonlinear and nonequilibrium systems.

This report provides a brief summary of the current state in our collective quest towards one of the grand questions of biology: how do genes encode functional geometry in living matter? To unfold the complexity of shape at the meso and macro scales in biology and characterize the role of growth and dynamics in living matter requires addressing two complementary questions: (i) how do we describe the geometry of cells and tissues, as driven collectively by cytoskeletal dynamics in a cell, and (ii) how do we create predictive mathematical frameworks that combine geometry, physics and chemistry to derive the robust qualitative aspects such as phase diagrams e.g. morphospaces and the state transitions therein. This will naturally also allow us to understand developmental dysmorphologies, pathologies with the failure of tissue maintenance that is often the basis for cancer, and perhaps harness our understanding to engineer tissue shape for repair and replacements.

2. Mathematical descriptions of shape

"The problems of form are in the first instance mathematical problems," wrote Thompson.[2] But what is shape? From a mathematical viewpoint, shape descriptors of an object are invariant to translation, rotation and scale, i.e. the elements of the similarity group. But this "definition" does not provide a unique characterization of shape. Indeed, one might choose to use any number of different geometries (or the equivalent symmetries) to describe biological shape e.g. the Euclidean description of planar objects such as (some) leaves and wings, the non-Euclidean description of (many) flowers and feathers, the conformal (or quasi-conformal) geometry of the surface of a mammalian brain, leaf, petal, etc., the projective geometry of an egg, bud, or bone, as shown in Figure 1. Thus the choice of geometry is often a matter of preference driven by notions of symmetry,[5] and driven by elegant mathematical and computational tools which may not necessarily reflect the underlying biophysical mechanisms of morphogenesis. On the positive side, a compressed description of shape that is agnostic to mechanism allows one to deploy a range of useful morphometric tools to compare phenotypes and look for relations that might shed light on the better characterized genotypes. Examples of this approach include the classic work of the paleontologist D. Raup who used Euclidean geometry to describe the helicoidal shapes of gastropod shells,[6] and recent attempts that have shown that conformal geometry describes leaf growth[7] and projective geometry describes avian egg shapes.[8] More broadly, this has led to thinking about the geometry of adaptive landscapes linking morphometry, morphogenesis and evolution.[9]

As the availability of shape data becomes more and more common, e.g. from 3D tomographic scans of skeletons[10] to 4D embryos developing in space-time,[11] one needs to go beyond deterministic descriptions since functional biological shape is variable, often in a scale-dependent way. This requires a statistical geometric description of shape, leading to probability distributions of the related continuously

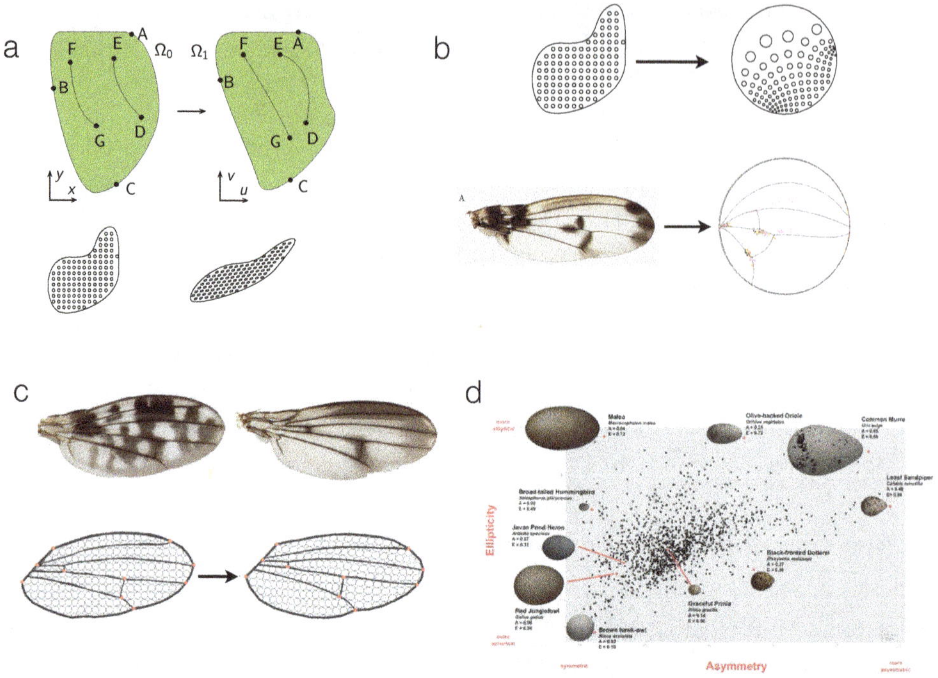

Fig. 1.   Morphometrics, the description and classification of biological shapes is often dictated by mathematical efficacy. (a) An important choice in comparative analysis of shapes requires choices about how to handle landmarks in the bulk and boundary. At the simplest level, one might consider affine maps, i.e. linear transformations that include shear, rotation and dilatation, as shown below. (b) Planar shapes can easily be transformed conformally, and transform circles to circles, e.g. a conformal map that takes *Drosophila* wings to a disk. The variability in the location of venation intersections is clearly seen (figure courtesy of G. Jones). (c) Conformal maps are unable to handle shear and differential rotation, both of which are biologically relevant. A generalization of conformal maps known as quasi-conformal maps allow one to handle dilatation, rotation and shear; a specialization known as Teichmuller maps take one wing onto another by transforming circles to ellipses of uniform eccentricity (figure courtesy of P.-T. Choi). (d) Closed objects such as eggs can be well described using projective geometry, here illustrated in terms of two parameters: ellipticity and asymmetry (adapted from Ref. 8).

variable geometric invariants, [12,13] still a nascent field in terms of its biological implications. In contrast, the field of quantifying genotype variations using statistical and computational methods is more mature as these need to only consider discrete differences between linear arrays written using a finite alphabet in such fields as sequence analysis. [14]

2.1. *Reparametrization invariance for classification*

Any morphometric study associated with the classification of shapes needs to be independent of representation, and thus invariant under reparametrization. Thus, if we choose some metric to measure the distance between two shapes f_1, f_2 defined as $E(f_1, f_2)$, and both shapes were subject to a (bijective) reparametrization so that the represenation $f_i \to f_i \circ \gamma$, then invariance under reparametrization implies that $E(f_1, f_2) = E(f_1 \circ \gamma, f_2 \circ \gamma)$ for all reasonable (bijective) transformations given by γ. Most current methods for shape classification do not respect this simple requirement, and instead choose a specific parametrization before deploying algorithms for geometric clustering. Recent advances in statistical geometry[15] combine ideas from differential geometry, statistics and pattern theory[16,17] to derive computational algorithms for shape classification that are reparametrization-invariant, replacing the usual L^2 metric with the more natural Fisher-Rao metric that considers geodesic distances between probability distributions characterizing shapes.[15] This allows for the simultaneous solution of the registration problem and the classification problem that has dogged the statistical description of shapes — and the time seems ripe for the deployment of these ideas to biological morphometrics.

2.2. *Probabilistic geometry of shape*

The invariant representations for shape depend on both the object and its embedding. Thus, to quantify one-dimensional objects such as protein backbones, neurons in a densely packed cortex, or plant tendrils or vines embedded in three dimensions, one needs to consider the curvature $\kappa(s)$ and torsion $\tau(s)$ of the center line as a function of the arc-length s that completely defines the geometry of curves up to rigid translations and rotations.[18] Equivalently, one can define discrete variants of these objects in terms of their natural counterparts in discrete differential geometry.[19] In a statistical setting as is the case in biology, one can define probability distributions of these objects as well as information-theoretic quantities such as $\int p(\kappa(s)) \ln(p(\kappa(s))) ds$. Similarly, for two-dimensional surfaces, one needs to consider the first and second fundamental forms $\mathbf{a}(\xi, \eta), \mathbf{b}(\xi, \eta)$ that reflect changes in length, angle and the variations in the normal to the surface as a function of surface coordinates (ξ, η), reflecting the fact that both their intrinsic and extrinsic geometries are important in a functional biological setting. Again this needs to be couched in the context of probability distributions on the invariants constructed from these fundamental forms, e.g. the trace and determinant of the respective tensor representations, and associated measures of information such as $\int p(\kappa_G(\mathbf{r})) \ln(p(\kappa_G(\mathbf{r}))) d\mathbf{r}$, where $\kappa_G(\mathbf{r})$ is the Gauss curvature at location $\mathbf{r}$. Finally, in three dimensions, the metric tensor $\mathbf{g}(x, y, z)$ and its invariants allow for a geometric description of bulk objects, but must also be couched in a probabilistic setting with associated measures of information. An interesting class of questions where these statistical geometric notions might be deployed is in the context of the fossil record[20] to allow for a study of morphological (phenotypic) diversity to complement our increasing knowledge of ancient genotypic diversity.

2.3. *Dynamic morphoskeletons for flows*

Biological shapes are dynamic, especially in a morphogenetic setting. Thus, one needs to move beyond shape analysis to understand how cells and tissues change with time. The ability to image large scale cell movements in space-time has advanced rapidly in the last decade with the advent of advanced microscopy methods,[11] combined with transgenic animals that allow for a comparative view of cell movements. Biological tissues can possess complex microscopic textures associated with variations in shape, size, number and position of cells that characterize the underlying texture encoded in the deformation and deformation rate (see Figure 2). But how does one glean information from these large dynamical data sets? Here, we need to combine ideas from statistical geometry with those from hydrodynamics

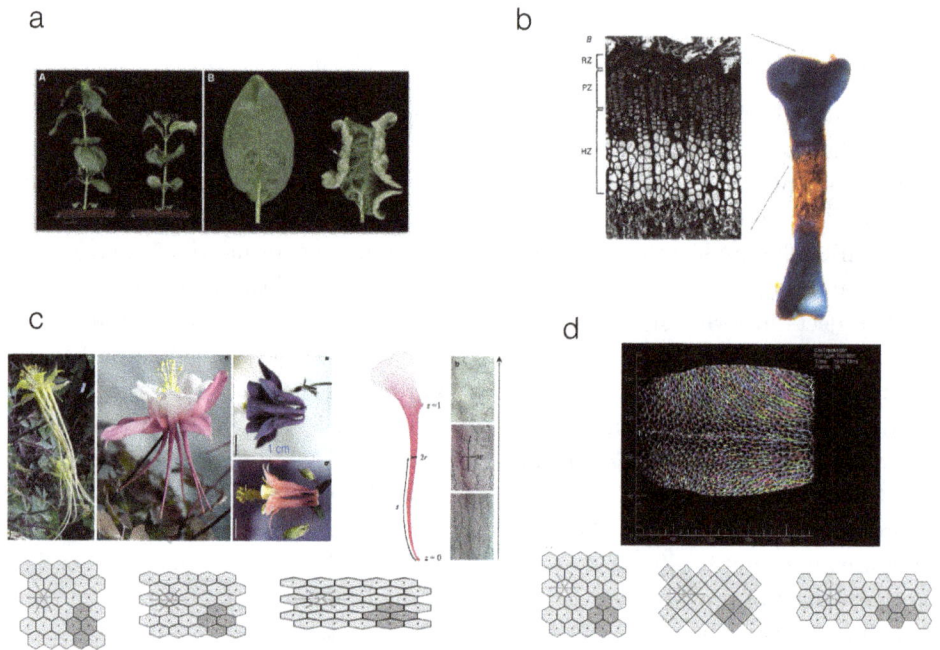

Fig. 2.   Cellular motifs underlying morphogenesis. (a) Cell number can be a function of space and time. In this example[43] a nominally flat snap dragon becomes potato chip-like because of excess cell proliferation along the margins of the leaf. (b) Cell size can be a function of space and time. In adolescence cells away from the joints expand in size in the hypertrophic zone, leading to disproportionate changes in limb length.[44,45] (c) Cell shape can be a function of space and time. In this example a square tissue can be transformed into a rectangular tissue with the same number of cells simply by making each cell switch from a regular hexagon to an elongated hexagon, preserving number, connectivity and overall area but changing shape. Seen here are variations in nectar spur length in *Aquilegia* flowers driven by this mechanism.[46] (d) In plant organs cell number, size and shape can be a function of space and time. However relative cell position cannot change because of the presence of cell walls. In animal tissues this constraint is lifted. This allows for a different way to transform a square tissue to a rectangular tissue, by the relative movement between cells, as evidenced in large scale cell intercalation.[22]

and continuum mechanics.[21] In general, cells can translate, rotate, change shape, change size and divide, while also moving relative to their neighbors.[22,23] In a dynamical context, all these quantities will vary and any description requires the use of frame-indifferent quantities that are invariant to transformations associated with translation, rotation and Galilean boosts. This immediately rules out the use of velocity (which is not Galilean invariant) and vorticity (which is not rotation invariant) representations.

Given a spatio-temporally varying velocity field $\mathbf{v}(\mathbf{r}, t)$, a natural variable to consider is the velocity gradient tensor $\nabla\mathbf{v}(\mathbf{r}, t)$ whose symmetric part describes both local changes in size (dilatation) and shape (shear), while its anti-symmetric part describes local rotations (vorticity), in an Eulerian (instantaneous) setting.[22,23] Recent advances in the analysis of spatio-temporal dynamical systems[24] suggests that fruitful dynamically invariant signatures of morphogenesis might be afforded by Lagrangian coherent structures that follow cellular trajectories. Since these trajectories involve information obtained by integrating the velocity fields to determine trajectories in space-time, they serve as memory traces and can be quantified in terms of the properly invariant (Cauchy-Green) deformation gradients[21] to yield the attracting and repelling manifolds that might serve as the structural organizers of tissue morphogenesis.

Characterizing these motions in epithelial morphogenesis has already led to our ability to distinguish how these different modalities can either together or separately induce tissue shape (Figure 2). Understanding how to lift this into the third dimension is a challenge that we need to meet now. Since tissues are made of densely packed cells, a second characterization of how strain correlations decay in space and time during morphogenesis is another critical aspect that needs to be addressed.

Together, the deployment of these statistical-geometric and dynamical approaches will serve to determine the morphospaces and the dynamical attractors of shapes during development and across evolution using data to provide a compressed geometric description of functional biological morphotypes.

3. Biophysical prediction of shape

To go beyond the mathematical descriptions of shape to biophysical predictions of shape, it is useful to first contrast biological morphogenesis from physical and chemical pattern formations.[25,26] In a range of chemical and physical systems that include reaction-diffusion, hydrodynamic and elastic systems, spatial patterns emerge via spontaneous symmetry breaking instabilities of a uniform and homogeneous state arising from simple interactions. The zoology of instabilities and patterns that arise even in these systems[25] is vast and is a consequence of both chance and determinism. In marked contrast from these systems, biological morphogenesis typically involves genetically preprogrammed "agents", i.e. cells, that can replicate (by division) but also change internal states (differentiate) in response to external (biochemical) signals, and can sense and act in response to external and internal signals. We refer the interested reader to textbooks[30,33] to review the bio-molecular, cellular and genetic and biochemical aspects of patterning from this perspective.

In a chemical setting, Turing's pioneering theoretical work[4] showed how differential diffusivity can lead to instability in a system that when well mixed would be stable, there began a search for viable morphogens — diffusible molecules that might serve as instantiations of the proposed mechanisms. However, the early hope that reaction-diffusion mechanisms might serve as explanations for the generation of patterns in biological systems[27,28] has not quite been realized. Indeed, the number of systems where the conditions for this remain small; the most prominent examples are associated with skin pigmentation patterns.[29] In the meantime, other chemical signaling approaches were proposed for "positional information"[30] within embryos and tissues, along with variants that showed how to solve the "scaling" problem in organisms, i.e. how to ensure that patterns scaled with body size. In its simplest form, the basic question raised is that of ensuring that the pattern wavelength scaled with the system size and thus leads to patterns that maintain the same proportions independent of absolute scale, rather than have an intrinsic wavelength that often happens in pattern forming instabilities associated with morphogen gradients. There have been a number of recent proposals to address this[31,32] but there is no unequivocal answer as yet.

We now turn to the physical and geometric aspects of morphogenesis, since morphology and physical shape cannot be understood without characterizing the physical forces and flows that underlic cellular rearrangements at the tissue level, which are themselves coupled to the biochemical and biomechanical processes at multiple levels.

3.1. *Cellular motifs*

The cellular basis for form in a developing organism arises from four spatiotemporal fields that relate cell geometry and topology to tissue, organ and organismal geometry and topology: (relative) cell number, size, shape and position,[23,34] as shown in Figure 2. The first three motifs are all that are accessible in plant morphogenesis where cells are incapable of moving relative to each other owing to the presence of very stiff cell walls. Nevertheless, even with these three spatio-temporal fields, plants have the ability to generate a range of remarkable tissue shapes.[35,36] In animal tissues, the additional complexity associated with cell movement leads to many redundant mechanisms for shape generation and the rescue of morphotypes.[22] These together drive overall tissue size, connectivity and shape and yield a number of tissue building motifs. Understanding how cells change their packing arrangements actively has some physical similarities to the passive physics of cellular foams,[37] but with the added complexity that cells can generate and respond to active forces driven by the contractility of acto-myosin composite networks.[38-40] Recent experimental and theoretical efforts to understand epithelial morphogenesis, the shape changes in sheets of cells that can deform in and out of the plane have taken this perspective by developing discrete[41] and continuum models of these sheets.[42]

3.2. *Tissue motifs*

There are three general tissue motifs which, with iterations and variations, define the range of possible shapes achieved in both plant and animal organs. The first corresponds to a simple change in size via the addition or growth of cells, or via the change in the shape of cells. For example, in early vertebrate morphogenesis, the embryo elongates via the addition of cells at the base of the pre-somitic meso-derm.[47,48] In plants, an example of tissue size and shape change without addition of cells via cell shape change.[46] A second tissue motif is associated with changes in topology and is seen in examples such as lumenization (the formation of a hole, or a lumen) and segmentation. In lumenization, seen in organs such as the brain, kidney, inner ear, etc.[49] holes are formed through a combination of tissue expansion and fluid secretion.[50,51] In segmentation, differential adhesion causes the tissue to

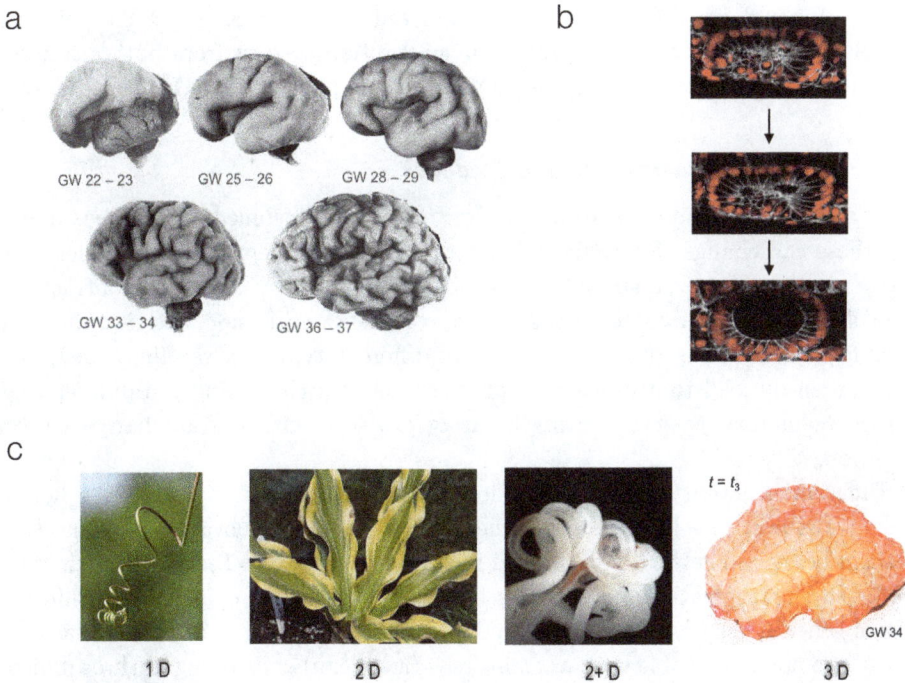

Fig. 3. Tissue-level motifs in morphogenesis. (a) Change in overall size as illustrated by brain volume increase. There is in addition a increase in the area of the cortex that leads to folding (adapted from Ref. 57). (b) Change in connectivity as illustrated by lumenization wherein tissue topology changes via the coordinated growth of a hole driven by fluid secretion.[49] (c) Change in shape or generalized buckling driven by differential growth, flow, activity, etc. which can occur in 1, 2, or 3 dimensions.

break up into periodic segments, e.g. during digit or vertebra formation.[52,53] The third motif in tissue morphogenesis is associated with tissue shape changes driven by a range of processes including differential growth,[2] differential adhesion,[54] differential activity,[55,56] and of course differential diffusion[4] among other processes. All these motifs can occur in multiple dimensions.

These topologic and geometric motifs are enabled by actively generated cellular and tissue flows and forces. Theoretical frameworks for these require us to revisit the basic tenets of statistical mechanics of passive materials to account for non-equilibrium behavior that can violate detailed balance, and the consequent hydrodynamic descriptions to account for additional broken symmetries, e.g. not requiring Galilean or rotational invariance, etc. Additionally, one has to augment the usual balance laws for mass and momentum (non) conservation by accounting for biochemical activity, positive entropy production rate, etc. While there is substantial activity in these areas,[56] what is needed is a transition from the physics of active matter to the biology of sentient matter, as theory and computation are still rather afar from biological experiments in most settings.

An exception is epithelial morphogenesis, typically associated with the change in form of monolayered tissues that deform in/out of the plane.[41,58,59] This involves actively moving cells, and there has been substantial progress from both an experimental and theoretical perspective, as described in the report by E. Wieschaus.

3.3. *Form from frustration and feedback*

Since growth and form involve changes in topology and geometry, one needs to ask how these can change. Spatially uniform change is an exception and only leads to changes in size. However, spatially non-uniform growth, flow, adhesion, contractility in a collection of otherwise identical cells, or differential diffusion, activity, rheology in multiphase systems that include cells of different types, extracellular matrices, etc., can easily lead to a bestiary of physico-chemical patterning instabilities that genetic regulatory pathways control during development, and are harnessed via evolution.

Differential activity, growth, adhesion, diffusion, and flow in an initially homogeneous tissue can lead to situations whereby cells in tissues move in a way that lead to them being geometrically frustrated in space, or unable to keep pace in time, or both. At a mathematical level, this is because in epithelial, epidermal or thin tissues in general, the first and second fundamental forms of the center-surface $\mathbf{a}(\xi, \eta)$, $\mathbf{b}(\xi, \eta)$ are not compatible with a strain-free (Euclidean) embedding in three dimensions.[18] The resulting incompatibility, that is due primarily to in-plane growth leads to tissues that buckle out of the plane into complex three-dimensional forms that are pre-strained[63,65,66] — indeed surgical experiments on both plant and animal tissues confirm this.[64,65] The stresses and strains associated with geometric incompatibility does not only lead to shape changes, but are correlated with variations in gene expression levels that control cell proliferation in epidermal patterning.[67]

In constrained epithelia such as skin, these strains driven by incompatible growth can cause cells to undergo apoptosis, and the tissue to change its rheological properties[60–62] — a form of tissue maintenance and homeostasis with a mechanical basis.[59]

This brings us to an important difference between the patterning and shaping of passive materials and active, sentient tissues — the role of feedback via chemical and mechanical cues. Chemical information can be shared either via diffusive processes or active transport. In the first, characteristic time scales as L^2/D and in the second, characteristic time scales as L/V, where L is a systemic length scale and D and V are diffusivity and active velocity, respectively. While these pathways have been shown to be efficacious at cellular and sub-cellular scales, on tissue scales, biochemical feedback may be too slow. A natural candidate for feedback on tissue scales is mechanical wave propagation[68] which takes a time L/V_e, where V_e is the elastic wave speed in a tissue, or L^2/D_p, where D_p is the poro-elastic diffusion constant.[69] These diffusion times can be orders of magnitude smaller than those associated with biochemical signaling, suggesting that mechanical feedback cannot be neglected. Quantifying the feedback mechanisms and the gains therein, and understanding the coupling between biochemical and biomechanical pathways[70] and regulation in morphogenesis remain one of the outstanding challenges in the field. Furthermore, there is constant feedback at multiple levels between cellular and tissue level events using chemical, electrical and mechanical signaling. Thus biological morphogenesis leads to very reproducible shapes, likely because of the multiple levels of feedback and the redundancy.[71,72]

To couch these issues in concrete terms, we will discuss two examples briefly, one each from plant and animal morphogenesis.

The meristem in plant shoots is its most actively growing region,[36] and over the past few years its morphogenesis has been the focus of biochemical and biomechanical investigations using a combination of experiments, theories and computations.[73,74] Using *Arabidopsis* as the model system, the molecular players that control cell growth and shape anisotropy are by now well established and include the plant hormone, auxin, and the cytoskeletal element, microtubules, as well as the dynamical cell wall that allows plant cells to sustain relatively large turgor pressures of the order of a few atmospheres. Mechanical stresses in plant cell walls cause microtubules to be oriented, leading to anisotropic stresses which changes auxin transport and thus modifies cell growth. Since differential cell growth leads to mechanical stresses, this closes the feedback loop. At the tissue level, these processes have begun to yield quantitatively testable predictions.[75] Recent studies suggest that plant shoots can integrate multiple stimuli in space and time[76,77] over their spatially extended growth zone. However, many questions remain, including, at one end, the experimental search for the molecular sensors and the circuits they feed, and at the other, the role of environmental forces due to gravity, fluid flow and contact with solids which act on macroscopic scales that might begin to explain the phenotypic morphospace seen in plants.

Many organs in vertebrates (and elsewhere) such as the brain, the lungs and the gut are biological solutions to a generic physiological problem faced in multicellular organisms: how can one (efficiently?) process matter, energy and information in a spatially constrained system. In each of these organs, packing constraints lead to folded, wrinkled and crumpled states on multiple scales. In the vertebrate gut, using the chick embryo as a model system, the biochemical and biomechanical morphogenetic mechanisms that underlie these processes have been investigated using a combination of experiments, theories and computations.[78,79] Differential growth of the constituent tissues has been implicated as the primary driver of morphogenesis in this system and helps quantitatively explain the looping of the gut, the formation of vili and the development of intestinal crypts that serve as stem cell niches in the adult.[80] Furthermore, molecular control of cellular processes such as division (under the control of the diffusible morphogen Bmp) and contractility (under the influence of a different morphogen FgF) have been implicated in controlling the loop morphology[81] and the elongation of the gut,[82] linking the molecular processes to the shape of the gut on multiple scales. Again, while physics has played a critical role

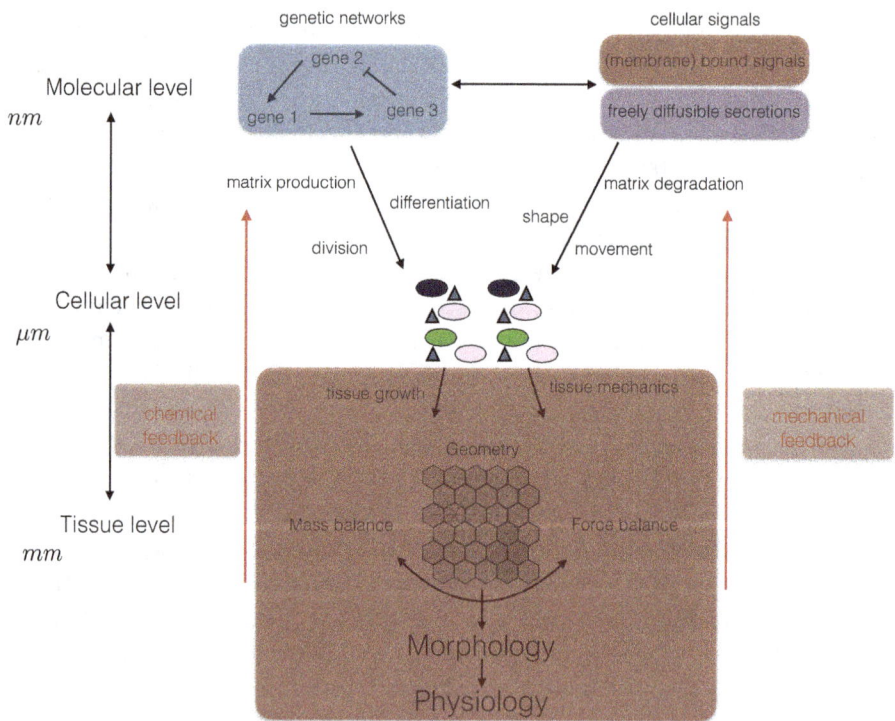

Fig. 4.   The prediction and control of tissue shape spans multiple scales from the molecular to the organismic. Here we show schematically the different types of biochemical and biomechanical interactions that drive cellular signaling, cellular differentiation, cellular division and cellular movement, all of which work to drive morphology which enables and constrains physiology.

in unraveling the mechanisms and sharpening questions, many challenges remain. At one end are the relative roles of (active) biomechanical and biochemical feedback in controlling gut growth and form remains open, while at the other, are the role of both molecular players and the ecophysiology of the organism as manifest in its diet in determining the overall adaptable morphology of the gut, which show extreme variability, as seen in organisms such as pythons and voles.

Answers to these questions will help move the field from the mechanistic questions that morphogenesis aims to explain, to the evolutionary questions that ultimately determine what we see in the wild.

4. Towards the Biophysics of Growth and Form

Morphometrics and morphogenesis attempt to quantify shape and its emergence from an interaction between biochemical interactions and biomechanical flows and forces over a range of scales in space and time. Classifying the range of shapes seen in biology poses a challenge that is simultaneously logistical, mathematical (involving geometry and statistics) and computational/algorithmic. Even as the logistic aspects of collecting and collating samples and shapes becomes viable, the question of what mathematical approaches to use to compress the description of shape and deduce the dimensionality of the occupied morphospaces is still open. The conceptual framework of developmental patterning till recently has focused primarily on molecular and cellular regulatory processes associated with chemical information

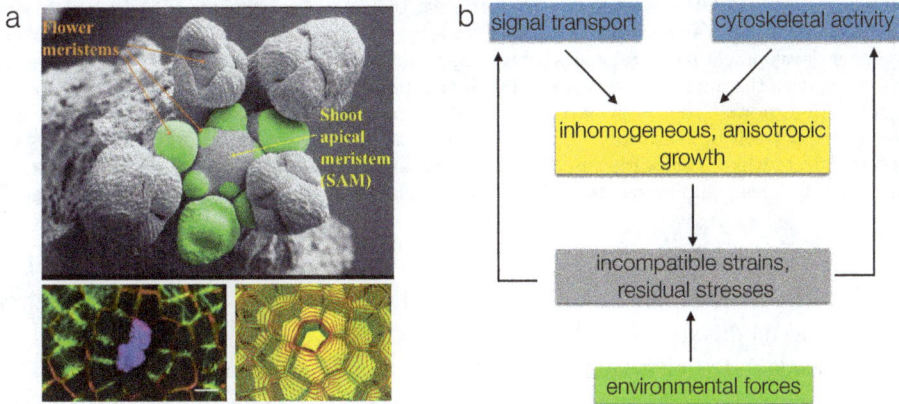

Fig. 5.   Plant growth and form — the shoot meristem. (a) Plant cells are physically akin to pressurized anisotropic balloons. Different views at the tissue and cellular level show how microtubules are oriented by principal stresses in the cell wall and reflect the localization of proteins along the cell wall.[36] (b) The biomechanical and biochemical interactions that drive plant tissue morphogenesis are shown here schematically and couple the scalar transport of a plant hormone auxin, that together with the vector orientation of microtubules, drive anisotropic tissue shape. The mechanical stresses induced by the growing tissue feed back on both auxin transport and microtubule orientation, and in turn are driven by environmental stresses such as those due to fluid flow and gravity.

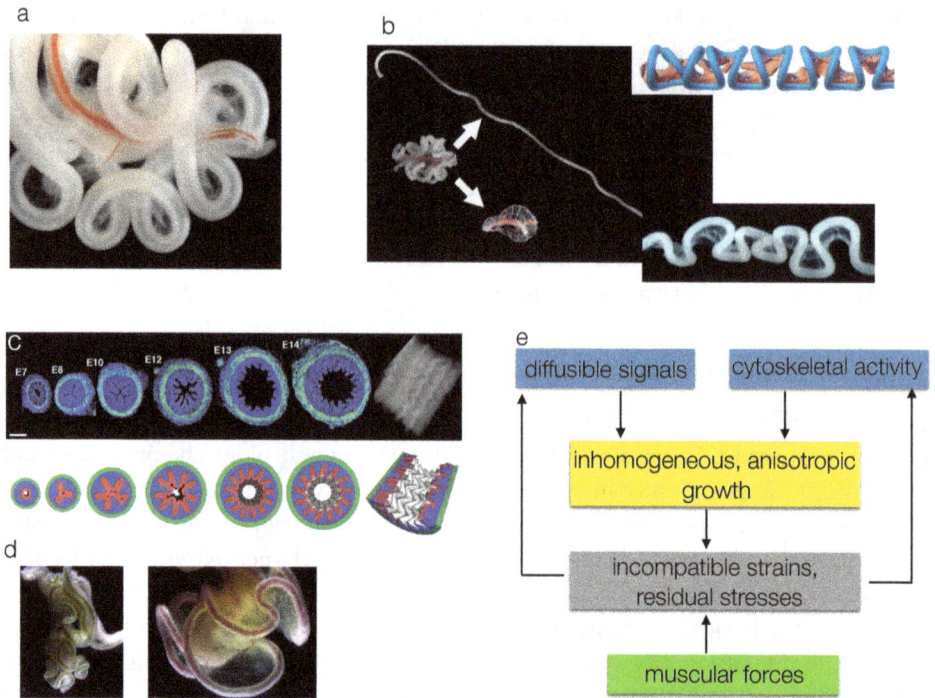

Fig. 6.   Animal growth and form — the vertebrate gut. (a) The chick gut has a characteristic looping morphology, as do most vertebrate guts. (b) Looping is driven by differential growth between the gut and the mesentery. Quantitative biological experiments, physical simulacra, computer simulations and scaling theory all converge to explain the gut loop size, all of which are under developmental control while being subjected to evolutionary selection.[78] (c) Differential growth between the multiple layers also drives lumen patterning into villi.[79] (d) Molecular control of differential growth can be achieved using diffusible morphogens and their inhibitors, which directly impact cell proliferation, e.g. Bmp2 and its antagonist Noggin. (e) A schematic of the process reflects what is shown in plant organ development in Figure 4, except that the molecular players are different, and cytoskeletal activity is associated with cell division and contraction.

processing, but describing morphogenetic processes also requires understanding how flows and forces associated with instabilities and patterns are harnessed by development and modified by evolution.

This leads to the view that the (discrete) logic of development together with the physics that underlies the (continuous) calculus of change, in the presence of feedback drives multicellular morphogenesis. However, the subjects of morphological (phenotypic) and regulatory (genotypic) diversity are separated by many orders in length scales. A fundamental challenge in morphogenesis is to connect these scales, deduce the dimensionality of these morpho-genetic spaces and provide a way to quantify the development of biological shape, using four themes:

1. **Measuring shape.** Information in space and time encoded in shape and its evolution. The link between form and function is intimate in connecting morpho-

genesis to physiology, development and evolution. The origin of form in biology requires us to think about the different ways in which how information is encoded in shape at every level of organization. Mathematically, shape is topology, geometry and dynamics. What would an information theory of shape entail? It would have to be scale-dependent, and be dynamic, i.e. time-dependent. An example of where this question can be addressed directly is epithelial morphogenesis,[23,83] and perhaps we are not too far from an answer.

2. Predicting shape. Complexity is easy to achieve temporally using chemical means, and spatially using mechanical means. However, the time scales for these are typically different, the first being diffusion controlled, and the second typically controlled by large-scale motion. How are space and time modulated chemically and mechanically during tissue shaping? Leaf and shoot/root shape are two large-scale examples in plants, and gut shaping and skin patterning are two large-scale vertebrate examples where we are close. Geometry plays a critical role in all these morphogenetic events but how does biology use dimensionality to control shape? Active matter is biologically inspired, but is it biological yet? What are the experimental constraints and tests? Examples which are very promising here include the order-disorder transitions associated with epithelial (ordered) to mesenchymal (disordered) tissues, and the morphogenesis of insect wings.[58]

3. Controlling shape. Feedback is an essential part of the process and can occur at multiple scales, either via chemical or mechanical means. While feedback is well understood in temporal systems, in spatial settings it is much less understood because sensing and actuation can often be spatially segregated which means that there are time delays that must be correctly accounted for. How is this done? Plants may be where we can first expect to get an answer.[35,36]

4. Evolving shape. Length and time scales in morphogenesis vary by orders of magnitude. At a chemical level, length scales might be set by the balance between production/degradation and diffusion, while at a mechanical level, they might be set by the balance between different modes of deformation e.g. bending and stretching, or internal and external viscous resistance, or even more simply by the system size. But how are these scales capable of coping with growing systems to realize robust patterns of geometry and topology? Are there natural ways in which mechanochemical coupling can lead to robust patterns that neither chemical or mechanical means alone will not? How is space converted into time and vice versa in morphogenesis?

No question in biology can be completely answered without couching it in an evolutionary setting, and so must it be for morphogenesis as well. The evolution of morphological complexity requires us to understand the range of morphospaces, and thus impinges on both the problem of mathematical and computational classification/compression and that of biophysical prediction. Why are certain shapes common/uncommon? How does physiology constrain and drive morphogenetic complexity and how much does developmental plasticity allow forms to be malleable? What are the relative roles of ancestry and convergence in these situations?

Understanding morphogenesis is as much a grand challenge for physics as it is for biology — for it brings together some of the oldest ideas of the subject with the some of the newest in a manner that is simultaneously easy to behold and difficult to comprehend. While this report has focused on aspects of multicellular morphogenesis, with a small number of changes, similar questions can be asked about the innards of a cell, with just as many mathematical and physical questions. But that is for a different time and place.

Acknowledgments

I thank Boris Shraiman for his comments and patience (beyond all reasonable limits) with this article. All errors are naturally mine alone.

References

1. I. Held, *How the Snake Lost its Legs: Curious Tales from the Frontier of Evo-devo* (Cambridge University Press, 2014).
2. D'Arcy W. Thompson, *On Growth and Form* (Cambridge University Press, 1917).
3. C. Waddington (ed.), Towards a theoretical biology, v. 1-3, IUBS (1968).
4. A. M. Turing, The chemical basis of morphogenesis, *Philos. Trans. R. Soc. London Ser. B* **237**, 37–72 (1952).
5. F. Klein, *Vergleichende Betrachtungen uber neuere geometrische Forschungen* (1872).
6. D. M. Raup, The geometry of coiling in gastropods, *Proc. Natl. Acad. Sci. USA* **47**, 602–609 (1961).
7. K. Alim, S. Armon, B. I. Shraiman, A. Boudaoud, Leaf growth is conformal, *Physical Biology* **13**, 05LT01 (2016).
8. M. Stoddard *et al.*, Avian egg shape: Form, function, and evolution, *Science* **356** (6344) 1249–54 (2017).
9. G. McGhee, *The Geometry of Evolution* (Cambridge University Press, 2006).
10. M. C. Milinkovitch *et al.*, Crocodile head scales are not developmental units but emerge from physical cracking, *Science* **339**, 78–81 (2013).
11. McDole *et al.*, In toto imaging and reconstruction of post-implantation mouse development at the single-cell level, *Cell* **175**, 859–76 (2018).
12. S.-I. Amari, *Information Geometry and its Applications* (Springer-Verlag, 2016).
13. R. Adler, J. Taylor, *Random Fields and Geometry* (Springer-Verlag, 2007).
14. D. Durbin, S. Eddy, and A. Krogh, *Biological Sequence Analysis: Probabilistic Models of Proteins and Nucleic Acids* (Cambridge, 1998).
15. A. Srivastava, E. Klassen, *Functional and Shape Data Analysis* (Springer-Verlag, 2016).
16. U. Grenander, Advances in pattern theory, *Ann. Statist.* **17**, 1–30 (1989).
17. D. Mumford, A. Desolneux, *Pattern Theory: The Stochastic Analysis of Real-world Signals* (CRC Press, 2010).
18. M. Do Carmo, *Differential Geometry of Curves and Surfaces* (Dover, 2016).
19. A. Bobenko, P. Schroeder, J. Sullivan, G. Zeigler (eds.), *Discrete Differential Geometry* (2010).
20. A. Goswami, J. Smaers, C. Soligo, D. Polly, The macroevolutionary consequences of phenotypic integration: from development to deep time, *Phil. Trans. R. Soc. B* **369**, 20130254 (2014).

21. C. Truesdell and W. Noll, *The Nonlinear Field Theories of Mechanics* (Springer-Verlag, 1965).
22. G. Blanchard *et al.*, Tissue tectonics: morphogenetic strain rates, cell shape change and intercalation, *Nature Methods* **6** (6), 458–64 (2009).
23. B. Guirao *et al.*, Unified quantitative characterization of epithelial tissue development, *eLife* **4**, e08519 (2015).
24. G. Haller, Lagrangian coherent structures, *Ann. Rev. Fluid Mech.* **47**, 137–62 (2015).
25. M. Cross, P. Hohenberg, Pattern formation outside of equilibrium, *Rev. Mod. Phys.* **65**, 851–1113 (1993).
26. C. Godreche (ed.), *Hydrodynamics and Nonlinear Instabilities* (Cambridge University Press, 1998).
27. H. Meinhardt, *Models of Biological Pattern Formation* (Academic Press, 1982).
28. J. D. Murray, *Mathematical Biology* V. 1, 2 (Springer-Verlag, 2002).
29. S. Kondo, T. Miura, Reaction-diffusion model as a framework for understanding biological pattern formation, *Science* **329**, 1616–1620 (2010).
30. W. Wolpert, C. Tickle, A. M. Arias, *Principles of Development* 6th edn. (2019).
31. D. Ben-Zvi, N. Barkai, Scaling of morphogen gradients by an expansion-repression integral feedback control, *Proc. Natl. Acad. Sci.* **107**, 6924–29 (2010).
32. S. Werner *et al.*, Scaling and regeneration of self-organized patterns, *Phys. Rev. Lett.* **114**, 138101 (2015).
33. S. Gilbert, J. F. Barresi, *Developmental Biology* 11th edn. (Oxford University Press).
34. A. Economou *et al.*, Whole population cell analysis of a landmark-rich mammalian epithelium reveals multiple elongation mechanisms, *Development* **140**, 4740–50 (2013).
35. E. Coen, A. B. Robocho, Resolving conflicts: modeling genetic control of plant morphogenesis, *Dev. Cell* **38**, 573–583 (2016).
36. A. Sampathkumar, A. Yan, P. Krupinski, E. M. Meyerowitz, Physical forces regulate plant development and morphogenesis, *Curr. Biol.* **24**, R475–R483 (2014).
37. F. Corson *et al.*, Turning a plant tissue into a living cell froth through isotropic growth, *Proc. Natl. Acad. Sci. USA* **106**, 8453–8458 (2009).
38. B. He, K. Doubrovinski, O. Polyakov, E. F. Wieschaus, Apical constriction drives tissue-scale hydrodynamic flow to mediate cell elongation, *Nature* **508**, 392–396 (2014).
39. L. LeGoff, T. Lecuit, Mechanical forces and growth in animal tissues, *Cold Spring Harbor Perspect. Biol.* **8**, a019232 (2016).
40. N. Noll, M. Mani, I. Heemskerk, S. Streichan and B. I. Shraiman, Active tension network model of epithelial mechanics suggests an exotic mechanical state realized in epithelial tissue, *Nature Physics* **12**:1221–1226 (2017).
41. R. Farhadifar *et al.*, The influence of cell mechanics, cell-cell interactions, and proliferation on epithelial packing, *Curr. Biol.* **17**, 2095–2104 (2007).
42. P. Haas and R. E. Goldstein, Nonlinear and nonlocal elasticity in coarse-grained differential-tension models of epithelia, *Phys. Rev. E* **99**, 022411 (2019).
43. U. Nath, B. C. W. Crawford, R. Carpenter, E. Coen, Genetic control of surface curvature, *Science* **299**, 1404–1407 (2003).
44. E.B. Hunziker, R. K. Schenk, Physiological mechanisms adopted by chondrocytes in regulating longitudinal bone growth in rats, *J. Physiology* **414**(1):55–71 (1989).
45. K. L. Cooper, S. Oh, Y. Sung, R. R. Dasari, M. W. Kirschner, C. J. Tabin, Multiple phases of chondrocyte enlargement underlie differences in skeletal proportions, *Nature* **495**(7441):373–378 (2013).
46. Puzey *et al.*, Evolution of spur-length diversity in Aquilegia petals is achieved solely

through cell-shape anisotropy, *Proc. R. Soc. B* **279**:1640–45 (2011).

47. B. Benazeraf, O. Pourquie, Formation and segmentation of the vertebrate body axis, *Annu. Rev. Cell. Dev. Biol.* **29**, 1–26 (2013).

48. P. McMillen, S. A. Holley, The tissue mechanics of vertebrate body elongation and segmentation, *Curr. Opin. Genet. Dev.* **32**, 106–111 (2015).

49. E. Hoijman, B. Alsina, Cavity morphogenesis: imaging mitotic forces in action, *Cell Cycle* **14**(18):2867–8 (2015).

50. T. Ruiz-Herrero, K. Alessandri, B. Gurchenko, P. Nassoy and L. Mahadevan, Organ size control via hydraulically-gated oscillations, *Development*, **144**, 4422–27 (2017).

51. C. J. Chan *et al.*, Hydraulic control of mammalian embryo size and cell fate, *Nature* **571**:112–116 (2019).

52. R. L. Johnson, C. J. Tabin, Molecular models for vertebrate limb development, *Cell* **90**, 979–990 (1997).

53. R. Sheth *et al.*, Hox genes regulate digit patterning by controlling the wavelength of a Turing-type mechanism, *Science* **338**, 1471–1480 (2012).

54. R. A. Foty and M. S. Steinberg, The differential adhesion hypothesis: a direct evaluation, *Developmental Biology* **278**(1):255–263 (2005).

55. C. Weber, C. Rycroft and L. Mahadevan, Differential activity driven instabilities in biphasic active mtter, *Phys. Rev. Lett.* **120**, 248003 (2017).

56. M. C. Marchetti *et al.*, Hydrodynamics of soft active matter, *Rev. Mod. Phys.* **85**(3): 1143–1189 (2013).

57. T. Tallinen *et al.*, On the growth and form of cortical convolutions, *Nature Physics* **12**, 588 (2016).

58. B. Aigouy *et al.*, Cell flow reorients the axis of planar polarity in the wing epithelium of Drosophila, *Cell* **142**, 773–86 (2010).

59. C. Guillot, T. Lecuit, Mechanics of epithelial tissue homeostasis and morphogenesis, *Science* **340**, 1185–1189 (2013).

60. E. Marinari *et al.*, Live-cell delamination counterbalances epithelial growth to limit tissue overcrowding, *Nature* **484**, 542–46 (2012).

61. A. Puliafito *et al.*, Collective and single cell behavior in epithelial contact inhibition, *Proc. Natl. Acad. Sci. USA* **109**:739–744 (2012).

62. G. T. Eisenhoffer *et al.*, Crowding induces live cell extrusion to maintain homeostatic cell numbers in epithelia, *Nature* **484**(7395):546–9 (2012).

63. Y. Klein, E. Efrati and E. Sharon, Shaping of elastic sheets by prescription of non-Euclidean metrics, *Science* **315**(5815):1116–1120 (2007).

64. Y. C. Fung, *Biomechanics* (Springer-Verlag, 1993).

65. H. Liang, L. Mahadevan, Growth, geometry, and mechanics of a blooming lily, *Proc. Natl. Acad. Sci. USA* **108**, 5516–5521 (2011).

66. E. Efrati, E. Sharon and R. Kupferman, The metric description of elasticity in residually stressed soft materials, *Soft Matter* **9**:8187–8197 (2013).

67. A. E. Shyer *et al.*, Emergent cellular self-organization and mechanosensation initiate follicle pattern in the avian skin, *Science* **357**(6353):811–815 (2017).

68. B. I. Shraiman, Mechanical feedback as a possible regulator of tissue growth, *Proc. Natl. Acad. Sci. USA* **102**:3318–23 (2005).

69. M. Biot, General theory of three-dimensional consolidation, *J. Appl. Phys.* **12**:155–164 (1941).

70. Y. Pan, I. Heemskerk, C. Ibar, B. I. Shraiman and K. D. Irvine, Differential growth triggers mechanical feedback that elevates Hippo signaling, *Proc. Natl. Acad. Sci. USA* **113**(45):E6974–E6983 (2016).

71. M.-E. Fernandez-Sanchez, T. Brunet, J.-C. Roper, E. Farge, Mechanotransduction's

impact on animal development, evolution, and tumorigenesis, *Annu. Rev. Cell Dev. Biol.* **31**, 373–397 (2015).

72. E. Hannezo, C.-P. Heisenberg, Mechanochemical feedback loops in development and disease, *Cell* **178**, 12–25 (2019).
73. V. Mirabet, P. Das, A. Boudaoud, O. Hamant, The role of mechanical forces in plant morphogenesis, *Annu. Rev. Plant Biol.* **62**, 365–385 (2011).
74. O. Ali, V. Mirabet, C. Godin, J. Traas, Physical models of plant development, *Annu. Rev. Cell Dev. Biol.* **30**, 57–78 (2014).
75. O. Hamant *et al.*, Developmental patterning by mechanical signals in Arabidopsis, *Science* **322**, 1650–55 (2008).
76. R. Bastien, T. Bohr, B. Moulia, S. Douady, Unifying model of shoot gravitropism reveals proprioception as a central feature of posture control in plants, *Proc. Natl. Acad. Sci. USA* **110**(2) 755–60 (2012).
77. R. Chelakkot, L. Mahadevan, On the growth and form of shoots, *Royal Society Interface* **14**, 20170001 (2017).
78. T. Savin *et al.*, On the growth and form of the gut, *Nature* **476**, 57–62 (2011).
79. A. E. Shyer *et al.*, Villification: how the gut gets its villi, *Science* **342**, 212–218 (2013).
80. A. E. Shyer *et al.*, Bending gradients: how the intestinal stem cell gets its home, *Cell* **161**, 569–580 (2015).
81. N. L. Nerurkar, L. Mahadevan, and C. J. Tabin, BMP signaling controls buckling forces to modulate looping morphogenesis of the gut, *Proc. Natl. Acad. Sci. USA* **114**, 2277–82 (2017).
82. N. L. Nerurkar, C.-H. Lee, L. Mahadevan, and C. J. Tabin, Molecular control of macroscopic forces drives vertebrate hindgut elongation, *Nature* **565**, 480–84 (2018).
83. N. A. Dye, M. Popovi, S. Spannl, R. Etournay, D. Kainmuller, S. Ghosh, E. W. Myers, F. Julicher, S. Eaton, Cell dynamics underlying oriented growth of the Drosophila wing imaginal disc, *Development* **144**(23):4406–4421 (2017).
84. K.D. Irvine, B. I. Shraiman, Mechanical control of growth: ideas, facts and challenges, *Development* **144**(23):4238–4248 (2017).

Discussion

U. Alon I pick up the last question about the evolution of tissue shapes and how can we have a functional morphology. So there are advances in the last few years in classifying morphospace, the space of traits. There are several tasks that a shape needs to fulfill, lets say a beak that need to eat seeds and pollen. However you cannot be optimal for both tasks with a single shape. That leads to fun situations that each task has an optimum which is, lets say, one of k points in the morphospace, and the best traits are in a polytope or a polyhedron which vertices are those archetypes. So there is a lot of empty morphospace and the observed sweep of variations has these sharp points. We do not know the tasks in advance but if you look at that shape and see what those organisms near the vertices are specialized to you can pick up what the tasks are, and find generalists based on functions and distances from those vertices. That is seen in a variety of morphological situations from peaks, to ammonite shell, to teeth shapes, to bones. Multi-objective optimality could be a guiding principle here.

T. Lecuit Other questions or remarks.

L. Mahadevan I think that surprises are always in the eyes of the beholder, so it depends on what else you know before that you will or will not be surprised. If I think about instabilities on their own, I think that you are right, there is very little reason to be surprised. I think that a more interesting question which I have skirted around is given how many different potential shapes are possible, even in very simple systems, how can you have essentially very few. So feedbacks seem to play a very important role in all these phases, all these ways, and all these examples. How much do we understand that? Actually I think we understand it very little, because if did not have strong feedbacks at multiple scales it would be very hard to explain why it is so reproducible because there are a lot of, for example, in this very little case I mention, shapes which are very close to each other. This addresses the first perhaps even the second question.

L. Mahadevan I think part of that is associated with plasticity of the morphological state itself. I mentioned very quickly, maybe I can elaborate a little. In some organisms gut length and villi length actually are very high variables. I told you that gut length is variable in rodents. A python is an extraordinary example of an animal which essentially eats very rarely, it can eat in fact sometimes just once a year, and once it eats within 24 hours every single villus, and there are probably 10 billions of these, will increase in length by a factor of 4. And then it will last for as long as it will take to digest the dear or whatever it is. The heart in the same organism, the python, can undergo hypertrophy. I do not remember the number on it, Jared Diamond has written quite a lot about it. So I think that the question that you are raising, and I do not know anything exactly at this point except stories and examples, depends on how plastic the environment demands the organism to be. So how much variability there is, and therefore how easy or how hard would it be to survive. I think all is pretty much anecdotally at this stage, there are many people that have other examples on this, unfortunately I cannot say much more of that.

T. Lecuit One question from Süel Gürol and then Rob.

G. Süel I have a question that may be potentially naïve. It is hard to get information about these evolutionary processes sometimes. But in the case of dogs I would have thought that this is the case. I am asking if something has been done, because there is a very beautiful, I am assuming, documentation of the breeding. If you look at a sort of chihuahua today, the fact that it came from a wolf is a pretty sad case of evolution driven by sort of humanity. But since there is such a dramatic change in morphology in these animals, and it must be documented because this is something that the breeders really paid attention to, the sequences in these shapes and skulls must be preserved somewhere. Has anybody actually gone in and seen how or what genes and so forth and how you can achieve such a dramatic change of morphology in such a short time? This is something we have to keep in mind, this not something over millions of years.

L. Mahadevan I do not know a lot of what exactly is going on with dogs, but Cliff Tabin, a colleague of mine, and his colleagues have shown that you can change beak morphology in chicks and make it resemble, for example, a dinosaur, or a juvenile dinosaur. There are ways, for example playing with Bmp4 signaling, where you can essentially tune or change morphology at least in early development (these things do not survive long enough). There is now a way to essentially start moving through this space of shapes, though I do not really know a lot about the problem that you are asking.

T. Lecuit Does anyone have a comment on this specific point of morphospace? After, we have two questions from Cristina and Rob. Nipam, did you want to say something?

P. Nipam I can actually answer the question about the dogs and tell you one example. A well studied example is short legs in dogs where it turns out that it is due to one single event that occurred in dog breeding. It is a retro gene that occurred, that is a reverse transcription of a signaling molecule that was re-integrated into the genome, it is expressed inappropriately in long bones and it keeps the legs from growing. So that is an excellent example where there is a huge morphological change but it is actually due to one single event that occurred in one dog and that has been inherited by all other dogs bred giving them short legs.

T. Lecuit We have a question from Cristina.

C. Marchetti I wanted to ask you, maybe this is just a way of resaying or restating something that you already said, but it seems that the notion that nature arises from interplay of geometry and shape is the possibility of incompatibility, geometric incompatibility. That means that a lot of these structures are intrinsically stressed even if you do not apply any load. Therefore, the description of these systems seems rather challenging and how do you start because you do not have, for instance, a reference configuration that you would need in elasticity.

R. Phillips I wanted to offer a little push back to Daniel and this notion of surprise. Perhaps it is a bit too philosophical but I think that one of the things that physics can bring to the table is that it can prove that we know what we think we know. With respect to Maha's paper, which I find incredibly impressive, what I like about it the most is that you can do the scaling analysis and all kind of stuff but if you look at the assays, they carefully looked at the radii and figured out what flexural rigidity was for different sizes, they measured the elasticity and actually tested it. I think I have not heard much about knowing things for real as supposed to thinking we know them. I salute that paper, Maha you already know it. This surprise thing really gets on my nerves.

T. Lecuit Thank you very much. I suggest now that we will have the first of the three prepared remarks. As you will see they will treat different aspects of morphogenesis, first in one dimension with Alexander Aulehla, then we will have with Ben Simons "Two-dimensional branching patterns", and then with Frank Julicher we will explore two-dimensional or three-dimensional exploration of shapes.

Prepared comment

A. Aulehla: Integration of time and space to control embryo patterning — the segmentation clock

My remark links to several themes discussed in this and also previous sessions, I will focus on two points.

- Time-Space: How do dynamic signals encode temporal and spatial information during embryonic patterning?
- Space-Time: How do embryonic growth/shape/mechanical cues feed back on signalling dynamics?

An excellent biological context to address the above questions offers the study of vertebrate embryo segmentation, which is controlled by a molecular oscillator, the segmentation clock. Since its discovery in 1997 by the Pourquie lab and the first real-time imaging by the Kageyama lab in 2006, the dynamics of this embryonic oscillator, which in mouse embryo includes oscillations of Notch, Wnt and Fgf-signaling activities with a period of approx. 2 hours, have been quantified in increasing detail and complexity. Most strikingly, oscillations occur to be phase-shifted between neighboring presomitic mesoderm (PSM) cells and hence produce spatio-temporal wave patterns that periodically sweep through the PSM in posterior-to-anterior direction. We have, in addition, established in vitro models, which recapitulate periodic wave patterns, period gradients and also physical segment formation, and hence serve as 2-D segmentation models with additional features and experimental possibilities. One central challenge remains to probe, directly, the role of signaling dynamics.

How do cells decode temporal information, do cells make first and second derivatives?

We have heard several times throughout the meeting about the importance and challenge to reveal function, in this case the function of signaling dynamics. To this end, I want to introduce an approach that employs universal physical concept — synchronization — to study complex biological systems, here the mouse embryo PSM.

We employ an entrainment/synchronization strategy to control endogenous segmentation clock by an external oscillator, i.e., periodic small molecule perturbations, which we implement using a microfluidic setup. Excitingly, we have evidence that entrainment of mouse embryo PSM cells follows very general synchronization rules described as Arnold tongues.

The ability to entrain endogenous embryonic oscillators serves at least two goals. First, it can be used to reveal the working principles of complex biological dynamical systems, i.e., the synchronization behavior of an oscillator ensemble can in itself serve to reveal fundamental properties allowing a general categorization.

Second, we can build on the ability to control signaling dynamics to decipher its function during segment patterning. I think of this also as an example for a top-down approach — we rely on universal synchronization rules to steer a complex cell/oscillator assembly towards a predictable outcome. This enables to test the function of signaling dynamics, without the need for a bottom-up approach, as long as entrainment is successfully used to control signaling dynamics.

- Time-Space: How do dynamic signals encode temporal and spatial information?

The simultaneous entrainment of both Notch- and Wnt- signaling oscillations enables, in addition, to experimentally control the endogenous, relative phase-shift between these oscillatory systems. Critically, this revealed that indeed, the relative timing between Wnt/Notch signaling oscillations encodes key information during segment patterning. We are at the beginning of deciphering the precise function of relative timing in this context.

Since oscillators are laid down as arrays in space and moreover, since relative timing between Wnt/Notch signaling oscillations changes as along the embryo axis, the possibility is raised that relative timing encodes positional information.

- Space-Time: How do growth/shape/mechanical cues feed back onto dynamics?

This brings me to the second important question also referred to by our rapporteur, i.e., how does space link to time and signaling dynamics? We do know that patterning is space dependent, for instance, the length scale between segments depends on and changes as a function of system size. In effect, we see that embryos of same stage but different overall size show proportional patterning, i.e., scale. We hence have to ask, how are growth dynamics, shape or also mechanical cues integrated and feed back on signaling and specifically signaling dynamics. Investigating this integrated feedback system will allow us to address a fundamental property i.e., the proportionality of developing embryos.

Discussion

T. Lecuit Thank you Alexander. Who wants to react to this presentation? Who else has any unprepared remark that could be related or unrelated to what we just heard?

T. Hwa I have a general remark about one thing that Maha brought up but it may also connect to biology. The really magic thing about biology is the connection between the molecular scale and the large scale. For various problems we can write down a theory at large scale that implies few parameters but the magic is that somehow these few parameters that are important to define shapes have to be directly connected to something at the molecular scale. In principle, we can have a lot of changes that affect these macroscopic parameters but then, in the case of a dog or whatever, changes of a few genes can change significantly shapes in a way that it is important for the physiology of the organism. And just to give another example, you know our facial features are important for us for many physiological aspects. How many genes does it take to determine them? Somehow there has to be a connection to the things that biology can change that directly affects the functions. What do people think about this?

S. Eaton I want just to point out that indeed we know, especially from well studied development systems like Drosophila, that there are many genes that can change shapes of things in a really reproducible way, in different dimensions, etc. But the key missing link has been that we do not know what that does to cell dynamics. That is the key thing that we have to understand. Maha, referring to you about the different sorts of things that cells can do to actually change tissue shape, we must really quantitatively describe these things. That is the thing that we have to do to bound these scales. Because there are a lot of things that control shapes.

T. Hwa To me a key challenge for this area is to know if there is some recipe to go from small scales through the dynamics that affects large scale. Ultimately, to have sufficient feature changes, biology needs to be able to manipulate things in the matter of few generations.

T. Lecuit I think we have a field of connecting molecular dynamics to cellular dynamics. Few quick remarks on this area before we move on. Maha first and Tony. Who has a point on that?

L. Mahadevan So I am not sure that I can address comprehensively all questions but can at least give it a shot. Bmp, which is a secreted signal which seems to be very important to cell proliferation, is very relevant in many of these different systems and is more or less directly (at the level of the physical description) changed. Bmp changes the growth rate at one location or the other. There is an experiment, that I did not mention which needs to be done in the context of the gut morphogenesis, where if you partially resect the gut from the rest of the body you should change the local stress and then change the amount of Bmp. Another example in the context of elongation processes is Fgf, which essentially changes motility. This seems to be present in outgrowth, like when your limbs form in body elongation or in gut elongation. So I think that a question, which comes back to something that Elowitz talked about yesterday, is how many different pathways there are and how many different pathways do you need. I do not know the answer but this question, I think, is one which definitely should be considered carefully.

T. Lecuit Thank you. Tony.

A. Hyman I just want to bring up an issue that relates potentially to yesterday's talk about pathways and messy pathways in eukaryotes versus prokaryotes. When we think about mammals we are always fascinated by the changes in size and shape because all seems to be set up in a mammal to evolve extremely quickly. It is a bit like the Darwin beak, a feature which is extremely evolvable in few generations. But if you would work, like I do, on nematodes you would know that there is no morphological changes. Over hundred millions years, you cannot tell a difference in a C. briggsae and C. elegans. In fact it is very dangerous to work on them in the lab because they get confused. Whereas what does change in nematodes are biochemical pathways, that is what is set up to evolve. The question is weather you have biochemical pathways that are set up to evolve, in the same ways that shape and size do in mammal, those are sets of different simpler definable pathways because they have to be evolvable as opposed to more complex pathways like mammals which are less evolvable in terms of the biochemistry.

T. Lecuit The issue of evolvability will be discussed later on and especially this afternoon. Let us move on, we have Holly and Boris

H. Goodson How general is this mechanism of the waves? Is it something seen also in fruit flies or is something that is laid on top?

T. Lecuit I think we can say that it is widespread. And, Albert, probably you have something to say about oscillations and waves?

A. Goldbeter I have some examples from mitogenesis because it is a beautiful example of transformation from time patterns to spatial patterns. It was not clear to me, Alexander, when you describe the importance of the phase differences how did you get the entrainment, what signal did you use? Is something known about the molecular mechanism at the origin of these oscillations? Some models indicate that negative feedbacks in each of the pathways involved, the Notch pathway, the Wnt pathways, the Fgf pathway, could produce oscillations. Is this the source of oscillations and what is the role of Fgf pathway? You showed Wnt and Notch, what about the third pathway?

A. Aulehla I will try to be very brief. To the first question the wave patterns indeed are quite ubiquitous, they are also seen in insects and Nipam would be the right person to talk with. This is quite general as a motif I would say. Of course the molecules involved are different but the motif is the same, which is very interesting. To Albert's questions. This is an entrainment approach that I really think to be a top down approach. The details of how we perturb this pathway periodically do not really matter, if you use an activator or inhibitor, it is a periodic perturbation and this is sufficient to entrain this complex machinery. So we do not have to have much knowledge about the downstream mechanism but you are able to control this complex dynamics in a top down fashion. This in itself is not sufficient to answer your question about what is the core oscillator. What we find is that they directly impact the dynamics that I was showing you, for instance, the wave velocity. These are controlled by the growth factor, and we are just starting to begin understanding how this growth factor integrates into the dynamics and to make the link between time and space.

T. Lecuit Boris.

B. Shraiman I will just take it back to Terry's and Suzanne's remarks. In the wonderful example of a face, there must be some notions presumably of the dimension of the phenotypic space. Police departments can synthetize a face with some number of continuous parameters (Daniel Fisher: 5). OK, it is 5 dimensional. On the other hand you can ask how many genes actually contribute to those 5 parameters. The answer is probably very many. Is it possible to make it into a tight statement? Is it possible to define some metrics on this phenotypic space? You can try to define *states* in the phenotypic space, map to the genetic space and now ask how easy or difficult, how many mutations it takes and how many ways there are, to go from one place to the other.

T. Lecuit Now we move on to the next prepared remark from Ben Simons.

Prepared comments

B. Simons: Branching morphogenesis as an emergent behaviour

In developmental biology, it can be easy to find consensus on the central questions confronting the field. The greater challenge lies in finding consensus on the correct language to articulate biological understanding. In an era dominated by functional genomics, it is natural to assume that mechanistic understanding can only be expressed in the language of molecules — genes and gene products — a tendency exacerbated by advances in sequencing technologies that now enable the molecular profiling of single cells.

In the realm of (terrestrial) physics, experience has shown that collective behaviour of complex systems may be difficult to understand from the bare interactions of the "elementary" degrees of freedom. Instead, mechanistic understanding of cooperative behaviour is more usefully articulated by tailoring theory to the appropriate level of abstraction: Cooperative phenomena are more usefully understood through the analysis of coarse-grained (hydrodynamic) theories involving just a few collective degrees of freedom. However, in condensed matter physics, attention is focussed typically on systems at, or near, equilibrium. By contrast, in biology, we are concerned with *evolved* systems driven far from equilibrium. As a result, the machinery of statistical physics — scaling, critical phenomena and universality — may not be relevant in biology. Nevertheless, a focus that emphasizes phenomenology (viz. functional behaviour) over cell state (viz. genomic/epigenetic signature) may provide a viable alternative route to understand developmental processes. To emphasize the point, Simons gave an example by way of a case study carried out in collaboration with Edouard Hannezo and several experimental partner labs.

To sustain life, organisms must exchange nutrients and metabolic wastes with the environment. In unicellular organisms, the surface of the cell is sufficient to meet these demands. However, in larger organisms, strategies must be developed to maximise the area of surfaces where such exchanges can occur. In volumnar organs such as the lung, exchange surfaces are maximised by packing them efficiently around intricate and ramified branched epithelial structures. But what is the basis for such complex structural organization? The situation is exemplified by the mouse mammary gland epithelium. In mouse, the mammary glands are specified as placode-like structures that invade an adipocyte-rich stroma. At birth, the mammary gland comprises a small rudimentary tree-like structure involving a minimally branched network. During puberty, cellular precursors — stem cells — drive a process of ductal bifurcation and elongation, leading to a complex ramified ductal network. How do stem cells integrate fate choice with collective cell migration to direct large-scale patterning of the ductal network? And are these mechanisms conserved in the specification of other branched epithelia such as pancreas and kidney? To address this question, most studies focus on the repertoire of genes and signaling pathways that regulate the fate of cells in the active tip. But are these the right starting variables?

Segmentation of the branched mammary network reveals a complex organization in which some ductal subtrees terminate early in development, while others expand, moving through as many as 30 rounds of sequential branching. Yet, during development, the average ductal length, width and proliferative index of active ductal tips remains constant. So where does the heterogeneity arise? Statistical analysis of the ductal network topology shows that the system conforms to a remarkably simple paradigm: In this model, active tips function as a niche environment supporting the renewal of lineage-restricted stem cells that act collectively to drive a stochastic process of ductal branching and elongation, which terminates when active tips encounter maturing ducts. Within this framework, a "branching-annihilating random walk", the network evolves as a soliton front of active tips that advance as a Fisher-KPP type wave, leaving behind a constant density of maturing ducts. This model, which depends on just one parameter, the ratio of the branching and elongation rates (fixed by the ductal density), predicts the statistical organization of the developing network, from the branching statistics and directional patterning, to giant ductal density fluctuations. Intriguingly, the same model describes quantitatively the statistics of other branched two- and three-dimensional networks, including mouse kidney and pancreas, as well as human prostate, pointing to a branching paradigm that is conserved across tissue types and organisms.

In the realm of cell biology, there is an unhelpful, if understandable, tendency to associate mechanism with transcription factor networks and signalling pathways. Leaving aside the fact that these descriptions are themselves an abstraction, surrendering information on binding kinetics, post-translational effects, time-delays, etc., it is evident that, for many questions, this level of resolution can mask the existence of "simple" guiding principles that can afford predictive, i.e. mechanistic, insights. As well as being of interest in its own right, a focus on the functional behaviour of biological systems, that places emphasis on cell states as fundamental biological units, may help to frame new questions into biological mechanism at the molecular scale.

Discussion

T. Lecuit This is indeed a beautiful example of how complex macroscopic structures emerge from local and simple rules. Who wants to react?

E. Wieschaus In the same organisms that have mammary glands, there are other branching organs, where the branching pattern is equally complex but very stereotyped. We know a little bit about the signaling pathways and individual events at branching that gives rise to these stereotype patterns. What I am wondering is, and this is a question of evolvability, whether these are two really distinct systems in that you do one or you do the other? Can you by relatively easy simple steps go from a self-organizing system, like you described for the mammary gland, to a system like the lung, where branching patterns are controlled at every individual step? How do you go back and forth between these two systems, or does the organism ever?

T. Lecuit I think this is a question about the contrast between deterministic and stochastic control of processes at any scales and self-organization. Who wants to have unprepared remarks or comments on that?

B. Simons Can I just comment on the comment? I am not supposed to answer the question because of the rules of our discussion. But my comment on the comment is that the latest stages of branching morphogenesis in mouse lung are precisely in this class and in human lung it does not show the stereotypic pattern that you see in mouse. So it is actually not clear what humans are really adopting.

T. Lecuit In the context of self-organization the amount of information that you need to describe and control the systems is actually very little as compared to the control at any scale (e.g. genetic programs of the lung or any other organism). So from the point of view of evolution, where you want to tune parameters to diversify structures, there will be an advantage to have something that self-organizes to some extent.

M. Desai Something related to what Boris was talking about in terms of five-dimensional face shape space and in general the idea that there is some specific small set of processes that controls shapes in any one of these systems. In reality there are a lot of things going on that are not necessarily the evolutionary preferred mechanism for controlling shapes but they actually do control shapes. For instance, the shape of our face is controlled by dietary preferences which presumably have nothing to do with the basic five-dimensional control of the shape space. I wonder from a point of view of evolution to what extent all these confounding factors go co-opted into new mechanisms for control.

S. Grill Just to pick up on that comment and on something that we heard from Maha. If I think about your gut system for example, of course the elasticity of the two layers is very important and there are going to be many molecular mechanisms that fit into setting an effective elasticity for a certain piece of tissue. So in some sense there is a kind of redundancy, maybe morphogenetic redundancy in terms of genetic pathways. But here maybe there is more of a physical redundancy and physical degeneracy because you have many molecular mechanisms that fit into one single parameter that is important for elasticity. Understanding this linkage is going to be important.

E. Wieschaus The issue in part for me is that if you take any of the mechanisms or any of the processes that we have talked about, the pathways if you will, they occur in a context where much is assumed in the context of physics and energy/osmotic landscape. All these things, if we change them, or if they are changed during the evolution can impact the process and change the morphology. But it seems to me that the more relevant question is actually, particularly in the context of developing a morphological change, what are the parameters that actually change, what are the features that actually change because traditionally those are the ones that we think are controlling the process rather than the totality of all the different parameters. The relevant thing is what in a particular organism and in a particular event is the feature, what is the parameter that is distinguishing between one choice and another.

T. Lecuit We will move on to the next prepared remark from Frank Jülicher.

Prepared comments

Frank Jülicher: Shaping a tissue by mechanical boundary conditions

How the shape of an organism is generated and how shape is encoded by genes is a fundamental question in biology. Together with Suzanne Eaton, we have recently studied the role of tissue mechanics in the process that shapes the fly wing. The characteristic shape of a fly wing can be easily recognized. Interestingly, the wing of a fly that is mutant in a single gene called dumpy exhibits a strong morphological phenotype. This single genetic perturbation leads to an abnormal wing shape.

What is the cause of this shape phenotype? The early wing tissue undergoes a mechanical contraction of its hinge area while its margin is mechanically attached via the extracellular matrix to a cuticle. The tissue contraction and shortening thus generates a shear stress in the tissue which induces cell deformations and cell rearrangements and thus tissue shear, forcing it to its final shape.

What is different in the dumpy mutant? The dumpy mutation leads to an abnormal wing shape because the mechanical attachments between tissue and cuticle are modified. The dumpy protein is a part of the extracellular matrix, a polymeric material that is formed outside the tissue. Its mutation affects the mechanical attachments between tissue and cuticle. As a result the boundary conditions change, the tissue margin can move, stress fails to build up and the tissue shear that is seen in wild type does not occur. This leads to an abnormal shape.

These results emerged from quantitative life imaging performed at high resolution over long times combined with theory.[a] Mechanical perturbations by laser ablation reveal the existence of anisotropic tissue stresses and show directly the relevance of mechanical connections at the tissue margins. A simplified continuum theory can capture the force balances and the magnitude of tissue shear, including the relative contributions of cell rearrangements and of cell shape changes to tissue shear.

In conclusion, the final shape of the wing is generated by an active mechanical process that is sensitive to mechanical boundary conditions. The dumpy genes can influence this process by changing the mechanical linkers that connect the tissue to the rigid cuticle. The change in mechanical boundary conditions results in a shape phenotype. This example highlights the role of mechanics in biological morphogenesis. Shape in biology emerges from physical processes that rely on material properties and involve active mechanical events. These are guided and choreographed with the help of genetic systems.

Discussion

T. Lecuit Remarks on boundary conditions or more? Clifford.

C. Brangwynne I am not sure this is necessarily a remark about boundary conditions. I just want to add some things that are not necessarily related to what Frank showed. We have this beautiful study that shows that we can understand to some extent the model and the mechanical forces and the way shapes emerge from these interactions. I just want to put it in a cautionary note that echoes some of the discussions that came up with Terry's, Rob's and Daniel's comments related to the nature of understanding. It is clearly a major accomplishment to build all sorts of perturbations and measure parameters and see changes and have some cautious use of the term predictability. But I also think that we should keep in mind that all these studies are sort of perturbations on biological substrates. I am referring to Feynman idea that if we really understand we should built it, and I think there is still such a huge gap in our understanding that in an engineering context there is no way we can build any of these things outside of biological substrates. This leads me to think that we really do not understand what we are doing in a deep way.

T. Hwa I come back to the issue of evolvability that Boris linked to the five-dimensional phenotypical space versus a much larger genetic space. These

---

[a] Etournay *et al.*, *eLife* **4**, e07090 (2015).

genes and pathways are highly redundant. They have to do many things and all these things are embedded in this dynamic process that needs to unfold: everything has to be done correctly until the end. I am not sure how long it would take by the process of "blind watchmaking" — random changes — to come up with the right design and answer. So what I am searching or wishing for is, if nature has figured out some kind of high order constraints so that within these constraints the changes that need to be made at the bottom are much fewer to reach the relevant changes at the top. If we understand that, then that addresses your question. Maybe nature has some tricks.

D. Fisher I want to pick up on some of the comments made by Cliff and others about being surprised or not. I love quantitative understanding as much as the next person, and the quantitative thing in going from one dog to another, that is really a quantitative trait. But I think that one of the roles of physics is to learn which are the things that we should not be surprised by because biology is manipulating a few things that can take advantages of really interesting physics and chemistry and so on to do things that would seem rather amazing. But somehow, in the end, I am much more interested in the qualitative understanding. I mean, I am interested in understanding evolutionarily how do we get things that are qualitatively different, not on how we change quantitatively things that are already there. And it seems to me that we are enormously further away from doing that, even as we are starting to understand the quantitative things. There are so many things that we do not understand even on the order of magnitude level and this is the case for almost everything in evolution.

T. Gregor I want just to pick up on the point that has just been made. I think one of the prime issues in this field is that even though we have made a lot of strides since those two defining books that were very descriptive, we are still in a very descriptive and anecdotal era of morphogenesis. Now we can make very nice quantitative measurements and be predictive but I think in order to understand we need to take different morphogenetic processes and try to understand or describe them with defining principles. I am wondering what needs to be done in order to get somewhere to a more coherent or more principle-oriented understanding.

J. Howard I just want to come back to the fact that it is very important to understand the molecular basis of these things. One thing to say is that hereditary information is encoded in some kind of pseudo crystalline substance. But, the elucidation of the structure of DNA provides such an extraordinary advance in our understanding of heredity. I think one has to distinguish between physical models that say "here is a plausible mechanism with some kind of rules and feedbacks and whatever". I do think that these models cannot be the end of the story because then it is just saying that this is a possibility of how this or that shape could rise. Instead we really need to understand the molecular basis behind that. That is the only way by which we can prove those things. For example, if there is a mechanical feedback between stress and biochemistry, than we have to find the molecules that are doing that. Actually that is a huge problem in mechanobiology because we know none of those molecules, we see it all over the place, but we know none of the molecular mechanisms. That is a huge gap in our understanding and my guess is that if we fill that gap we will get a huge understanding at higher level.

A. Perelson I just wanted to say that this notion of using 5 dimensions to describe shapes for faces resonates with me. Quite a few years ago, George Oster in Berkeley and I used the idea of shape space to understand molecular recognition in the immune system. And we thought that antibodies and T-receptors have shapes, and there are complementary shapes that are the ligands interacting with them. And by some theoretical arguments that I don't want to go through here we deduced that this shape space should be something like five-dimensional. Later experiments, looking at large panels of molecules interacting with each other and binding, used multi-dimensional scaling to find out what the lowest embedding dimension is in order to make sense of what we call antigenic distance or difference in shapes as seen by the immune systems, showed that the space is very low dimensional, something between two or five dimensions. So maybe there is something in biology that is more universal. Somehow to do recognition we do not need to know all molecular details and all coordinates and everything, there are some common features from which biological systems evolved to sense and utilize that help drive evolution, so it can operate in this low dimensional manifold.

S. Chu I just want to throw perhaps a topsy-turvy way of looking at this. What about the fact that if you look at fly or insect wings there are variations, there are communalities but they are different. How those shapes actually alter their aerodynamics, the maneuverability, how much mass they have to lift up? When I was a small child I was told that we do not know how bumblebees fly, that they shouldn't be in the air anyway. In addition to molecular feedbacks that actually confer the shapes there must be something about the aerodynamics, the mass of the insect, and maneuverability. It determines how good or strong is the eventual shape? People must know about this.

T. Lecuit I think that this resonates with the comment from Maha. Functional feedbacks from the perspective of evolution are essential.

T. Lecuit I suggest that we pause for coffee. I just try to summarize a question that has been phrased by different people. There is a huge space of genes and interactions and at the other extreme there is a huge space of shapes and morphologies that is also multidimensional and very complex. In terms of understanding, we want to try to find a reduction of dimensionality in between, where we want to have possibly few physical parameters and few control parameters that emerge from the interactions among genes. The question is what approach can be used to identify the number of relevant physical parameters and how we discover the genetic and biochemical parameters that connect with the few physical parameters. So that we understand the core morphology, the theme of morphology and the variations around the theme that evolution is exploring. Let us go for coffee and resume afterwards with the second rapporteur, Eric Wieschaus.

Rapporteur Talk by Eric Wieschaus: Epithelial Morphogenesis during Early Embryonic Development

Although morphogenesis in general involves behavior on different length scales (molecular, cellular and tissue-wide scales), I will focus on changes in shape and arrangement of groups of cells, generally two-dimensional epithelial primordia of 10^2 to 10^4 cells. Such changes can often be modeled using simple 2D surfaces, but the deformations that occur during morphogenesis ultimately need to be considered in three dimensions. Morphological change requires localized differences in forces and other physical properties within the epithelial sheet. In the best-studied cases, the forces are generated by local activation of contractile actomyosin networks. The pattern of this activation and the distribution of other physical properties reflect the underlying positional information within the sheet and depend ultimately on differential expression of genes that control development.

1. Introduction

Function follows form — the complex biochemistry and the mechanical functionality of life presuppose a similarly complex spatial organization of underlying molecular components. At the single cell level, this spatial organization is the key to life itself, and cells spend energy to maintain that organization. Although cells can undergo significant changes in shape, they maintain the defining functional features of their spatial order throughout their lifetime and pass that organization on to their daughter cells when they divide. This continuity in spatial organization is the essential feature of life and underlies the 19th century realization that life begets life and all cells are derived from previously existing cells. [1,2]

At a more complex level, spatial organization is also a central feature of multicellular organisms. Our adult bodies function because our heads are in the right place [mostly], our skin is on the outside of our bodies and the tubules of our kidneys are localized in a precise pattern relative to the overlying mesenchymal matrix. The spatial patterns that characterize multicellular organisms differ in one important respect from that characteristic of single cells. They are built de novo each generation. During embryonic development, individual cells maintain and pass on their life essential internal organization as they divide. The obvious continuity is less apparent in the super cellular changes in morphology that characterize development. There is no obvious continuity between the complex changing morphology of the embryo or morphological features of the adult and the simple cytoplasmic distributions in the fertilized egg. This means that new spatial patterns are generated continuously during embryonic development.

In the embryo, each transition requires changes in gene activity at the RNA and protein levels. An attractive feature of studying cellular mechanics during these early stages is that over the past thirty years, much has been learned about the genetic patterning of cell fates in the early embryo. Genes have been identified in laboratory model organisms that control pattern and have been shown to operate in the mother that makes the egg as well as in the embryo itself. In many of these organisms, the lists of relevant genes are sufficiently complete to build circuit diagrams describing the pathways and developmental strategies that control cell fate. Ultimately such molecular patterns of gene activity must be coupled to physical properties like force, resistance and viscosity that control changes in cell shapes and position. Our detailed knowledge of the genetics of early development therefore provides a rare opportunity for the experimentalist to combine physics and biology in approaches that connect quantitative modeling, molecular biochemistry and physical measurements.

I have focused the following presentation on morphogenetic processes that occur in specifically in the embryo, not just because of their central important to multicellular life, but also because certain simplifying features of embryonic development facilitate thinking about the actual mechanics and physics. I will begin by discussing embryos in general, reviewing some of the simplifying features of embryonic epithelia. In the middle section, I will discuss the relationship between the localized forces that drive morphogenesis and the preceding genetic events that program regions of the embryo to specific cell fates. The third section discusses how physical parameters of embryonic tissues like viscosity and viscoelasticity affect the outcome of localized forces.

2. Morphological Changes in Early Development — Origami vs. Modeling Clay

Any movie of early development emphasizes one of the characteristic features of embryonic development, namely the speed at which the morphology of the embryo is transformed. In the Drosophila embryo, for example, only three hours are required to transform the single fertilized egg cell into a blastoderm in which 6000 epithelial cells are positioned over a large centrally located yolk cell[3] (Figure 1). The cycles of DNA replication and nuclear division that achieve this transformation occur at ten-minute intervals. The formation of a mitotic spindle and the physical separation of chromosomes into sister nuclei (i.e., the process of mitosis) take less than three minutes per division.

The rapidity of the transformations that follow once the blastoderm has cellularized and the embryo has begun gastrulation is even more impressive. The mesodermal cells formed on the ventral side of the embryos will be internalized by a furrow that forms along the length of the embryo. During the formation of this furrow, individual mesodermal cells constrict apically, elongate and then shorten, transforming their morphology from columnar to trapezoid in shape (Figure 2). These cell shape changes drive the internalization of mesoderm over a period of 15 minutes. Once in the interior of the embryo, the cells re-initiate cell division, divide once and then begin to spread out over the ectoderm.

An important simplifying feature in embryo development is that morphological changes most often occur in 2D epithelial sheets. The 2D character is evident in the early stages of Drosophila where cells are organized in ordered monolayers of single cells. A similar epithelial character is observed in mammalian development where

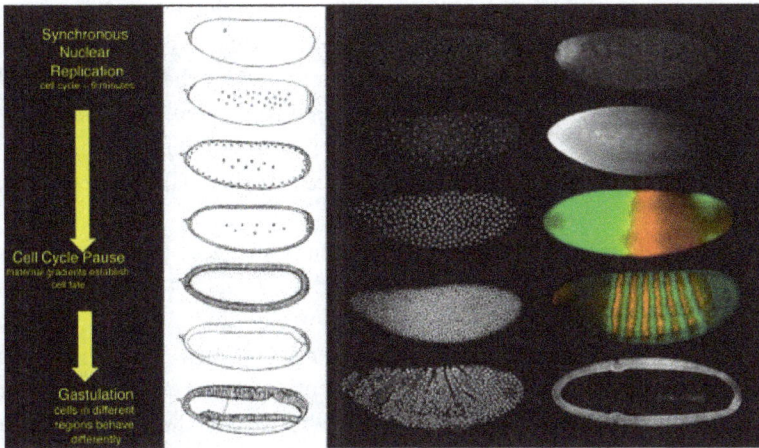

Fig. 1. Early development of Drosophila. After fertilization, the embryo undergoes a rapid sequence of nuclear replications without intervening subdivision into cellular units. These replications pause after thirteen rounds, and the nuclei at the surface are separated into individual cells by invagination of membrane from the surface. During this "cellularization" cells are assigned position specific fates reflected in specific patterns of gene expression. The patterns are established by specific RNAs localized in the egg during oogenesis in the mother. In the example shown in the figure, RNA from the bicoid gene are translated into a gradient of Bicoid protein that activates transcription of different gene in different regions of the embryo. Differential gene expression then results in different cell behaviors at gastrulation.

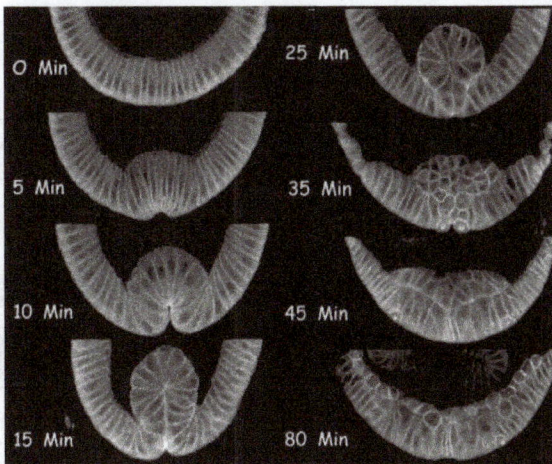

Fig. 2. Changes in cell shape drive internalization of mesodermal cells in Drosophila. At the blastoderm stage, cells on the ventral region of the embryo have been programmed to mesodermal fates. In the first step of their internalization, the apices of these cells undergo constriction, driving an elongation of the cells and a basal redistribution of the cytoplasm and nuclei. Once the cells have completed their apical constriction, they begin to shorten and their bases expand. The shift from columnar to trapezoidal shape produces a torque in the sheet that drives the internalization of the ventral furrow. In the panel at 10 minutes, note the stretched appearance of the cells on either side of the forming furrow.

the early morphogenetic events that give rise to the definitive embryo occur in the epiblast, a disc of cells that forms from a small residue of cells once the majority of the blastocyst has given rise to external membranes of the placenta.[4,5] The first morphogenetic events visible in this disk are a visible infolding of cell along the "primitive streak" marking the site where the embryo will form. Even in organisms where morphologies are more complex and three-dimensional, the organization can often be broken down into layers.

The 2D epithelial nature of embryonic tissues suggests that morphogenesis in embryos is more like origami than modeling clay. Cells retain their organization as coherent sheets and the three-dimensional form of the embryo arises through a sequence of complex folding patterns and internalizations. The analogy between origami is attractive because it focuses our attention on two fundamental problems the embryo must solve — it has to decide where to put the folds and then must define mechanisms that localize force to produce folds in those areas. The situation in the embryo, however, is a bit more complicated than origami. Unlike paper, the epithelial sheet in the embryo behaves as a 2D structure that is fluid and stretchable. The transition to a more fluid character is obvious in mammalian embryos once the cells move through the primitive streak and establish an underlying mesodermal layer. Once internalized, these mesodermal cells spread under the epiblast in a pattern of swirling flows as cells move around the future posterior of the embryo and toward the anterior (Figure 3).[6,7] Similar flow patterns and redistributions are observed in Drosophila[8] in ectodermal cell layers that remain epithelial in character. Within the sheet, such flows involve a fluid-like movement of cells relative to each other unlike anything that occurs in origami paper.

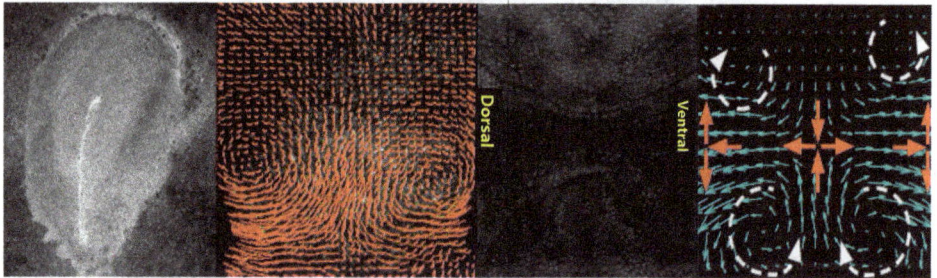

Fig. 3.   Cellular flow patterns in sheets in mouse and Drosophila embryos during early gastrulation. In mouse embryos the patterns of cellular flow begin in the mesodermal layer after the cells have been internalized at the primitive streak. In Drosophila, the earliest flows occur in the ectodermal layer. In both cases, cellular flows are characterized by saddles and vortexes that result from displacement of cells along the future anterior-posterior axis. Images modified from Refs. 6–8.

In many cases, these flows can occur with little or no change in cell shape or surface area as individual cells. Flows can also create regions of compression, or even loss of surface along one axis and an expansion along the other, changing the overall aspect ratio of the cell. If the cells maintain constant volume, decreases in local surface area must result in corresponding increases in cell height. Groups of cells undergoing apical constriction or compression therefore form thickenings in the epithelial sheet called placodes. Such placodes often mark the first stage in the internalization of primordial. Well-known examples are optic and ear placodes in vertebrate development and the tracheal and salivary placodes in Drosophila.

Despite this treatment of morphological change as a 2D phenomena, morphogenetic events produce "out of the plane" 3D transformations that can be described as folds or buckles. During normal development, such invaginations occur in a defined direction relative to the surface of the embryo, producing tissue internalization or out pocketing. The inside/outside directionality may be imposed by features outside the sheet itself (e.g., interactions with the substrate or underling yolk mass). More often it reflects some internal directionality with the 2D sheet itself. Such directionally may come from the inherent cell biology of the systems. Epithelial cells are polarized with a clear apical/basal axis perpendicular to the sheet (Figure 4). Each cell is tightly coupled to its neighbors by band of adherens junctions immediately below its apical surface. The basal regions of the cell have a different cytoskeletal organization, and are less adhesive to each other and often closely associated with an underlying extracellular matrix. In a way that we do not understand, these features allow localized constrictions to drive invaginations in one direction or the other.

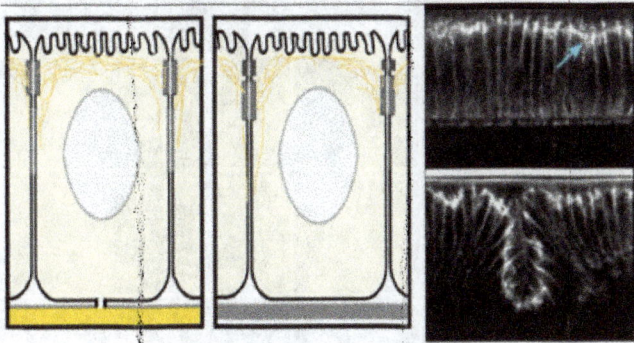

Fig. 4.   Epithelial cells in both flies and mammals are polarized with a clear apical/basal axis perpendicular to the sheet. In the apical most region of the lateral surface, cells in fly embryos (left) and mammalian embryos (right) are held to their neighbor by adherens junctions containing the transmembrane adhesion molecule E Cadherin. Adherens junctions also contain components that anchor cytoskeletal elements like actin and myosin. In Drosophila, contraction of the actomyosin cytoskeleton drives apical constriction and the cell shape changes are associated with internalization of mesoderm described in Figure 1. Repositioning of these adhesive sites on the dorsal side of the gastrula is associated with the formation of folds in the dorsal epithelium. Adapted from Ref. 40.

3. Force Generation in the Epithelial Sheet

Any movement and morphological change in epithelial sheets requires forces that are unequally distributed on the surface of the epithelium. Many experiments suggest a predominant role for the motor protein Myosin in this force generation. Myosin interacts with cytoskeletal actin meshwork, producing contraction by hydrolyzing ATP. A remarkable feature of the actin myosin cytoskeleton, initially observed in mesodermal cells during Drosophila gastrulation but subsequently detected in many morphogenetic events in many invertebrates and vertebrates is the pulsatile nature of the contractions.[9] In Drosophila mesoderm, a central apically located mass of myosin aggregates and disperses in repetitive 90-second pulses (Figure 5). Each pulse is associated with stepwise reduction in apical surface area, such that over a period of 8 to 10 minutes the apical surface achieves its maximal reduction in surface area. Perhaps more surprisingly, all the other morphological changes occurring in the mesodermal cells during this period are also pulsed. These include apical/basal elongation of the cell, its nuclear displacement, increase in surface area and changes in junctional position morphology. Their tight correlation with the timing of the myosin pulses argues that directly or indirectly, all these diverse morphological changes are driven by myosin contractility.[10] This possibility is consistent with the pulsing myosin network being the most obvious force-generating machine observed in the embryo during this period. Myosin's localization to the apical extreme of the cell presents a challenge, however, in that it requires that forces driving apical

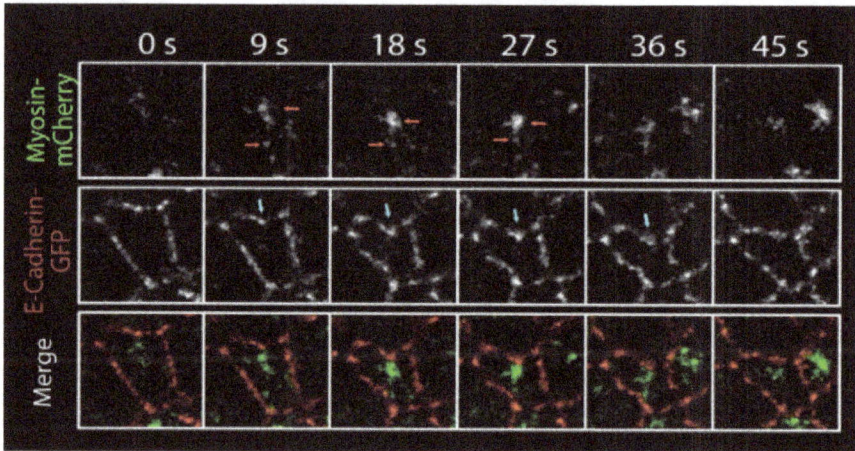

Fig. 5.   Stepwise cell shape changes in ventral furrow are driven by contractile pulses in the actomyosin cytoskeleton. The cell shape changes that internalize mesoderm in the Drosophila are driven by asynchronous stepwise reductions in apical surface area occurring in individual mesoderm cells. Each step is associated with a transient accumulation of myosin in the apical surface of the cell. The contracting actomyosin appears to pull on the cadherin based adhesive junctions on the cell surface, distorting its outline. The interval between myosin pulses averages about 90 minutes but is highly variable. Over the 10 minutes required for the apical surface of the mesodermal primordium to undergo complete constriction, each cell undergoes about 6 contractile pulses.

constriction impact the changes occurring across the entire volume of cell. The simplest models build on the observation of constant volume and invoke myosin contractility as driving a basal displacement of cytoplasm and the associated stress and expansion of other regions of the cell.

Any predictions of morphological change based on myosin must also take into account myosin's different subcellular distributions. In the Drosophila embryo, for example, there are different pools of myosin. In addition to the apical medial myosin responsible for apical constriction of mesoderm and endodermal cells, all the blastoderm cells have a distinct pool of myosin localized on their basal surface that arose during cellularization and is thought to play a role during cellularization.[11,12] Contraction of both these networks is isotropic and results in stresses on neighboring cells located in all directions from the contraction cell.[8,13] A third pool of myosin arises in the ectoderm at the onset of gastrulation. Unlike the apical-medial and basal pools, this myosin is highly polarized and is enriched on anterior posterior interfaces.[14] This enrichment is thought to drive anisotropic cell rearrangements leading to germ band extension and elongation of the embryo.

4. Later Development and Longer Timescales: Growth as an Alternate Force Generating System in the Embryo

On longer time scales, the transcriptional patterns that determine cell fate can also control behaviors like growth. There are many developmental systems where differences in growth can account for the observed morphological changes. For example, in plants, as leaves grow, they often lose their initial uniform shape and assume more complicated lobed morphology. The final leaf shape is controlled by *AcNAM*, a transcription factor that is expressed in regions between each future lobe and inhibits growth.[15,16] Development of the looping pattern of the gut in chick provides another example of impact of growth on morphology.[17] In this organism, the gut grows more rapidly than the attached mesentery, causing the gut to curve. In experiments where growth rates are artificially adjusted, looping patterns and loop radius change accordingly. Differential growth is thus a powerful tool for shaping embryos and organs. The difference between local myosin-based force generation and local growth are the time scales available for morphogenesis. For growth to matter, the relevant time scales are always longer.

5. Spatial Patterns of Gene Expression and Cell Movement

To drive functionally meaningful morphologies, changes in shape must be coupled to the same genetic mechanisms that control cell fate decisions over longer terms. These patterning events have been well studied in the past 40 years and a broad outline has emerged that allows comparing animals of different species. Cell fate decisions are generally transcriptional in nature and depend on the expression of selector genes whose on/off expression determines specific cell fates.[18,19] In embryos like frog and flies, selector gene expression is prefigured by graded distributions of maternal RNAs and proteins already present in the unfertilized egg.[20] Mammalian embryos appear to be an exception to this general rule. Mammalian eggs are small, development is long and associated with significant growth. The earliest morphogenetic events concern the development of trophectoderm of the placenta and do not directly address cell fate distinctions and patterning in the embryo itself. The initial events of patterning in the mammalian embryos are not thought to depend on maternal gradients but instead are thought to be generated by interactive networks and self-organization properties that are as yet poorly understood.[21]

The same features that affect the differing mechanisms of cell determination in flies and mammal also have an impact on how the cellular components governing mechanics are provided to the embryo. In embryos like flies frogs and fish that develop rapidly to a larval stage outside the body of the mother, there is little growth or uptake of nutrient until the individual hatches and is able to feed for itself as a larva, tadpole or fish hatchling. In all these cases, eggs are relatively large. During the initial stages, embryos are subdivided into cells by a sequence of rapid mitosis that produces a blastula or blastoderm. In addition to all the nutrients the embryo will need until hatching, the eggs also contain the entire cell biological machinery, the cytoskeletal components and adhesion molecules necessary to transform the blastoderm into a more complex morphology.[22] These maternal proteins are distributed to all cells, irrespective of cell fate and regardless of the specific levels required in particular cells. Their activity and their impact on morphology is controlled by the cell fate choices made in each region of the blastula.

Due to the longer time scales and the increases in cellular volume, mammalian embryos require increasing amounts of new gene product from the early stages of growth. Even though the mammalian embryo continues to be supplied with nutrients via the placenta, proteins and RNA do not pass through the placenta once it is formed. The mammalian embryo is therefore reliant on its own transcriptional activity to supply cytoskeletal and motor proteins from the earliest stages of development.[23] Transcription therefore supplies the fundamental cell biological components that are maternally supplied in other organisms, as well as playing a role in cell fate decisions themselves.

Regardless of their different mechanisms for establishing the initial patterns of gene expression, in both mammals and species that pre-pattern eggs, the final steps in cell fate determination involve the expression of transcription factors that define cell fate. A major challenge has been to relate the expression of these transcription factors localized in the nucleus to the visible changes in cell structure and morphology that predominantly occur in the cytoplasm and in many embryos that occur immediately after the transcription factors pattern are definitively expressed.

Transcription factors that control cell fate produce visible effects on morphology by regulating the expression of target genes whose products interact with the maternally supplied cytoskeletal and adhesive components. In Drosophila embryos

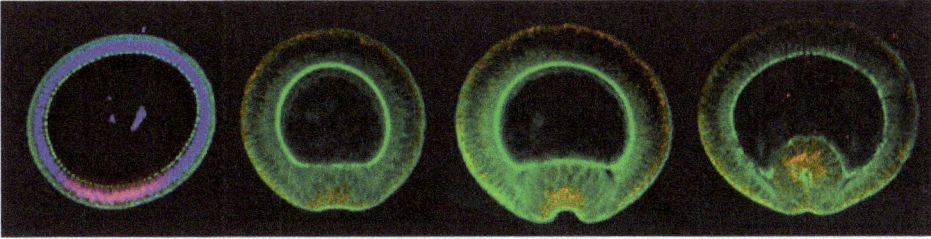

Fig. 6. The mesodermal determinant Twist induces expression of folded gastrulation gene in Drosophila. During the late stages of cellularization, the transcription factor Twist is expressed in the central region of the Drosophila embryo, here seen in the cross-section. Twist expression induces expression of the cell signaling protein folded gastrulation, which together with a second zygotic product T48 drives myosin accumulation and apical constriction in mesodermal cells. Twist (red), unstained nuclei (blue), Myosin (green), folded gastrulation (orange).

mesodermal cell fate is established by the expression of the transcription factors Twist and Snail, which in turn drive the local expression of cytoplasmic factors like Folded Gastrulation and T48 that more directly control activity of the cytoskeleton (Figure 6). [24,25] Identifying such cytoplasmically active target genes for other morphological processes has been more difficult. [26] Molecular strategies like transcription profiling reveal that most major cell fate regulators control the expression of large batteries of downstream genes, only a fraction of which may play a role in immediate morphological changes. [27,28] Genetic analysis suggest that target genes may play overlapping redundant roles such that elimination of any single target gene may only subtly affect morphological transition and may not produce easily measurable effect on viability. This has made genetic screens difficult since mutations in single downstream components may have only minimal effects on morphology. Consistent with this view, most of the mutant lines effecting overall embryology in Drosophila are in genes encoding transcription factors and upstream the cell signaling factors that control cell fate. [29] It has been relatively difficult to identify mutants in single genes with large-scale effects on morphology.

6. Time Scales and Feedback Regulation

To drive functionally meaningful morphologies, changes in shape must be coupled. It is interesting to compare the times scales for morphological transformations with the time scales required for gene expression. Based on the average size of most genes, the measured rates of RNA polymerase along DNA and the time required for nuclear transport and translation, about 10 minutes are required between the initiation of transcription and the presence of protein activity in the cytoplasm.

In organisms like mammals that develop slowly, cell fate decisions and changes in morphology occur on time scales sufficient for multiple rounds of transcription and cross regulation. The longer time scales might allow cell fates and cell morphologies to be established by interacting signals and transcriptional circuits that only gradually achieve their final precision. Positive and negative feedback circuits might also play a significant role in cell shape changes, enhancing or reducing the contractile stresses or local differences in growth to achieve precise shapes.

Feedback circuits that rely on transcription are more difficult to imagine in rapidly developing organism. The entire sequence of ventral furrow formation cell in Drosophila (apical constriction cell elongation and shortening, basal expansion) require about 15 minutes, about the time required for one round of transcription. One possibility is that such systems are highly tuned to give reproducible morphological changes from simple "feed forward" circuits that drive a sequence of events.

In all these examples, differential accumulation of Myosin II appears to act as a primary driver of morphological change in early embryo. Myosin interaction with the actin cytoskeleton will produce contraction of the cytoskeleton and will generate local stress. Whether that stress is translated into movement and thus morphogenesis depends on the properties of the surrounding tissue and whether those properties are uniform. Local differences in stiffness or elasticity will affect how the localized forces play out in terms of morphological change. Even with uniform surface tension (or presumably uniform myosin distribution), a local relaxation can produce tissue flows away from that relaxation point that mimic what would be produced by forces localized elsewhere. Such relaxation is thought to help position the cytokinetic furrow during cell division and local relaxation has been invoked to explain various morphogenetic events in vertebrates and invertebrates.[30] Given that it is hard to measure local differences in stiffness, documented examples are more rare than what may actually occur in nature. Local relaxation can also be coupled to adjacent force generation and it is not easy to tease apart the contribution of the two opposing mechanisms. Moreover, the flows associated with relaxation can produce local accumulations of myosin, which can concentrate even further through positive feedback. These problems are difficult to address experimentally with the currently available tools.

A similar problem arises in the interpretation of cell shape change that occurs outside the regions of obvious myosin accumulation and contractility. As the apical surface in Drosophila mesoderm constricts for example, the adjacent cells appear to be stretched and pulled toward the invagination (see for example the stretched cells adjacent to the ventral furrow at 10 minutes in Figure 2). The process of epiboly during fish gastrulation provides another example of such indirect effects, where a ring of contractile cells in the margin of the ectoderm appears to expand and pull an initially small ectodermal cap over the entire surface of the embryo. It is possible that the observed expansion here (and potentially also that observed in the cells that neighbor the ventral furrow in Drosophila) may also involve some local autonomous active cell shape change in the ectoderm itself. Ectodermal cells might actively flatten or expand their surface to push epiboly forward or close the ventral furrow in Drosophila. Under the constraints of constant volume, it is not easy to sort out what effects are active and local and which are passive responses to nearby contractions.

7. The Physical Properties of Cells: Viscous versus Elastic Responses

Localized forces produced by local myosin may generate local stress, but even if those stresses could be measured, morphological outcomes cannot be predicted without knowledge about resistance supplied by the cytoplasm and the cells in the tissue. Most models assume that resistance is uniform across the surface of the embryo and thus the differences in local forces generate all differentials in behavior. How the tissue responds to those forces depends on the property of the tissue itself, whether it is viscous or elastic. All biological tissues are elastic until they become viscous. The displacements that occur when forces are exerted initially results in internal stresses that tend to drive a reversion of the displacement. This spring-like behavior can be short lived such that the potential energy stored in the resisting system is soon dissipated by restructuring chemical bonds. Based on the recovery of fluorescence following photobleaching of the cytoskeleton, the time scales at which this restructuring occurs in biology are generally thought to be very short, on the order of seconds.[31,32] Based on these and other experiments, biological systems are generally thought to behave as viscous fluids when subjected to localized forces.[33,34] This conclusion is supported by the measured behavior of cytoplasmic flows during morphogenetic movements, which are consistent with predictions from the Stokes equations that describe the motion of viscous fluids at low Reynolds numbers (Figure 7).[35]

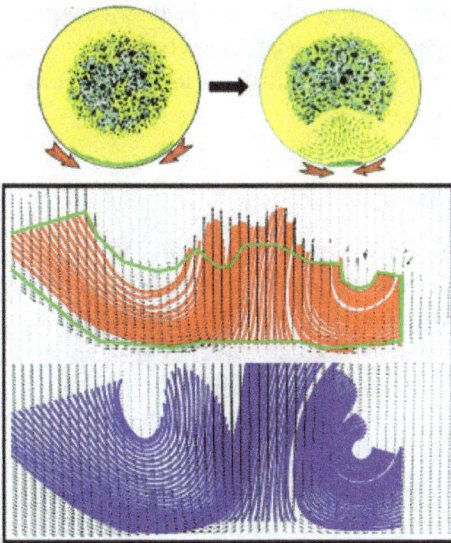

Fig. 7. Redistribution of cytoplasm during Drosophila gastrulation follows the flow patterns predicted by Stokes Equation. Although cytoplasm in the Drosophila embryo during gastrulation is packaged into distinct cells, the redistribution of that cytoplasm during cell shape changes the results in flow patterns that do not appear to be affected by the lateral membranes that separated adjacent cells. In these experiments, cytoplasmic flows (red) were tracked in living embryos using injected fluorescent beads and compared to the flow patterns predicted by the Stokes equation (blue) using the velocity distributions at the boundary of the region of the actual flow analyzed (green). Figure from Konstantin Doubrovinski.[35]

Assuming a viscous cytoplasm of uniform resistance allows various fruitful theoretical approaches. One can in principle extract the changing distribution of forces during morphogenesis from the morphogenetic movements themselves. This approach has been effectively applied to the ventral furrow in Drosophila.[36] There are cases however when strictly viscous models may not capture all features of a morphogenetic sequence. During ventral furrow formation, the apical localizing of constricting myosin appears to drive a basal-ward flow of cytoplasm and an associated cell elongation of the apical basal axis. As indicated above, this initial pattern is compatible with a strictly viscous response of the cytoplasm.[35] What drives the subsequent internalization of the cells however is a cell shortening and expansion of the base of the cells. During the internalization, the cells return to their original length, in behaviors that are visually analogous to that of a stretched spring. Incorporating spring-like behaviors allow the morphogenetic sequence to be described solely in terms of chemical work of the apically localized myosin and an independent slow loss of the basal stiffness. In this model[37] some fraction of the work of myosin is transiently stored as potential energy in the stretched lateral surfaces and it is that energy that drives shortening and internalization. The challenge for this model is that this requires biological springs that maintain their elasticity (that is, do not restructure and thus remember resting lengths) for up to eight to ten minutes, much longer that the turnover times for the molecular component of known cytoskeleton components. The challenge has been addressed in a model using active rearrangements of the cytoskeleton which could in principle maintain approximate internal force balance.[38] Physical distortions of the cortex of living embryos using ferrofluids and magnetic probes also suggest that long-lived elastic elements may exist in biological systems,[39] but it is doubtful whether extrapolation from the cortical elasticities characterized in those experiments to the properties of the lateral membranes is justified. The model is intriguing however because it allows the temporal sequencing of biological process (elongation and shortening) to be generated by a single force-generating machine, coupled to spatially distinct localized changes in passive features like stiffness of the cell surface.

8. Morphogenesis in Non Epithelial Tissues

This presentation has focused on epithelial tissues in early embryos where morphogenetic processes are simple and rapid. This simplicity is especially apparent and advantageous during the Drosophila gastrulation where the initial condition (the blastoderm) consists of a single cell layer held together by a band of adhesive junctions. Extrapolating from this simple pattern to more complex epithelia with multiple layers is interesting but challenging. Part of the challenge reflects our limited knowledge of the way that multilayered epithelia are held together and how forces are transmitted between the layers in the absence of well-structured bands of adhesive junctions.

Morphogenesis in embryonic tissues like mesoderm that seem to have lost their epithelial character entirely provide greater challenges. This presentation began by comparing morphogenesis favorably to origami rather than building structures by modeling clay. Perhaps, non-epithelial morphogenesis is in fact more like the alternative, modeling a pliant tissue like clay into complex three-dimensional structures.

The difficulty of that analogy is that clay is modeled by the application of external forces, at the hands of the potter or sculptor. Although in practice this is also true for the origami artist, I have tried in this review to develop the possibility that locally generated forces within the epithelial sheet itself might drive an origami-like transformation from two to three dimensions. How to apply that view to shape changes in an initially solid 3D mass of adherent potentially unpolarized cells is unclear and remains an exciting frontier in developmental biology.

References

1. R. Virchow, *Cellularpathologie* (Hirschwald, Berlin, 1871).
2. H. Harris, *The Birth of the Cell* (Yale University Press, New Haven and London, 1999).
3. E. Wieschaus, C. Nüsslein-Volhard, Looking at embryos. In *Drosophila: a Practical Approach*, ed. D. B. Roberts (Oxford, UK: IRL Press, 1986) pp. 199–226.
4. J. Rossant, P. P. Tam, Blastocyst lineage formation, early embryonic asymmetries and axis patterning in the mouse, *Development* **136**(5):701–13 (2009).
5. O. Voiculescu, F. Bertocchini, L. Wolpert, R. E. Keller, C. D. Stern, The amniote primitive streak is defined by epithelial cell intercalation before gastrulation, *Nature* **2007** 449: 1049–1052 (2007).
6. E. Rozbicki, M. Chuai, A. Karjalainen, F. Song, H. M. Sang, R. Martin, H. J. Knölker, M. P. MacDonald, C. J. Weijer, Myosin-II-mediated cell shape changes and cell intercalation contribute to primitive streak formation *Nat. Cell Biol.* **17**(4):397–408 (2015).
7. J. Firmino, D. Rocancourt, M. Saadaoui, C. Moreau, J. Gros, Cell division drives epithelial cell rearrangements during gastrulation in chick, *Dev. Cell.* **36**(3):249–61 (2016).
8. S. J. Streichan, M. F. Lefebvre, N. Noll, E. F. Wieschaus, B. I. Shraiman, Global morphogenetic flow is accurately predicted by the spatial distribution of myosin motors, *eLife* **7**:e27454 (2018).
9. A. C. Martin, M. Kaschube, E. F. Wieschaus, Pulsed contractions of an actin-myosin network drive apical constriction, *Nature* **457**(7228):495–9 (2009).
10. M. A. Gelbart, B. He, A. C. Martin, S. Y. Thiberge, E. F. Wieschaus, M. Kaschube, Volume conservation principle involved in cell lengthening and nucleus movement during tissue morphogenesis, *Proc. Natl. Acad. Sci. USA* **109**(47):19298–303 (2012).
11. P. E. Young, T. C. Pesacreta, D. P. Kiehart, Dynamic changes in the distribution of cytoplasmic myosin during Drosophila embryogenesis, *Development* **111**(1):1–14 (1991).
12. A. M. Sokac, E. Wieschaus, Zygotically controlled F-actin establishes cortical compartments to stabilize furrows during Drosophila cellularization, *J. Cell Sci.* **121**:1815–1824 (2008).
13. A. C. Martin, M. Gelbart, R. Fernandez-Gonzalez, M. Kaschube, E. F. Wieschaus, Integration of contractile forces during tissue invagination, *J. Cell Biol.* **188**(5):735–49 (2010).
14. C. Bertet, L. Sulak, T. Lecuit, Myosin-dependent junction remodelling controls planar cell intercalation and axis elongation, *Nature* **429**:667–671 (2004).
15. T. Blein, A. Pulido, A. Vialette-Guiraud, K. Nikovics, H. Morin, A. Hay, I. E. Johansen, M. Tsiantis, P. Laufs, A conserved molecular framework for compound leaf development, *Science* **322**(5909):1835–9 (2008).
16. A. Maugarny-Calès, P. Laufs, Getting leaves into shape: a molecular, cellular, envi-

ronmental and evolutionary view, *Development* **145**:161646 (2018).

17. T. Savin, N. A. Kurpios, A. E. Shyer, P. Florescu, H. Liang, L. Mahadevan, C. J. Tabin, On the growth and form of the gut, *Nature* **476**(7358):57–62 (2011).

18. A. García-Bellido, Genetic control of wing disc development in Drosophila, *Ciba Found Symp.* **0**(29):161–82 (1975).

19. R. S. Mann, S. B. Carroll, Molecular mechanisms of selector gene function and evolution, *Curr. Opin. Genet. Dev.* **12**(5):592–600 (2002).

20. C. Nüsslein-Volhard, Gradients that organize embryo development, *Scient. Amer.* (1996).

21. H. T. Zhang, T. Hiiragi, Symmetry breaking in the mammalian embryo, *Annu. Rev. Cell. Dev. Biol.* **34**:405–426 (2018).

22. P. H. O'Farrell, Growing an embryo from a single cell: A hurdle in animal life, *Cold Spring Harb Perspect Biol.* **7**:a019042 (2015).

23. D. Jukam, S. A. M. Shariati, J. M. Skotheim, Zygotic genome activation in vertebrates, *Dev. Cell.* **42**(4):316–332 (2017).

24. M. Costa, E. T. Wilson, E. Wieschaus, A putative cell signal encoded by the folded gastrulation gene coordinates cell shape changes during Drosophila gastrulation, *Cell.* **76**(6):1075–89 (1994).

25. V. Kölsch, T. Seher, G. J. Fernandez-Ballester, L. Serrano, M. Leptin, Control of Drosophila gastrulation by apical localization of adherens junctions and RhoGEF2, *Science* **315**(5810):384–6 (2007).

26. E. Wieschaus, From molecular patterns to morphogenesis: The lessons from Drosophila, from *Nobel Lectures, Physiology or Medicine 1991–1995*, ed. Nils Ringertz (World Scientific Publishing Co., Singapore, 1997).

27. S. De Renzis, J. Yu, R. Zinzen, E. Wieschaus, Dorsal-ventral pattern of delta trafficking is established by a snail-Tom-neuralized pathway, *Dev. Cell.* **10**(2):257–64 (2006).

28. T. Sandmann, C. Girardot, M. Brehme, W. Tongprasit, V. Stolc, E. E. Furlong, A core transcriptional network for early mesoderm development in Drosophila melanogaster, *Genes Dev.* **21**(4):436–49 (2007).

29. E. Wieschaus, C. Nüsslein-Volhard, The Heidelberg screen for pattern mutants of Drosophila: A personal account, *Annu. Rev. Cell. Dev. Biol.* **32**:1–46 (2016).

30. K. Murthy, P. Wadsworth, Dual role for microtubules in regulating cortical contractility during cytokinesis, *J. Cell. Sci.* **121**:2350–9 (2008).

31. M. Fritzsche, A. Lewalle, T. Duke, K. Kruse, G. Charras, Analysis of turnover dynamics of the submembranous actin cortex, *Molecular Biology of the Cell* **24**(6), 757–67 (2013).

32. L. Blanchoin, R. Boujemaa-Paterski, C. Sykes, J. Plastino, Actin dynamics, architecture, and mechanics in cell motility, *Physiol. Rev.* **94**: 235–263 (2014).

33. M. Mayer, M. Depken, J. S. Bois, F. Jülicher, S. W. Grill, Anisotropies in cortical tension reveal the physical basis of polarizing cortical flows, *Nature* **467**(7315):617–621 (2010).

34. M. Behrndt, G. Salbreux, P. Campinho, R. Hauschild, F. Oswald, J. Roensch, S. W. Grill, C. P. Heisenberg, Forces driving epithelial spreading in zebrafish gastrulation, *Science* **338**(6104):257–260 (2012).

35. B. He, K. Doubrovinski, O. Polyakov, E. Wieschaus, Apical constriction drives tissue-scale hydrodynamic flow to mediate cell elongation, *Nature* **508**(7496):392–6 (2014).

36. G. W. Brodland, V. Conte, P. G. Cranston, J. Veldhuis, S. Narasimhan, M. S. Hutson, A. Jacinto, F. Ulrich, B. Baum, M. Miodownik, Video force microscopy reveals the mechanics of ventral furrow invagination in Drosophila, *Proc. Natl. Acad. Sci. USA* **107**(51):22111–6 (2010).

37. O. Polyakov, B. He, M. Swan, J. W. Shaevitz, M. Kaschube, E. Wieschaus, Passive mechanical forces control cell-shape change during Drosophila ventral furrow formation, *Biophys. J.* **107**(4):998–1010 (2014).
38. N. Noll, M. Mani, I. Heemskerk, S. J. Streichan, B. Shraiman, Active tension network model suggests an exotic mechanical state realized in epithelial tissues, *Nat. Phys.* **12**:1221–1226 (2017).
39. K. Doubrovinski, M. Swan, O. Polyakov, E. F. Wieschaus, Measurement of cortical elasticity in Drosophila melanogaster embryos using ferrofluids, *Proc. Natl. Acad. Sci. USA* **114**(5):1051–1056 (2017).
40. Y. C. Wang, Z. Khan, M. Kaschube, E. F. Wieschaus, Differential positioning of adherens junctions is associated with initiation of epithelial folding, *Nature* **484**(7394):390–3 (2012).

Discussion

T. Lecuit Thank you Eric.

One of the two themes that we have seen is the tension between control and self-organisation in 2D or 3D morphogenesis. A theme that we will have to discuss also, is connected with the first part of this morning session: should we treat continuously morphogenesis that happens in 3D like mesenchymal cells for instance that have high motility, 2D sheets which behave as viscoelastic fluids on longer time scales, and 2D and 3D morphogenesis in plants in particular, that Maha discussed this morning. Are these completely discrete kinds of dynamics or should we use a continuous and a homogeneous mechanical models to explain these phenomena?

J. Howard I just wanted to make the point, and went back to something Rob Philipps made. What is really important is the measurement in these systems, to measure elasticities, viscosities and forces. The reason why this is important is that we know a lot about forces, elasticities and viscosities at the molecular level. So if we can make measurements at this mesoscopic level then we can make connections to the motor proteins, for example. In Eric's case, we can make connection to a number of active motors giving rise to these things. I think that it is a crucial connection to make to truly understand this.

E. Wieschaus I agree that the challenge is in the measurement. We can measure the general elastic properties of the whole cortex, but what we need are technologies for measuring local tensions either optically or mechanically. These are technologies that we do not have yet, although we can imagine them.

S. Mayor I think that some of the issues are about translating the mechanical information into some kind of a program which unfolds in a stepwise fashion. In some level, the translation of that may occur through mechanical signals being converted into chemical signals which in turn influence the cell that is experiencing this mechanical input. And therefore you see a transformation which is not purely mechanical but which is occurring through this interface of the chemistry that is being changed. I think understanding how mechanochemical processes are occurring in cells is crucial, as Joe brought that up as well this morning.

J. Lippincott-Schwartz How robust are these changes that you are seeing, the morphogenetic changes, to mutations in some of these genes. If you knock out one of the genes, that is the key transcriptional element of that cascade, what is the impact on the whole morphogenic process?

E. Wieschaus If you eliminate the program in the upstream transcription factor, you eliminate specific local behaviors. Because the embryo is complex and there are other regions, you may still see morphogenic movements. What might be more interesting is to use genetics to modulate the levels of downstream components. We know, for instance, that these components are supplied maternally but their activities are regulated through supplied co-factors. When you lower the level of myosine, you prolong and slow down the process in interesting ways but the process still occurs. Those kinds of experiments, I think, these subtle modulations that are possible in genetic systems are potentially the way to really understand mechanistically what is going on.

R. Philipps I wanted to give you a bacterial analogy. I thought that it was very important that you noted the issue of time scales in elasticity. In a growing bacterium, there is roughly 10^4 new lipids being inserted into the membrane every second. We have been interested in the tension in the membrane, but there is 5000 nanometer square area being donated every second. So the issue of growth and elasticity in time scales is very complicated and really interesting.

T. Lecuit I suggest that we move on to the prepared remark from Boris.

Prepared comment

B. Shraiman: Molecular mechanism of mechanical feedback on growth

We have been talking about mechanics, and growth control, and feedback, and molecular mechanisms, so Thomas has asked me to say something about the molecular mechanism of mechanical feedback controlling growth. Specifically, the feedback to control uniformity of growth. Quite generally, in order to build a feedback control system, one needs a read-out for the property one wishes to control, which in this case is the non-uniformity of local rate of growth within epithelial layer. Luckily, physics naturally provides such a readout signal. Time scales are important. If tissue can maintain elastic response on the timescale of cell growth and proliferation, then non-uniform growth will generate internal (elastic) stresses. Uniform growth — when the layer dilates uniformly — is the only way of maintaining a stress-less state. It is then easy to imagine a feedback system, where local rate of growth depends on local stress in such a way that uniformity of growth is stabilized: specifically local tension should promote growth, while compression should inhibit it.[a] However, as Joe Howard said here: "Why should we take this seriously, without a molecular mechanism?" Well Ken Irvine's lab at Rutgers has recently identified one particular pathway of growth control — the Hippo pathway — and one particular protein in this pathway — called Ajuba — as a likely conduit of information about the mechanical state of

---

[a]B. I. Shraiman, Mechanical feedback as a possible regulator of tissue growth, *Proc. Natl. Acad. Sci. USA* **102**:3318-23 (2005).

the cell into the growth control circuitry.[b] Ajuba localizes into the cortical cytoskeleton in a way that depends on local myosin tension and the effect of that — I will skip a bunch of molecular details in the interest of time (they are published) — is to activate growth in response to tension. So, more myosin and hence more tension beget more growth. This has correct sign, as the tension should promote growth, but you may ask: where is the feedback? How is myosin a *read-out* of a mechanical stress in the tissue? The answer is: through another layer of mechanical feedback, that recruits myosin into the cortex in response to externally applied tension. This makes local myosin level a read-out of tissue tension. Most importantly, experiments in Irvine's lab have demonstrated that small clones induced to grow faster than surrounding tissue end up with *lower* level of cortical myosin than the surrounding tissue.[c] This is what one expects on the basis of physical considerations (for non-uniform growth) and the effect has the correct sign for the desired mechanical feedback: reduction in myosin level releases Ajuba from the cortex and down regulates growth. Many molecular details are still missing, but still the mechanism may be coming into focus.

Discussion

T. Lecuit Any reaction to what has been said?

A. Murray In that situation do you know mechanically how extra myosin is recruited in response to stress? So if it is a feedback mechanism everything would be perfect, I expect you would have to explain how that last part works.

B. Shraiman You add another layer of molecular mechanisms that has to be filled in. And then there are probably more.

T. Lecuit We will continue on mechanics with Stefan Grill, with a prepared remark.

Prepared comment

S. Grill: Guided mechanochemical self-organization

I will stick to the theme of mechanics and regulation, and would like to discuss some work that we have been doing together with Frank Jülicher to understand feedback modules that generate a self-organized system combining both mechanics and regulatory processes. A particular focus is to

---

[b]C. Rauskolb *et al.*, Cytoskeletal tension inhibits Hippo signaling through an Ajuba-Warts complex. *Cell* **158**:143–156 (2014).

[c]Y. Pan *et al.*, Differential growth triggers mechanical feedback that elevates Hippo signaling, *Proc. Nat. Acad. Sci USA* **113**(45):E6974-83 (2016).

understand how such self-organized systems can be guided to arrive at the correct pattern in the right place and time.

We have been working on the process of PAR cell polarity establishment in the early *Caenorhabditis elegans* embryo. Here, three molecular structures are relevant, two species of PAR complexes, anterior and posterior, as well as the actomyosin cell cortex underneath the membrane. PAR complexes assemble on the membrane and thus on the surface of the embryo, and an interesting feature is that actomyosin flows transport PAR complexes in order to establish a cell polarity pattern.[a] We have performed precise quantifications of the spatiotemporal dynamics of the PAR polarization process, and have now available the concentration fields of both PAR species over the surface of the embryo as a function of time, together with both the concentration and the flowfield of non-myoscle-myosin-II as a function of space and time. All molecules exchange with the bulk, and the module of self-organization is as follows: The two PAR species "kick" each other off the surface, their normal residency time on the surface is of the order of 100s, and this time is reduced to less than a second in the presence of the opposing species. In addition, the anterior PAR species controls the dissociation rate of myosin: the myosin residency time is approximately 10s, but this residency time is modulated by the anterior PAR complex, where an increased local concentration of the anterior species leads to a longer residency time and thus to a higher concentration of myosin. Inhomogeneous myosin distribution generates active tension gradients that drive flows.[b] The actomyosin flowfield in turn transports both PAR species as well as myosin, and thus affects all fields. So this is a system of feedback that does not allow for a separation of mechanical and regulatory processes. The system is subcritical, there is an instability present but the parameters are such that we are far away from the instability. The pattern is established by guiding cues that control the pattern forming process. One such guiding locally removes myosin at the tip of the embryo, this leads to an active tension gradient and flows that immediately transport the anterior PAR complex away from the point of action of the guide, giving rise to self-amplification and positive feedback.

On the side of theory, we capture the underlying dynamics in a reaction-diffusion system coupled to a thin film active viscous fluid, and we can with the measured guiding cue distributions recapitulate the full spatiotemporal

[a]E. Munro, J. Nance, J. R. Priess, Cortical flows powered by asymmetrical contraction transport PAR proteins to establish and maintain anterior-posterior polarity in the early *C. elegans* embryo, *Developmental Cell*, 7(3):413–424 (2004); N. Goehring, P. Khuc-Trong, J. S. Bois, D. Chowdhury, E. M. Nicola, A. A. Hyman, S. W. Grill, Polarization of PAR proteins by advective triggering of a pattern-forming system, *Science* 334, 1137–1141 (2011).
[b]M. Mayer, M. Depken, J. S. Bois, F. Jülicher, S. W. Grill, Anisotropies in cortical tension reveal the physical basis of polarizing cortical flows, *Nature* 467, 617–621 (2010).

dynamics of the associated concentration and flow fields at a quantitative level, also under perturbed conditions. So how does polarization proceed? In this system mechanochemical feedback from the PAR species to myosin generates two basins of attraction, one around the unpolarized and one around the polarized state, separated by a boundary line. Guiding cues drive the system away from the unpolarized state and far across the transition line. Guiding cues are then released, and the system self-organizes to reach the polarized state.

To conclude, we have for this system characterized how the handover from a pre-pattern of guiding cues to mechanochemical self-organization is orchestrated.

Discussion

T. Lecuit Thank you Stefan. Who wants to either discuss biomechanics and morphogenesis or have other themes to present?

H. Goodson I was just wondering, can you ignore all of the turning that is going on inside of the cytoplasm? The model is beautiful but is that also playing a role?

S. Grill The surface flow drives intra-cellular cytoplasmic flows. This is understood by the work we have done, but also work that a group in Japan has been doing. The key aspect of the cytosol here is the exchange of molecules between the surface and the inside. Inside, they diffuse rather rapidly, the flow speed inside the cytosol are microns per minute and with these rapid diffusion coefficients, the cytosol still remains quite well mixed in terms of protein distribution.

H. Goodson So the microtubule cytoskeleton is not playing a role there. Just there is a lot of motion of large material inside there.

S. Grill There is almost no microtubules. In fact, the microtubules trigger the process at the tip. This cytoplasm does not have much structure at this point in time.

T. Lecuit I guess one of the questions that we are discussing, stemming from one of the concluding points from Eric, is what is it that transcription specifies in an embryo. We know that we have in mind positional information but what do we actually mean by that in terms of connection to mechanics. Obviously the information provides regionalization. What Eric presented in the mesoderm is not happening exactly the same in the ectoderm, which is just the abutting tissue. There is a regionalization of mechanical properties. How much tension you build up, for instance. Another one is polarization, the symmetry breaking in terms of tension is obviously important to set different kinds of dynamics. Temporalization, to follow up on the theme developed this morning by Alexander, is how I think we can understand how transcription factors could specify these kinds of information. If you look now at cellular dynamics exactly following what you just said, there are pulses, there are flows, there are waves. By no means, are these specified transcriptionally because the time scales are different. Maybe a theme to be discussed here is: you have the subcellular dynamics, that cannot be controlled transcriptionally, nonetheless, are they specified transcriptionally and so how?

R. Neher I do not have an answer. I have a naive question instead. We have seen this very rapid dynamics in early embryos that are clearly set up by maternal gradients, there is no transcription going on. Everything is basically boiling down to a certain predefined program. And then there are these things that happen much later which are coupled to the genome. There is transcription, there is a whole different level of feedback that comes in. Is there a fundamental difference between these stages? Is there something concrete that one could say about what is organized differently here and there?

S. Grill Maybe I can also remark on this point. I think there is a lot of synergies we can draw from looking at these systems at different scales. In what Eric was telling us, large scale sheet-like structures that contract but they also behave on some time scale as viscous fluid. I was telling you about an intra-cellular actomyosin layer that has a viscoelastic relaxation time of just 5 seconds in comparison to 8 minutes. I am going to say that a force balance is a force balance, so the physical picture that we should be thinking about to describe morphogenetic changes, if it is at least restricted to plants, might not be so different in between these situations. Then protein localisation, in what I was telling you, amounts to activation of transcription factors in different cells in different regions at larger scales.

J. Lippincott-Schwartz We heard yesterday from Jitu about the role of the plasma membrane and its intimate relationship with actin, the cortical actin system. I am curious both for Stefan and Eric. Have you looked at flows of particular lipid species or just the pathways that are feeding the plasma membrane and the exocytic pathways to what extent are they important for the actin remodeling and the dynamics that you are seeing in these embryos.

S. Grill I can answer that because we have looked at mobility of lipids. So the flow speed of actomyosin are microns per minute, so with the typical diffusion coefficient of a single lipid the Peclet numbers are such that they are not very much redistributed.

J. Lippincott-Schwartz I am not talking about movement of particular lipids. I am talking about lipid species like sphingolipids, cholesterols, proteins that can be associated with particular domains that might cross-talk with actin cytoskeleton.

E. Wieschaus One of the biases that has emerged at several times over the course of this meeting is that lipid bilayers represent passive responses to the structure of the cell, structure of organelles. It is an interesting question to ask how much do we think in terms of active cytoskeleton and in terms of lipid bilayers as responding to tension easily. And that bias might be true.

J. Lippincott-Schwartz We know for instance that apical domains are enriched in cholesterol. It could be very helpful in turning the dynamics of the actomyosin system.

T. Lecuit We had a question from Daniel.

D. Fisher Maha talks about what happens when you change various things like the gut and so on. Eric talks about ways if you stop the cellularization what sort of things change. Then there has been some discussions about time scales and the importance of those. But the time scales are one of the easiest things for biology to manipulate, by biochemical rates, strength of a transcription control and so on. I wonder if there are ways of trying to get at how many possible things can one get the development to do instead of what it would normally do by changing around some of the time scales, or the prospect of actually trying to look at that experimentally, either in development but also in cell biology.

T. Lecuit This issue of time scale is really important and I think we will address that. We will take a comment from Maha and then Thomas Gregor has a specific one on time scales.

L. Mahadevan So, I think an important question which really has not come up, I may have briefly alluded to it. You can control time scale by playing with the amount of cells relative to the extra-cellular matrix. In the gut, certainly it is true, in the mesentery they are very few cells and a lot of extra-cellular matrix. It could be that a way, at least mechanically, to play with things is to change the amount of extra-cellular matrix and then change how much or how little of it is degraded.

T. Lecuit Thomas Gregor will have a prepared comment on time scales.

Prepared comment

T. Gregor: Fundamental lack of understanding of time scales during developmental processes

A fundamental pillar of biological systems is their composition by unitary building blocks: molecules, cells, tissues, organisms, ecosystems. These building blocks span spatial scales from nanometers to the size of the earth, and their associated reaction timescales and lifespans range from sub-second to the age of the earth. Our understanding of how to bridge or transition between these different scales, however, is scarce to inexistant. How do biosystems measure time and space? How do they operate at different temporal and spatial scales? What parameters were dialed throughout evolution to tune temporal and spatial scales for different systems? Focussing here on time, I argue that we need to think about simple reaction coordinates that can be tuned to act globally on all molecular players in order to understand how fundamental life processes such as cytokinesis or somitogenesis can happen at vastly different time scales but with essentially identical molecular players. A possibility might be found at the metabolic level where, i.e., availability of ATP could be tuned to globally reduce or increase reaction speeds.

A first morphogenetic process, somitogenesis (i.e. the process that is at the origin of the formation of our vertebrae), exemplifies the problem as we compare it across different species, that all use the same molecules to implement the clock that generates somite after somite at periods that change by almost an order of magnitude: the interval of two successively generated somites ranges from 30 minutes in fish to 5 hours in humans. The molecules that constitute the network responsible for the clock are nearly identical in all species, and the question is how their properties can be tuned such that the period can be adjusted to the particular size and developmental speed of a given species.

The second example is cytokinesis (i.e. the final step during cellular division that brings about the separation of the two daughter cells) that happens at very different time scales within the same organism. In fly development, cytokinesis during early stages takes 3 min, but later on lengthens to 15 min and more.

Discussion

T. Lecuit Thank you Thomas. Who wants to discuss this issue of timing of cellular developmental processes? We have Uri and Joe.

U. Alon This comment is about having an oscillator that in different species shows a ten-fold different frequency. Also we heard about different frequencies across the tissues, but keeping a similar amplitude. This touches a recurring theme in this meeting, the need to classify dynamical phenomena. There we talked about switches, here we have oscillators. I just want to mention, for example, work by James Ferrell comparing two kinds of major oscillatory models, a delay oscillator and a relaxation oscillator based on bistability. It turns out that in the delay oscillator if you try to change the frequency, you change the amplitude too. That is built in the way it works. In a relaxation oscillator based on bistability, you can tune the frequency by differences in just one parameter, it could be the transcription rate of the slow variable, and then the amplitude is set by those jumps of the switch. By making cross species comparison, like you did, and by experimental manipulations and classifications of our phenotypes, the phenotypes in this case are the oscillator properties, entrainment properties, noise resistance properties, we can hope as a field to have a clear naming of what it is we are studying. This naming, like naming the demon, is extremely powerful. And can then narrow us down to classes of mechanisms. We will always have many molecular arrows, we will see both feedback loops, positive and negative, but this will tell which ones are likely to be important. Is it a couple of positive and negative as in bistability or just a cycle of negative as in the delay oscillator, so go and help us clarify at the molecular scale what to measure and what experiments to do. So, classification.

T. Lecuit Thank you Uri, we have Joe now.

J. Howard I will just make the point that molecularly it is easy to change the time of things. A very good example is this myosin II, there are fast myosin II and slow myosin II and the role in the very related proteins. In our muscles, we have combination of these proteins. They can contract or move at time scales that are different by a factor of 100. It may not be so hard to tweak the time scale of things, also taking into account what Uri said.

S. Eaton I just would like to point out two things that would change the rates, even in the same cell and with the same molecules in the process, that would be the temperature which in a segmentation clock also changes the period of these things, but also just the rate of metabolism, most likely because you can get a fly to develop twice as fast overall, based on the nutritional conditions. So I imagine most processes are like this, that are going on, must themselves just speed up as the result of nutrient supply or ATP production.

A. Goldbeter Another factor that may be involved in explaining the differences in period may be post-translational modifications, like phosphorylation, dephosphorylation reactions. In the circadian clock, you obtain eventually a 24-hour period but there is no single step which controls the 24-hour periodicity. You build multiple steps into that. What is also very important is the rate of protein synthesis and degradation. That was shown also in the segmentation clock to play a key role in setting the period.

T. Lecuit Who else wants to say something about timing processes? No one. We move on with Nipam.

Prepared comment

N. Patel: The morphology of entire organisms

I would like to step back a bit and tell you about the evolution of the morphology of entire organisms as opposed to what we have been focusing on which is the morphology of cells in tissues. In particular I like to illustrate this with an example from anthropod evolution where the overall morphology of the organism is very much controlled by the identity of the individual segments of which the body is made of.

You heard a little bit already about Hox genes, they were mentioned a couple of times, which are the transcription factors which are found in all animals, generally organized in a cluster. They control the fate of segments along the anterior or posterior axes of the body. In flies there are three genes expressed in the more posterior part of the body called ultrabithorax (Ubx), abdominal-A (abd-A) and abdominal-B (Abd-B). They control identity from the middle of the thorax to the abdomen. But in flies there is not a whole lot of difference between those segments, in fact the abdomen does not even have any appendages on it.

When we teach our students about the basic rules of how to think about these genes, we have a simple paradigm which definitely exists in drosophila which is called posterior prevalence: the more posterior gene overrides the action of a more anterior gene. So when you look at the phenotypes from mutating the Hox genes in flies, at least for these three genes, the rules are pretty simple: segments that lose expression simply take on the identity of a more anterior segment.

But what I am illustrating here in this slide is a different anthropod, an animal called Parhyale, a beach hopper. It is closely related to animals you are used to as you eat them, so it is related to lobsters, shrimps and crabs.

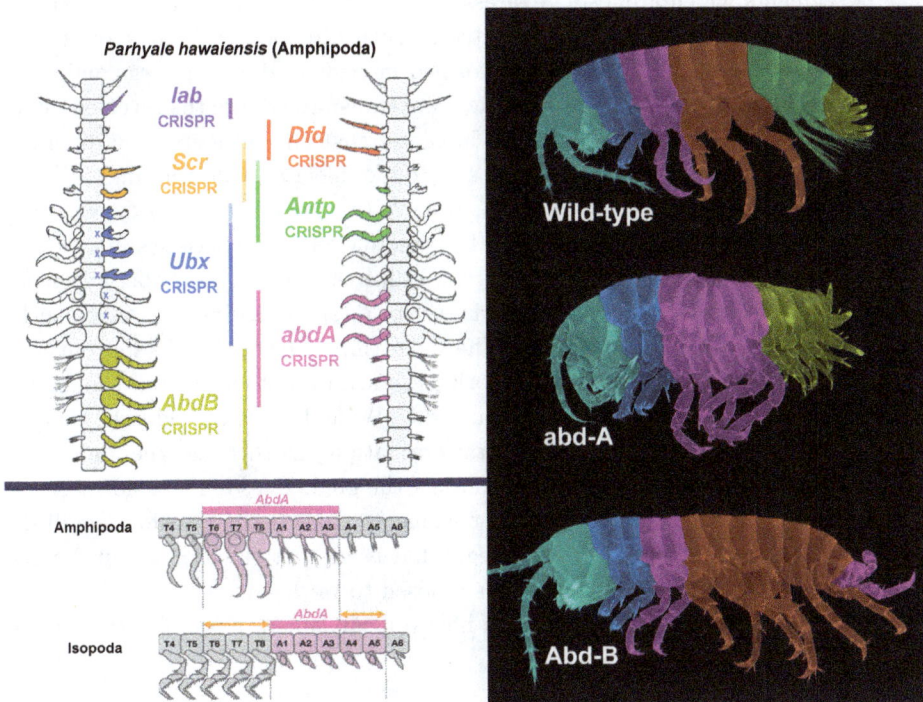

Fig. 1. The effects of knocking out Hox genes in Parhyale.

There is an interesting issue here because they have a pair of appendages on every segment. And not only do they have appendages on all segments but in fact they have a lot of different appendages. So if you ever tried to tackle a lobster that did not have rubber bands on its claws, you know this: it can simultaneously attack you, run away and feed because it has specialized appendages for all these functions. So what we wanted to address here is how can you have so many different types of appendages and how you would pattern them.

So of course an obvious possibility is that you would have more Hox genes, that you would have more of these transcription factors. But it turns out that this is not what this organism does. Instead what it does is that it breaks the rules that you established from drosophila. I illustrate that here, if you look at the top right, the wild-type animal. The appendages we are dealing with, the dark blue ones, have claws on them. The purple ones are for walking forward, the red ones are for jumping backwards, the green ones are for swimming and the yellow ones are for anchoring in the substrate. So we are trying to use those three genes in explaining how you can have all those kinds of appendages.

The key experiment comes from knocking out the middle of those genes, abd-A, which is shown in the lower embryo. What you can see is that the segments which are losing expression are not simply adopting a more anterior fate. They are adopting fates in both directions. What should have been the jumping legs are transformed into forward walking legs and what should have been the swimming legs are transformed into the reversed direction, into the anchor legs. This indicates that the rules are very different. We have gone on into making all sorts of double and triple mutants to really test this. The bottom line is that to have this greater complexity, you have not changed the number of genes but you rewired the genetic system. So you have done it in such a way that you broke the rule of posterior prevalence and now you have a combinatorial set of interactions instead, that can be used to pattern the segments. The other thing that really violates our idea about how these genes should work, is shown from knocking out the Abd-B gene, shown in the embryo down right. What I want to highlight there is that you also get a non-linear transformation. In this case the abdominal segments, in green and yellow, most of the green ones are transformed into the jumping legs but the very last segments are actually transformed all the way into a much more anterior leg. That is another violation of our "sacred laws" of how the Hox genes are supposed to work.

This provides an example of how you can generate complexity by actually rewiring genetic systems.

Discussion

T. Lecuit Thank you Nipam. Who wants to comment on that?

J. Lippincott-Schwartz It would seem to me that there must be a real intimate connection between the functions. What those appendages are actually doing relative to each other, force-wise etc., that is fundamentally driving that whole systems. I am curious to whether you have any clues to what it is that would be progressively sensed. Do the appendages couple each other based on what the organism is doing?

N. Patel It is a great question. Obviously as you also think about evolving these systems, you have to keep them functional all the time. One of the things we have been looking at also is how the musculature is coordinated with the morphology. Interestingly enough, you only have to change the Hox genes expression in the ectoderm and the muscle will simply follow along. Another challenge for us in thinking about this, is also how you wire it to the nervous system. Not only how you change the mechanics in what it does but how the brain would know that it is actually altered. Those are definitely challenges that we are trying to address. Now we have the tools to do that.

E. Wieschaus One of the things that strikes me, Nipam, is the difficulty in deciding what a fundamental law is. What we see are the properties of the specific organism that we study, the properties that explain some puzzling things. Until the "law" was formulated, posterior prevalence was just a way for thinking about genetic circuits in flies. I still believe that there are fundamental laws, generalizations across all species that use Hox genes to pattern. The trick is to figure out what are the ones that are really fundamental and which are not fundamental. Is that even a good question to ask "what is fundamental"?

T. Lecuit The time is running, I would like to try to attempt to summarize and outline what I have perceived as the fundamental questions that were discussed this morning.

The first one is, in the face of diversity and in the search of general principles, when looking at morphologies and shapes, one of the important themes that we discussed is the search for relevant parameters, and if possible few parameters, to describe shapes, to control — and I mean by that to develop shapes — and the third one is to evolve the shapes, which we will discuss this afternoon. Anything about the parameters that we need for description, control and evolution of shapes, we have in mind physical parameters, we talked about mechanics in particular, and regulation and by that we mean biochemical regulatory processes. The subquestions that emerged are: How many parameters do we need? What is the scale of these parameters? If you think about cellular processes, the parameters you might come up with may not be the same as parameters when you look at more coarse-grained descriptions in tissues, for instance. Or is it the case that you can find the same parameters which will be relevant irrespective to the scale that you consider. Active Matter provides a theoretical framework to describe and adequately predict the dynamics of structures in a way that transcend this scale boundary, which I think is an important example of that.

The second important question that we discussed this morning is the interplay between control, at one extreme high-level wired control, transcriptional control which Eric tried to push at the end of his presentation, and the other one presented by different people is self-organization. It seems to me that on the one hand you need a lot of information to provide exquisite control of a process having in mind robustness and precision. But evolution can pay the cost for that. And the question of cost which was discussed by Joe Howard in a very explicit way yesterday comes to mind. At the other extreme, you have self-organization which is very cheap in the sense that few parameters can provide control over a number of states and dynamics. I give the example of pulsatiles, waves, or flow patterns that exist at subcellular scales and at tissue scales.

The third important question that we discussed was rates. How do you control rates of processes? Again whichever the scale that you might consider, three important notions were discussed: temperature, energy (metabolism which Suzanne reminded us), and visco-elastic properties. In plants, growth occurs on scales of hours and days when with the same number of cells, you would achieve the same extent or rate of deformation over few minutes. So visco-elastic properties of wall cells versus animal cells with a cortex of active myosin are different.

The fourth and last point, that was not really discussed but that I would like to bring in, is the notion of the built-in irreversibility, or reversibility in some instances, of processes. What I have in mind is morphogenetic processes, some of them are reversible. Eric told us that cells contract and then relax. So here is a case of reversibility. But the tissues in which these cells are embedded irreversibly invaginate. What is the origin of this irreversibility at the tissue scales?

And the last aspect which will be discussed I guess this afternoon, is the reversibility or irreversibility of evolutionary processes. If you go now not from cells to tissues but to organisms, is the case such that we can understand the fact that shapes can evolve or de-evolve and if so, how?

Maybe this is a transition to this afternoon session. And I thank you all for being very active participants of this morning session.

Session 5

Evolutionary Dynamics

Chair: *Daniel Fisher*, Stanford, USA
Rapporteurs: *Richard Neher*, Biozentrum, Basel, Switzerland and *Aleksandra Walczak*, ENS, Paris, France
Scientific secretaries: *Tom Lenaerts*, ULB/VUB, Belgium and *Remy Loris*, VUB, Belgium

D. Fisher I have not quite figured out yet whether it is an honor or punishment to chair the last session. But it is certainly appropriate that it is on evolution since that ties everything together. Dobzhansky's dictum that nothing in biology makes sense except in the light of evolution is certainly the most quoted in all of science. But it also seems to be the most ignored. There have been already discussions earlier in this meeting on trying to use evolution as an Occam's razor in biology and I think that is something we will come back to later in the general discussion. But this session is actually on evolutionary dynamics and I had a couple of slides just to start off.

So, the facts of evolution, at least until recently, have been entirely based on snapshots, or on phylogenies, with no dynamics being observed directly. The laws of evolution and the basic theory have been known for an enormously long time. The problem is that these laws are so general, and almost like string theory, in principle everything could follow. Once you have these laws anything can evolve. But we have no idea how long various things take and we also have no idea about the dynamics because most of the data does not contain that information. The dynamics we are asking about, and the timescales of evolution. In particular, one of the questions is how much depends on rare events and how much is pseudo-deterministic (I wouldn't use predictable) and of course the roles of the various processes in those.

To what extent do we not understand the numbers? Let us think about the rare events. So, if this room would be completely filled with bullshit, there would be about 10^{20} bacteria in the room. If the earth would have only 10^{20} bacteria in total, but there was a similar diversity in environments, would the evolution of bacteria become as diverse as now, and would they become as sophisticated in some ways? The actual number of bacteria on earth is more like 10^{30}. I would say that there is a factor of 10^{10} of uncertainty in our understanding of numbers in evolution. Even small numbers like how long it takes to evolve antibiotic resistance (which is a few years) we do not understand.

The other really big question seems to me is trying to understand diversity on all scales, and whether that is a general consequence of sufficient complexity or somehow depends on niches or other things we read about in textbooks. On that, unfortunately, we are not going to have much discussion. One of the big hopes for progress, of course, comes from DNA sequencing. We can do an awful lot with genomic evolution and the diversity there, but it is so much harder to get as much data at the level of phenotypes. And this is the opposite of what it was traditionally with evolution where everything was about phenotypes.

Before asking what we are trying to do here overall, I thought it was useful to make a simple analogy with physics. There are the basic laws, which for this level of physics is quantum electrodynamics, and Darwin and Mendel together constitute the corresponding laws for evolution. And then there are basic processes like a selective sweep or neutral variation which corresponds to population genetics. And single protein evolution is like simple molecules in physics. Then on the other end there is the fossil record, evolution of macro-organisms, the traditional evolutionary theory and that is really in some ways analogous to geology and geophysics. But what we are missing, is an enormous gap in the middle, namely things that are analogous to condensed matter physics. And we know that to get at anything from the fundamental laws level we need to understand basic processes up to understanding geology and geophysics and there are many layers in between. One of the important things one learns from condensed matter physics is that different people care about different things and have opinions on which things are important and which things constitute understanding. And it is really important to have that whole range and not to try to convince each other of what is important. I think, for example, that what we understand about "mechanisms" varies a lot from person to person, and that is healthy. So, I think that at this stage, we are trying to formulate good questions as we do not know what the good questions are. We are moving forward but how do we make good questions?

So this session is divided into two parts. The first is going to be on micro-evolution and things we can now follow the dynamics of. On this, I think we are developing a lot of understanding. After the break, we will deal more with things about macro-evolution and turning to bigger questions and then flow into the general open discussion section at the end.

Rapporteur Talks by Richard A. Neher and Aleksandra M. Walczak: Progress and open questions in evolutionary dynamics*

Evolution has fascinated quantitative and natural scientists for decades as a realization of a random process that generates a diversity of forms and structures. Historically, we seem to have come full circle: from collecting species, equating diversity with genetic diversity and the hope of sequencing, to realizing that while genomic information is invaluable it does not necessarily completely determine the evolutionary trajectories. We review theoretical models and data on genetic and species diversity has been gathered in the field, more recently using in lab evolution experiments, longitudinal sampling of simple systems, microbial communities and complex systems such as immune repertoires. One of the messages that emerges from all of these systems is that the evolution of a given organism does not occur in vacuum but is strongly driven by both its biological and physical environments, showing the limits of current theoretical approaches.

1. Introduction

Over the centuries, scientists have been cataloging the diversity of life, and discovered a hierarchical organization. This hierarchical structure prompted the idea that organisms evolve by passing on their phenotypic traits to their offspring, with possible modification — so-called *descent with modification*. On much shorter time scales, plant and animal breeding showed that phenotypic traits can be dramatically altered by consistently selecting the desired types. In the 20th century mathematical frameworks were developed to describe genetic diversity within species (population genetics and quantitative genetics) and between species (phylogenetics). However, until recently, molecular data to test specific predictions and assumptions have been lacking. The advent of relatively cheap high-throughput sequencing has changed this situation profoundly. The elegant theoretical models, however, often fail to describe the now abundant data sets in a quantitative manner.

These data sets are either observations of natural populations or are obtained from laboratory evolution with microbes or other rapidly reproducing organisms, e.g. flies. Laboratory evolution experiments with microbes have the great advantage over observational studies in animals or plants that population turnover and the spread of mutations happens on observable time scales. Such time dependent datasets are much more powerful at differentiating between models than snapshots of natural populations. The only natural 'measurably evolving' populations where observations beyond snapshots have been made use are pathogenic bacteria or RNA viruses (influenza virus and HIV) and certain cancer types. But time resolved data of commensal and environmental bacterial populations are starting to emerge.[1–3] Strong selection and rapid mutation are also at play during somatic hypermutation of B-cell receptors. Genetic diversity of entire microbial populations or immune

---

*Note from the editors:** This is a joint write up of the rapporteur talks by Richard Neher and Aleksandra Walczak.

receptors can be characterized by high-throughput sequencing and allows genome wide tracking of mutations at high resolution.

These experiments and observational studies have shown that microbes adapt rapidly to new environments, even if these environments are kept as simple and stable as possible. Furthermore, there are ample opportunities for adaptation and populations are typically diverse with many similar trait modifications competing against each other. Such rapid and diverse responses are at odds with classical population genetics theory, which assumes that adaptive changes are rare and occur sequentially with periods of stasis between so-called selective sweeps.

Furthermore, models typically assume an externally specified environment in which every member of the population competes with everybody else for a common resource. However, even in the simplest experiments that try to approximate this idealized model, the populations rapidly split into different types that feed on secondary metabolites or specialize in growth at different nutrient densities. In other words, ecology rapidly develops. Ecological theory, on the other hand, typically ignores evolution and assumes that species are static monomorphic entities that interact with each other.

Here, we review theoretical models of population genetics and discuss recent data that suggest that the assumptions of many established theoretical models are not met. We discuss recent theoretical developments that aim at addressing some of the conflicts, and present an overview of the wide open questions that arise when evolution and ecology meet.

2. Traditional Population Genetics

Nineteenth century scientists realized that the diversity of life is the product of heritability, mutation, and natural selection. Population geneticists and quantitative geneticists in the 20th century attempted to develop these qualitative ideas into a quantitative description of evolution. The models studied in the 20th century can be broadly classified into (i) deterministic dynamical systems, (ii) phenomenological quantitative genetics models, and (iii) stochastic models of frequencies of within species variants. On one hand, quantitative genetics models describe the response of diverse populations to the selection for particular phenotypes without reference to the genetic determinants of the traits.[4] Patterns of genetic diversity of variant frequencies, on the other hand, require probabilistic models of the processes that introduce and remove genetic variants.

The two prominent models for genetic diversity are the backwards in time coalescent[5] or a forward in time diffusion approach.[6] The Kimura diffusion equation describes the distribution $P(x,t)$ of the frequency x of a single genetic variant A subject to genetic drift and selection in a (haploid) population of size N. If individuals with and without variant A have on average $1 + s$ and 1 offspring, respectively, and the variance in offspring number is σ^2, the distribution of the variant frequency evolves according to

$$\frac{\partial P(x,t)}{dt} = \left[-s\frac{\partial}{\partial x}x(1-x) + \frac{\sigma^2}{2N}\frac{\partial^2}{\partial x^2}x(1-x) \right] P(x,t). \tag{1}$$

The first term on the right accounts for selection with strength s; the second term accounts for demographic stochasticity that results in non-heritable, undirected fluctuations in variant frequency. This stochastic contribution is known as *genetic drift*. Genetic drift with strength $\frac{\sigma^2}{N}$ describes the stochastic dynamics of variant frequencies only when fluctuations are uncorrelated in time (non-heritable) and across individuals. In most populations, however, many additional processes affect variant frequencies in an undirected manner but their effects are correlated over many generations. If these correlations are weak enough and fluctuations of allele frequencies have sufficiently fast decaying distributions, they can be accounted for by introducing an effective population size N_e. The latter is often orders of magnitudes smaller than estimates of the census population size. However, such correlations can qualitatively change the stochastic dynamics and different models outside the universality class of the Kimura diffusion equation are needed.. While formally possible, it is difficult and unhelpful to generalize the diffusion approach to complex populations with more than three mutations.

Another popular framework to analyze neutral diversity ($s = 0$ for all variants) is the Kingman coalescent, which models the genealogy of the population backward in time: Any pair of lineages merges at random with rate $k(k-1)\sigma^2/2N$, where k is the number of lineages remaining. A typical tree generated by the Kingman process in shown in Figure 2. The time to the most recent common ancestor (T_{MRCA}) of a large population is on average $2N/\sigma^2$, where σ^2 is again the variance in offspring number. The coalescent can be used to predict diversity at many genomic loci and is easy to simulate under a range of demographic models. Diffusion and coalescent descriptions are two dual ways of looking at the same processes, but some questions are more conveniently formulated in one framework than the other.

The basic coalescent model assumes that all individuals are equivalent, i.e., all variation is neutral and there is no population structure. These assumptions can be relaxed by introducing a structured coalescent where individuals have types that restrict who can merge with whom.[7] While Kimura's diffusion equation readily accommodates positive selection at a single locus, it is challenging to incorporate positive selection into the Kingman coalescent.[8]

A different framework that has been explored at some depth is that of a monomorphic population evolving in a static landscape that assigns fitness to all possible genotypes. This approach assumes that mutations are rare enough and fitness differences between neighboring genotypes large enough that the population is monomorphic most of the time. This dynamics results in a Boltzmann distribution with the inverse population size and fitness playing the role of energy and temperature, respectively.[9–11]

3. Confronting Theories with Data

While sequencing a single bacterial genome was a major undertaking 20 years ago, a small lab can now sequence hundreds of bacterial genomes in a few days. Thousands of animal, plant, or parasite genomes have been sequenced and observed patterns of genetic diversity in these populations have been compared to predictions of patterns of genetic diversity.[12] Using these data, scientists have estimated demographic parameters such that as the time to the most recent common ancestor, historical population size changes and migrations as well as patterns of conservation. The observed patterns are mostly broadly consistent with predictions in the sense that large populations tend to harbor more genetic diversity and that there are many more rare than common variants.[13–16]

However, essentially all sequence data from eukaryotes are static snapshots and our ability to learn dynamical properties from such data is limited. In the best case, inference from static data yields estimates of model parameters relative to an intrinsic time scale of the system, e.g. the above mentioned coalescent time scale (or effective population size). More often, static data is unable to differentiate between models and seemingly good fits of simple models hide the actual dynamics. Any inferences from static data are necessarily an average over history. Different ways of calculating expectation values of the same parameters average over different past intervals and can give contrasting results. For example, Bergland *et al.*[17] surveyed genetic variation in Drosophila populations over several annual cycles and found oscillations of large amplitudes suggesting strong selection pressures (10% or more) that vary with seasons. By contrast, inference from static data suggests selection coefficients of 1% or less.[18] This discrepancy is just one example of how inference from static data can lead to misleading results because the dynamics are not identifiable from static data alone.

To study evolutionary processes in real-time rather than inferring them from snapshots, high throughput sequencing technologies have been applied, for example, to the long term evolution experiment (LTEE) with E. coli conducted by Richard Lenski and colleagues,[19] the tracking of millions of lineages over short times intervals,[20] and sequencing of serially sampled HIV populations,[21] or the global surveillance of influenza viruses.[22]

Evolution experiments have the advantage over observational studies that the environments can be controlled and replicated, but the environments used to propagate the populations are typically simple and artificial. As a result, the majority of adaptive mutations that are observed are loss of function mutations, that is inactivation of proteins and pathways that are not necessary in the lab environment. This mode of adaptation is not necessarily representative of evolution in the wild.

Outside the lab, rapid evolution can be directly observed in populations of pathogenic RNA viruses such as HIV or influenza. These viruses continuously evade a co-evolving immune system. The Global Influenza Surveillance and Response System (GISRS)[22] collects thousands of influenza virus samples. The common ancestor of all circulating A/H3N2 viruses is typically only three years in the past and in any given year two randomly sampled HA sequences differ on average at ∼10 positions. This rapid turnover and diversification is driven by increasing human immunity against circulating viruses. New variants with different antigenic properties emerge and reinfect previously immune individuals.[23]

Similar dynamics can be observed in HIV populations in infected humans. Shortly after infection, the adaptive immune systems target the virus and escape mutations quickly spread.[24] Typically, a few strongly selected mutations evade cytotoxic T-lymphocyte responses spread during the first three months of infection, followed by a multitude of more weakly selected escape and reversion mutations. While segments of the influenza virus genome are propagated asexually, HIV recombines by crossing over (template switching of the reverse transcriptase).[25] Recombination is rapid enough that different regions of the genome show profoundly different dynamics. Surface proteins exhibit rapid turnover similar to global influenza virus dynamics, while the enzymes often slowly accumulate diversity over many years.[21,26]

Good *et al.*[27] recently sequenced 120 samples of each of the 12 lines of the LTEE that evolved in a constant environment for about 60,000 generations. They discovered that multiple lineages co-existed in these cultures for many years. Instead of being in a simple environment with a single fitness optimum, the bacteria partitioned the system into niches creating an environment in which multiple types can coexist.

These three examples, together with many other recent observations of microbial evolution, have demonstrated that diverse populations, in which many variants compete vigorously are the rule rather than the exception. Furthermore, the environments in which microbes evolve change, either because they coevolve with a host or because they generate their own ecology. A quantitative description of these populations requires theoretical frameworks beyond single locus diffusion theory, neutral coalescent models, or static fitness landscapes.

4. Traveling Waves Models of Rapid Adaptation

The earliest models of diverse populations under selection were developed to describe plant and animal breeding. In sexual populations with large genomes, quantitative traits typically depend on many loci resulting in an approximately Gaussian distribution of the trait in the population, see Figure 1a. This model is known as the infinitesimal model with many effectively unlinked loci. The response to selection will then be proportional to the trait variance and the heritability of the trait the breeder selects.[4]

Outside of plant or animal breeding programs, natural selection operates on fitness. Individuals at the high fitness end of the fitness distribution increase in frequency of lineage, while less fit lineages die out. The dynamics of the bulk of the fitness distribution can often be described by deterministic growth and shrinkage of lineages. This deterministic dynamics describes the short term response to selection,

(a)

$$v = \sigma^2 - \Delta_\mu$$

(b)

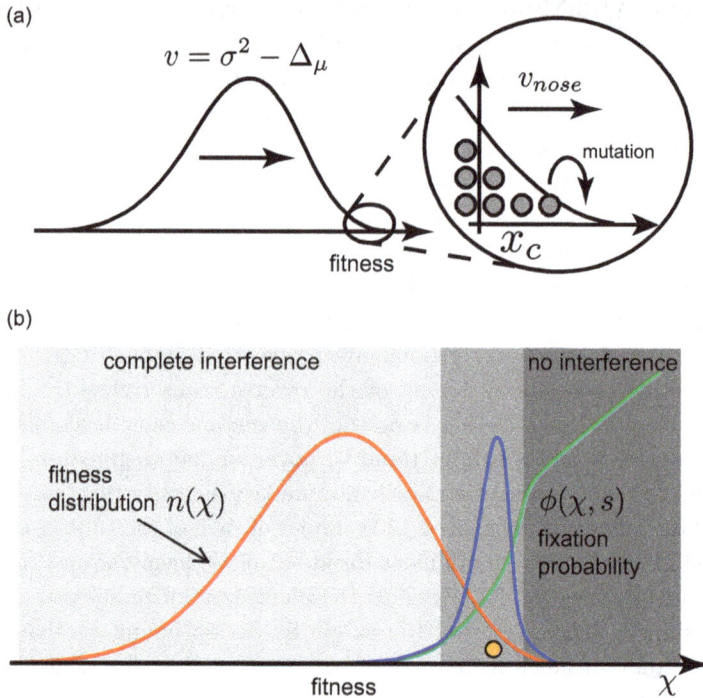

Fig. 1. **Traveling wave models**: (a) The mean fitness in the population increases with rate $\sigma^2 - \Delta_\mu$, where σ^2 is the variance in fitness and $\Delta_\mu = -u\langle s \rangle$ is the mutation load be generation. The fitness variance σ^2 is determined by matching the speed of the mean fitness to the speed at which the "nose" of the fitness distribution advances. The latter is determined by the stochastic dynamics of high fitness individuals. (b) The fixation probability (green line) is very small in the bulk of the fitness distribution. In contrast, fixation of a mutation that increases fitness of an individual beyond the nose of the distribution is almost as likely as if there was no competition. The product of the population fitness distribution and the fixation probability defines a narrow zone around x_c where the ancestors of future populations originate from.

but it cannot explain how fitness variation is maintained over longer times. De-novo mutations that contribute to future fitness variation arise in the high fitness tail of the distribution where only very few individuals reside and stochastic dynamics is important. Fluctuations and stochastic events among these high fitness individuals are amplified exponentially and dominate the bulk of the population after a delay. Almost all individuals that contribute to future populations originate from a narrow zone in the high fitness tail (see Figure 1). Hence a probabilistic description of the high fitness 'nose' is essential.

Tsimring *et al.*[28] studied this problem in a simple model of an adapting RNA virus population: Each individual suffers from many small effect mutations, resulting in a diffusive motion in fitness. The deterministic diffusion/selection equation describes the dynamics of the bulk of the distribution. The stochastic instability in the nose is handled by a cut-off beyond which exponential amplification does not operate. This ad-hoc modification of the model is justified by the idea that "fractional" individuals do not reproduce and no exponential amplification can take place in this part of the distribution. This modification results in a consistent solution that predicts that the speed at which the population adapts depends only logarithmically on the population's size.

Tsimring *et al.*[28] assumed frequent small mutations such that the fitness distribution always remains smooth and no large monomorphic clones exist. At steady state, the variance of the fitness distribution is approximately given by

$$\sigma^2 \sim D^{\frac{2}{3}} (\log ND^{\frac{1}{3}})^{\frac{1}{3}} \tag{2}$$

where $D = u\langle s^2 \rangle$, u is the total mutation rate, and $\langle s^2 \rangle$ is the second moment of the distribution of fitness effects of novel mutations. The rate of adaptation, i.e. the rate at which fitness increases, is then $v = \sigma^2 + u\langle s \rangle$ where $u\langle s \rangle$ is the net effect of mutations. The fitness increase will typically be offset by a co-evolving or environment.

If mutation rates are small, yet large enough that many variants compete, other models are a more natural starting point.[29–31] A commonly used model assumes that fitness can increase in discrete steps of size s (stair-case model) and the rate of such beneficial mutations is u. In this case, the speed of increases with N, u and s as

$$v \sim 2s^2 \frac{\log Ns}{(\log u/s)^2} . \tag{3}$$

In both models the speed of adaptation depends logarithmically on the population size and the nose of the wave is $\sim \sqrt{\log N}$ standard deviations above the mean. Newly arising mutations have a negligible chance of spreading through the population unless they emerge in the very tip of the fitness wave, see Figure 1b.

While the above models were restricted to asexual populations, traveling wave models have been generalized to sexual populations. In the limit of very rapid recombination (compared to the time scale of selection), different loci in the genome completely decouple and selection operates on individual alleles rather than individuals or genotypes.[32] Once the recombination rate drops below typical fitness differentials in the population, clonal subpopulations begin to form and the dynamics starts to resemble asexual traveling waves, albeit with much higher rates of adaptation.[33–36]

5. Genetic Diversity in Rapidly Adapting Populations

Most of the early studies of traveling wave models calculated the speed of adaptation, i.e., the rate at which populations accumulate beneficial variants. Adaptation, however, is difficult to measure and even when measured has limited power to differentiate models. Much more accessible and more informative are patterns of genetic diversity such as the site frequency spectrum (SFS). The SFS is the density of mutations found at different frequencies in the population. Different models make qualitatively different predictions for the SFS.

Since neutral mutations accumulate uniformly along lineages, neutral diversity is determined by the genealogical tree: the length of the branches is proportional to the number of segregating mutations and the branching pattern determines how common a mutation has become since its origin. The simplicity of the Kingman coalescent allows for analytic expressions for many neutral diversity statistics, including the SFS.

Recently, similar progress has been made for traveling wave models of adapting populations. Brunet *et al.*[37] showed that a class of models in which a population moves as a front in fitness space gives rise to a coalescent process known as the Bolthausen-Sznitman coalescent (BSC). This process is exchangeable, and emerges naturally if the number of offspring individuals that contribute to the next generation is drawn from a distribution that has a power-law tail $\sim n^{-2}$ (see Ref. 38). This same process was later shown to describe populations in traveling wave models of rapid adaptation.[39,40] A large class of models seems to converge to the same coalescent model after coarse-graining in time.[40]

The trees generated by the BSC differ qualitatively from those generated by Kingman coalescent trees (Figure 2). While Kingman trees tend to branch sym-

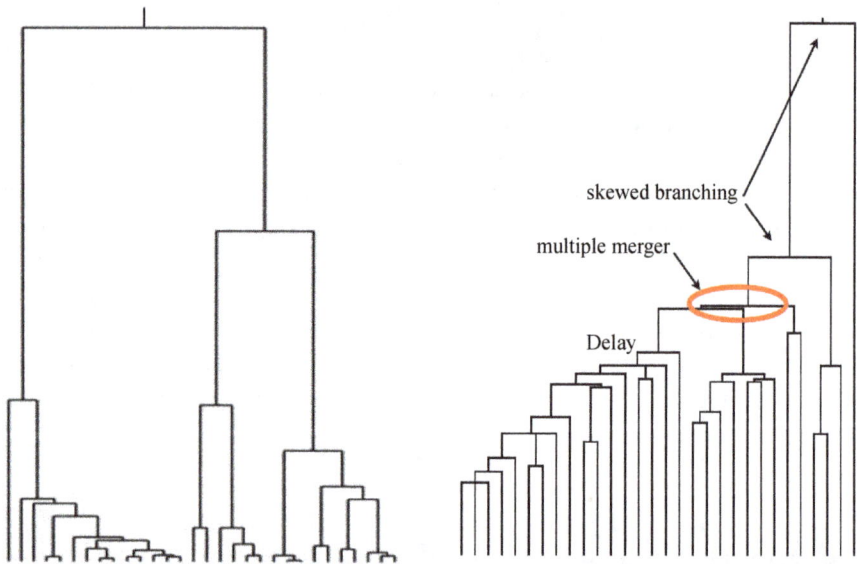

Fig. 2. **Coalescent trees in neutral and adapting populations**. The left panel shows a typical Kingman coalescent tree. Most merging happens close to the present and the remaining lineages tend to split the populations into similarly sized families. The tree on the right was sampled from a simulation of an adapting population. At first, little coalescence is happening until ancestral lineages have moved to the high fitness tail. Here, competition between exponentially growing clones results in approximate multiple mergers and skewed branching.

metrically (given k lineages, the distribution of leaves is uniform on the $k - 1$ dimensional simplex), BSC trees are often very skewed. In Kingman trees, most merging happens in the recent past since the merger rate is proportional to square of the number of lineages. In the BSC, coalescence happens deeper in time and often occurs in terms of multiple mergers, which correspond to exceptionally fit individuals in traveling wave models.[37,41]

How come genetic diversity in traveling waves of adaptation is described by an exchangeable, neutral process? The rapid exponential expansion of exceptionally fit clones occasionally results in one lineage taking over a macroscopic fraction of the population and gives rise to an $\sim n^{-2}$ tail of the distribution of the number of offspring after a characteristic time. Over this same time scale, the fitness of lineages competing against each other in the high fitness nose decorrelates. In essence, heritable variation in fitness (resulting in small systematic differences in offspring number) is exponentially amplified into heavy-tailed iid fluctuations on intermediate time scales.

In contrast to the SFS predicted by the Kingman coalescent, the SFS of rapidly adapting populations is non-monotonic with specific asymptotic behavior both for rare and common variants.[40,42] Qualitative features of the SFS predicted by the BSC have been observed in systems we expect to be under strong positive selection, for example HIV-1 populations of B-cell clones during somatic hypermutation.[21,43]

The correspondence between the BSC and models of adaptation requires a large number of mutations that contribute similarly to fitness of the organisms. More typically, a moderate number of mutations contribute to most fitness variation, which limits the quantitative agreement between model and data. Nonetheless, the BSC as a universal limiting case of rapidly adapting polymorphic populations is a very useful abstraction.

6. Repeatability and Predictability

In addition to comparing diversity to model predictions, reproducibility and predictability of evolutionary dynamics can be used to estimate parameters, test models, and interrogate the redundancies at different levels of the genotype-phenotype map. At the level of phenotypes the response to selection is extremely reproducible. Selection on quantitative traits in animals and plants, for example, invariably leads to a strong and consistent response. Anti-microbial resistance evolves rapidly. But reproducibility of the accompanying changes in the genotype vary. Strong very specific selection, for example for resistance to drugs with a well defined target, can lead to parallel mutations in the same nucleotide. This happens frequently in viruses like HIV. In bacteria, however, it is more common that drug resistance selection is reproducible at the level of the gene or the biochemical pathway. Toprak *et al.*,[44] for example, observed repeated mutations in the same genes of E. coli in response to antibiotic selection.

One of the biggest systematic studies of repeatability and consistency of molecular evolution was undertaken by Olivier Tenaillon and colleagues[45] who selected many initially identical *E. coli* populations to reproduce at elevated temperatures. This study revealed a striking parallelism in the genes and pathways that accumulated mutations, while repeatability at the individual nucleotide level remained limited. Only in cases with very specific selection pressures, like selection for drug resistance, or in organisms with very high mutation rates, are repeated mutations the norm.

While the repeatability of pathways hit by mutations has been observed in other instances of microbial adaptation, our ability to predict *a priori* the genomic responses to particular environmental challenges is essentially non-existing. Instead of predicting where and how adaptation can occur, one could settle for the more modest aim of predicting which of the existing genotypes will prevail and take over. In this case the challenge moves from predicting molecular changes to predicting the changing population composition — ideally from a one-time observation. Several such approaches have been developed to predict influenza viruses that likely dominate future influenza seasons.[46]

Łuksza and Lässig[47] developed a model for virus fitness based on scores of mutations that have been historically associated with antigenic novelty and for mutations that are expected to be detrimental, for example, because they destabilize protein structure. Neher *et al.*[41] used an alternative approach that infers fitness of different parts of the tree given the tree topology and branch length. This inference is based on a traveling wave fitness model in which lineages diversify and adapt by many small effective mutations. The distinguishing feature of the latter approach is that it only requires a single time point without any influenza specific input or historical data, while the former is easier to complement with additional biophysical or antigenic phenotypes. Both approaches predict the composition of future influenza populations with a much higher accuracy than choosing a random recent virus.

7. Complex and Variable Environments

So far we have assumed that populations adapt towards a well defined and fixed goal, for example rapid growth in evolution experiments or immune evasion in host-pathogen systems. More generally, however, the biotic and abiotic environment will change on many time scales and the extent to which lessons learned from studying systems in stable environment are applicable to natural situations is unknown. To address this question researchers studied the evolution of bacteria in the mammalian gut. The rate of molecular evolution was found to depend strongly on the bacterial strain: experiments conducted with a pathogenic hospital strain showed that adaptation to the mouse gut proceeds ∼5 times slower than in experiments starting with the lab adapted K-12 strain (isolated in the 1920s),[48] suggesting that clones with lower initial fitness to the given environment improve faster — a result that has also been observed in the lab. In analogy to in vitro experiments, evolution in the mouse gut exhibits convergence among replicate lineages, mainly at the phenotypic, pathway, and gene level.[48-50] While strains with mutator phenotypes have an initial advantage and adapt faster, they accumulate many secondary mutations that are deleterious after the environmental change and eventually loose out.[51]

Experiments in vitro as well as in the mouse gut suggest that drastic environmental changes often result in early mutations in global regulators. Such mutations are very common in early mutations in laboratory experiments but were not observed in the gut with a pathogenic hospital strain.[48] In experiments colonizing mice guts with lab adapted K12 strains such mutations do occur.[50]

Furthermore, interactions of the microbial population with the host are important. Molecular evolution of E. coli was found to be much slower in immune compromised mice than in healthy wildtype (WT) mice, even though the mutation rates of the bacteria are the same in both environments.[52]

Theoretical investigations of evolution in fluctuating environments have shown that environmental fluctuations can amplify demographic fluctuations, which results in more rapid coalescence, lower genetic diversity, and a higher probability that deleterious mutations fix compared to a constant environment.[53-55] Fluctuations in the environment translate into large fluctuations in the sizes of the different clones that make up the populations. For example, the long tailed clone size distributions observed in adaptive immune repertoires, measured as the abundances of the B or T-cell receptors in a given individual,[56,57] can be explained in terms of strong fluctuations in the pathogenic and self-protein environments the repertoire experiences, which translates into strong fitness fluctuations felt by each receptor clone.[58]

Organisms can develop strategies to mitigate the effects of fluctuating environments.[59,60] These strategies have been studied by considering a population of size $N_t(\sigma)$ at time t, where individuals are described by their phenotype or genotype σ. The environment changes in time between different phenotypic states denoted by x_t. In each generation individuals can switch their phenotype or keep their parents' but they cannot go extinct. Different phenotypes are better suited for different environments, with the best phenotype guaranteeing more offspring and larger fitness. Different phenotype switching strategies result in different long term population growth rates, $\Psi[x_{0:t}] = \ln \frac{\sum_\sigma N_t(\sigma)}{\sum_\sigma N_0(\sigma)}$, and the optimal strategy depends on the environmental fluctuation statistics.[59,61] In extended approaches, differentiating between phenotype and genotype leads to considering different modes of inheritance and sensing.[62] In all cases, the optimal strategies are strongly dependent on timescales of environmental fluctuations. For example, the diversity of vertebrate and microbial forms of immunity can be rationalized using this theoretical framework in terms of the differences between the timescales of the lifetime of the host compared to the pathogen: vertebrates typically have much longer timescales than their pathogens, while microbe lifetimes are similar to those of their invaders.[63] The different immune strategies (such as CRISPR-like vs bet-hedging or adaptive vs innate immunity) then would result from differences in the frequencies of the invading pathogens.

Lastly, it is also worth noting that these frameworks based on considering long term fitness make direct links with stochastic thermodynamics and information flow.[60,64–67] The long term fitness can be decomposed into ratios of probability distributions that correspond to knowing the environment and the uncertainly coming from the environment. While these terms are very hard to evaluate in detail since they account for path integrals over random trajectories, they correspond to well known fluctuation relations in stochastic thermodynamics, such as the Jarzynski equality.[68–70] Despite certain attempts to link with experiments,[71,72] this body of work remains mainly theoretical, accentuating the rare event nature of evolution and connecting evolution to ideas in sensing and information flow at the formal level.

8. Fitness Landscapes and Constraints on Diversity

At any given time, the paths along which a population can adapt depend on the environment and the organism's biology which is a product of billions of years of evolution. The effect of individual mutations, as well as which evolutionary paths are accessible, are therefore contingent on this history. Since it is easier to destroy protein function or a working pathway than to improve function, the majority of mutations in functional regions of a genome are expected to be deleterious. Historically, fitness landscapes have often been depicted in two dimensions as shown in Figure 3a. These illustrations serve as a useful metaphor, but do not capture the structure of the genotype to phenotype map. Genotype-phenotype maps can also be represented on a fitness graph hypercube with edges corresponding to the nucleotides or amino acids at a particular position. Each pair of sequence that differ by one point mutation are connected by an arrow that points to the genotype of higher fitness. The peaks of the fitness landscape can be identified as sequences with only incoming arrows and are often highlighted in color. One such hypercube for four specific mutations in the fungus Aspergillus nigris[73] is illustrated in Figure 3b.

High-throughput technologies have revolutionized the way scientists can interrogate the effects of mutations and large scale comparisons between organisms, conditions, and proteins are now possible. Deep mutational scanning (DMS) experiments[74] generate many possible mutants and measure their relative performance with respect to a specific function (e.g. binding of a ligand, enzymatic activity, stability), linking genotype to phenotype. An example of the output of a DMS experiment that represents amino acid preferences at specific positions is shown in Figure 3c.

In some cases, the effects of specific mutations in chosen environments are of intrinsic interest. Specific mutational effects can inform molecular biology investigations, treatment strategies, and interventions. In other cases, the focus is on statistical properties of fitness landscapes that might be comparable across environments and organisms.

(a) Metaphorical fitness landscape

(b) Fitness hypercube

(c) Deep Mutational Scanning

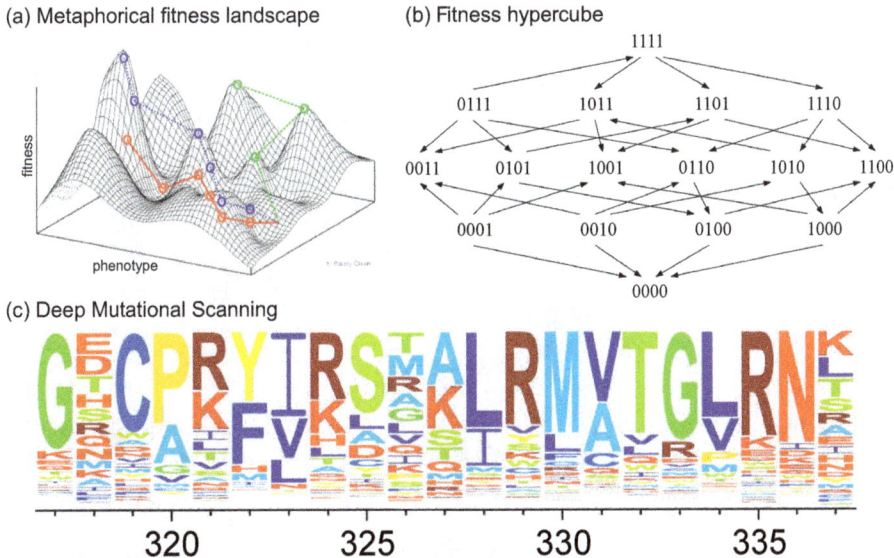

Fig. 3. **Fitness Landscapes**: (a) Fitness landscapes are often drawn in two dimensions where axes represent genotype or phenotype in some abstract away. Mutational paths, illustrated by colored lines, tend to maximize fitness. Depictions of this nature are more metaphorical than faithful descriptions of reality (modified from Randy Olsen, wikipedia). (b) Fitness graph hypercubes offer a more informative description of the genotype-phenotype map. A four-dimensional hypercube of a fitness landscape in aspergillus niger (reproduced from Ref. 73). (c) Deep mutational scanning (DMS) experiments quantify amino acid preferences at each site. Rather than a complete landscape of all combinations of mutations, DMS explores the landscape around a wildtype sequence.

9. Lessons from Proteins

Not all mutational paths in a fitness landscape are possible: some mutants are dead-ends of evolution. This observation is true across all evolutionary scales, from the evolution of proteins and regulatory networks to the evolution of large organisms. At the smallest scale, proteins have proven very informative in identifying the constraints stemming from function and molecular physics. From protein families — statistically meaningful ensembles of proteins coming from different organisms that can be aligned to each other since they share an evolutionary history — we can learn how protein structure and function constrain evolution. By analyzing alignments of these families we see that mutations at certain sites are more common than at others (Figure 4). Additionally mutations at certain sites are correlated: mutating one amino acid at a certain position may always be accompanied by a mutation at another position.[75] In certain protein families correlations between amino acids in their sequences prove extremely important for their structure and function: a substantial fraction of artificially engineered proteins that constrain the same pairwise correlations as naturally occurring WW domains were shown to fold and have a similar function to WW domains found in nature.[76,77]

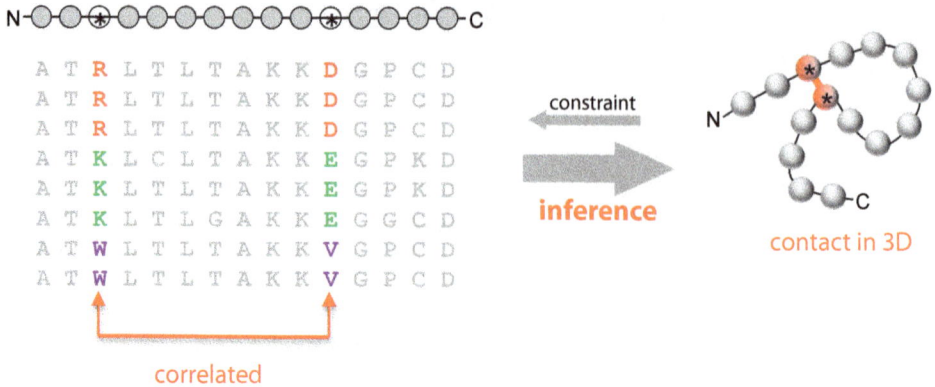

Fig. 4.   Interacting residues in a protein result in correlated amino-acid substitutions and hence correlated alignment columns. Such correlations in large alignments have been used to infer protein structure and interaction surfaces. (Figure from Ref. 89).

These promising results started a whole field of understanding to what degree contacts between amino acids and protein structure can by inferred from pairwise correlations, $C_{ij}^{ab} = f_{ij}^{ab} - f_i^a f_j^b$, between the occurence of amino acid a at position i and amino acid b at position j.[78–81] This point f_i^a and pairwise f_{ij}^{ab} frequencies can be estimated from data, or learned from a model that describes the probability of the whole protein sequence $\vec{\sigma}$, $P(\vec{\sigma})$. It turns out that a division into strong direct interactions and weak collective interactions successfully predicts the fitness of all non-viable mutants[81] in the antibiotic resistance conferring bacterial beta-lactamase TEM-1. Overall, these models and other analysis[82–85] of ordered proteins show that protein stability mediated by amino acid interactions is an important determinant of fitness.

A model that maximizes the entropy, $S = -\sum_{\vec{\sigma}} P(\vec{\sigma}) \log_2 P(\vec{\sigma})$ of the possible protein sequences, σ, while constraining the experimentally measured correlation matrix, C_{ij}^{ab}, and imposing normalization of the probability to see a given protein sequence, $\sum_{\vec{\sigma}} P(\vec{\sigma}) = 1$ leads to a prediction of the form of the probability to observe a given protein sequence, $P(\vec{\sigma}) \sim \exp[\sum_{i,j,a,b} J_{ij}(\sigma_i^a, \sigma_j^b)]$. σ_i^a denotes amino acid a at position i and $J_{ij}(\sigma_i^a, \sigma_j^b)$ is the 20×20 matrix encoding the interactions between all possible amino acids at positions i and j. This framework is based on pairwise localized interactions. An alternative approach has also been developed based on a delocalized spectral decomposition of the correlation matrix weighted by positional conservation, $\tilde{C}_{ij}^{ab} = \sum_k |k\rangle \lambda_k \langle k|$, into modes $|k\rangle$ termed "sectors" ordered by their eigenvalues λ_k.[86,87] Positions contributing to dominant "sectors" are contiguous in the tertiary structure, and show common biophysical and biochemical properties and form functional units with independent patterns of sequence divergence. These delocalized and localized frameworks have been unified by showing that they are both limiting cases of a more general decomposition in terms of a Hopfield-Potts model, $P(\vec{\sigma}) = \frac{1}{Z} \exp\left[-\sum_{ij} \left(\sum_{\mu=1}^{K} \lambda_\mu \xi_i^\mu \xi_j^\mu\right) \sigma_i \sigma_j\right]$, where ξ_i^μ correspond to a set of delocalized modes that can be mapped onto sectors, while their summation reproduces the effective interaction matrix.[87,88] For a full description of protein sequence diversity both small and large scales are needed.

Knowing that correlations are important motivated researchers to map out the mutational landscape and link specific mutations, and their interactions to phenotypic variables, such as the fitness of the *E. coli* bacterium carrying a given beta-lactamase TEM-1 mutant,[81] or the affinity of antibodies — proteins involved in an immune response — against a specific antigen.[90] These studies, which are examples of deep mutational scans, find that most mutants are detrimental — they reduce the fitness or binding affinity compared to the wild-type protein.

10. Evolutionary Paths

Deep mutational scanning experiments probe the neighborhood of a particular wild-type genotype, but typically have little to say about plausible evolutionary paths that are several mutations in length because of epistatic interactions between successive mutations. Weinreich *et al.*[91] addressed this question by constructing all possible intermediate mutants between inhibitor sensitive and resistant variants of the TEM-1 beta-lactamase. Five mutations in this enzyme are enough to increase resistance by ~ 100000 fold compared to the wild-type (WT) TEM-1. The authors constructed all $5! = 120$ mutational paths between the five-point mutant and the WT. Out of the 120, only 18 were found not to lead through strongly deleterious genotypes and most of these 18 paths were unlikely — half of the weight of all possible evolutionary trajectories followed two paths. Additionally, the order of mutations was important for their effect: mutations that were beneficial in one mutational background become deleterious in another background.

The experiment by Weinreich *et al.* probed paths between *a priori* known successful endpoints. Possibly more representative evolutionary trajectories are generated by repeated diversification and selection. Similar to DMS, directed evolution experiments generate a diverse library of genotypes and select the best molecules for a given function, such as binding a specific ligand.[74] The pool of sequences that emerges as "winners" can be analyzed to identify successful mutations. This artificial selection process is very efficient, but it is unclear whether this combination of very high mutation and strong selection results in similar trajectories as evolution in the wild.

Affinity maturation in B-cells — important actors of the immune system that recognize pathogens — is a natural process that is similar to directed evolution experiments. Affinity maturation starts from a diverse source of immune cells that get selected for their ability to recognize an invading pathogen. The cells that bind more strongly to the pathogenic molecules proliferate faster, undergoing Darwinian evolution: the offspring cells acquire somatic mutations, which then get selected upon in the next round. This way, affinity maturation consists of several rounds of mutations and selection, forming evolutionary lineages just like viruses or microbes, and the affinity of B-cells to pathogenic molecules increases 10–100 fold.[92] At the end, a fraction of the evolved cells are kept in a subcompartment of the immune system called the memory repertoire, which responds rapidly upon re-infection by a similar pathogen, conferring fast protection. Recent experiments have shown that when B-cells are stimulated with a synthetic molecule, a hapten, affinity maturation looks very much like the directed evolution experiment: the B-cell binding affinity for the hapten increases and the diversity of B-cells decreases during the process. However, when B-cells are stimulated with pathogen derived molecules, binding affinity still increases but diversity remains high.[93] Interestingly, the cells that are kept in the memory pool at the end of affinity maturation are not necessarily the best binders. So it seems that the real life "directed evolution" experiment is more complex than the in-lab version.

Does evolution also have a well defined target like in the directed evolution experiments? Bloom and Arnold allowed the cytochrome P450 enzyme to accumulate mutations that were neutral for one specific substrate.[94] When they tested the activity of the mutant enzymes to five different substrates they found that some of these mutants showed up to four-fold higher activity to the new substrates than to the original one. The enzyme evolved to what is known as cross-reactivity and is also an important feature of how the adaptive immune system works — one immune cell receptor can recognize many pathogenic molecules and one pathogenic molecule can be recognized by many receptors.

11. Co-evolution

The TEM-1 experiment showed that evolutionary paths within isolated proteins are strongly constrained. Additionally, evolution occurs in the presence of other interaction partners, constraining the fitness landscape of interest by other fitness landscapes. A number of studies describe how two mutating partners influence each other's evolutionary paths, creating a mutual fitness landscape.[95–97] Anderson *et al.* reconstructed all possible paths in an ancient transition between two hormone ligands and two binding sites, closed evolutionary paths were opened by permissive mutations occurring in the partner. These mutations allowed the first partner to tolerate mutations that would otherwise be deleterious.[97] Similarly, certain paths were closed by co-evolution. The evolutionary history of the pair depends strongly on the partner's history and shows that co-evolution has strong implications on the space of evolutionary paths, in turn, are trajectories conditional on their surroundings.

The hormone-binding site study shows an example of the positive effects of co-evolution opening otherwise certain forbidden paths. However, co-evolution can also involve more than two interacting partners that infer strong negative selection pressures upon each other. This is the case in the co-evolution between pathogens and the adaptive immune system. The adaptive immune system consists of a diverse set of B and T-cells that are endowed with specialized receptors able to bind pathogenic molecules, successfully recognize them as foreign to the host organism and trigger an immune response aimed at eliminating them. The diversity of different cells is needed, given the different pathogens that a host may encounter. However, as the immune system recognizes pathogens, it exerts pressure on the pathogens that mutate and escape the immune system. In turn this forces the immune system to evolve and chase the pathogens, resulting in an arms-race like co-evolutionary process. The evolution of broadly neutralizing anti-bodies against HIV is a good example of this process. [98]

It is very hard to sample all the possible pathogens that we are exposed to, but thanks to recent high-throughput sequencing experiments we can now sample a significant fraction of the immune receptors in a given organism. [99–105] These experiments combined with an advanced statistical analysis show that non-related healthy people that do not necessarily share the same lifestyle, share as many unique immune receptors as expected by chance, if we use the structure of the repertoire for the random estimate. [106,107] Outliers from these estimates can be linked to functional selection pressures. [108] Even in such diverse ensembles we can thus identify reproducible patterns and ask how is diversity constrained, given an appropriate statistical analysis.

12. Ecology

Each species evolves in a dynamic environment shaped not only by abiotic factors but also by other species. Thanks to the advances in sequencing technologies, [109] we can now sample the diversity of microbial communities in the wild, on fermented food or cheese, and in various bodily orifices. We can first ask about the diversity of species living in a given community. The study of different cheese rinds reveals that most of the large bacterial diversity originates from environmental sources, such as the air, as opposed to starter cultures, and that abiotic factors (e.g. salinity, pressure, moisture) shape the community composition. [110] The environment favors certain strains and the growth of these strains changes the environment for the community. Theoretical descriptions of such interactions go back to the 1970s [111] and are described by generalized consumer-resource, or Lotka-Volterra like models. The differences in the assumptions matter, but in general, one considers the change in time of a subpopulation of size N_i, due to effective growth $f_i(\vec{N})$, community interactions with other substrains and itself, α_{ij}, and noise, $\xi(t)$: $\partial_t N_i = N_i(f_i(\vec{N}) - \sum_j \alpha_{ij} N_j) + \xi_i(t)$. Both the effective growth and community interaction terms can additionally depend directly on environmental factors. These approaches allow us to ask questions about timescales, the role of interactions and the environment, and have produced a very large body of work. [112–114] However until recently [115,116] they remained mostly disconnected from experiments.

Characterizing diversity gives us a static picture of a community. We can also ask how diversity evolves. From studies of the marine plankton bacterium Prochlorococcus we learn that the changes in the environment can be transmitted to the community through changes in the metabolism of its members.[117] Prochlorococcus lives at different depths in the ocean and its subpopulations have adapted to the levels of light and nutrients available at each immersion depth. The whole population responds to modifications in the external environment, such as light exposure, by a niche constructing a chain of adaptation. New ecotypes change their metabolism, chemically modify their environment, which exerts pressure on their co-inhabitants and globally change the diversity of the whole ecosystem[118]. Theoretical studies of interacting Lotka-Volterra-like models with random interactions show that these systems have a limited capacity for how many different species they can support[111] and diversity is predicted to exactly surf the allowed bound[119]: a community with initially more species will decrease the number of species until the maximum allowed number survive. This diversity constraint means the system is marginally stable, which is known in spin glass physics to correspond to a highly rugged landscape with many minima. Even small perturbations of the system will lead to large, nonlinear rearrangements of the landscape and a new community composition. These experimental and theoretical results open up a series of questions about the timescales for community rearrangements, their dynamics and whether the communities we observe are stable or transient.[120] How should we interpret the fact that random interaction models often give reasonable expectations?

The adaptation of Prochlorococcus to different environmental niches is also visible at the level of genetic variation. The genetic diversity is huge: in 1ml of water there are hundreds subpopulations and the sequencing of each new organism adds 160 new genes, which corresponds to $\sim$6–8% of the known pan genome.[117] The genetic variation can be linked to niche adaptation. On top of the backbone genome adapted to light, nutrient and temperature variability and shared between organisms in a given niche, each bacterium has many flexible genes, resulting in this very large genomic diversity. Despite large population sizes of $N = 10^{13}$, a relatively average mutation rate $\mu = 10^{-7}$ but extremely strong environmental selection, no selective sweeps have been reported and the diversity is huge. These observations defy the predictions of traditional population genetics. Why is there so much diversity given the strong selection pressures? Many ideas have been put forward, including co-evolution with predators (phages), recombination, which is extremely common in bacteria in the wild, the potentially complicated role of selective environments, as well as the role of the population structure, which suggest that we should study the whole federation of organisms living in a given ecosystem as selectable units. At this point the list of possibilities means we do yet have an answer.

Cyanobacteria are part of the Prochlorococcus federation. But they also live in hotsprings, where similarly to Prochlorococcus they show huge genetic diversity despite strong selection.[121] However, unlike Prochlorococcus, which despite the within-niche diversity does show a niche structure, these cyanobacterial biofilm genomes form a freely recombining gene pool, where allele frequency correlations between individuals decay on very small length scales along the genome. Selection seems to structure the genomes on short genomic scales, while recombination dominates on longer scales. Only an advanced statistical analysis combined with appropriate sequencing experiments was able to reveal this picture of a community without coherent strain that correspond to genome-wide stable ecotypes.

Analysis of E. Coli genomes in the wild find even more interesting signatures of the interplay of the underlying evolutionary forces with the surrounding habitat. These studies leave us with the impression that simply gathering data will not answer the question about the structure of diversity and the relevance of different forces. Only analysis methods driven by a theoretical understanding of the evolutionary processes can help us make sense of the vast datasets and understand the structure of microbial communities.

13. Conclusions

Our understanding of short term evolution in controlled or stereotypic environments has improved dramatically over the past decades. At the same time, quantitative high-throughput methods to survey fitness landscapes and eco-evolutionary dynamics of populations expose the complexity of biology and highlight the short-comings of the theoretical models of evolution.

Detailed studies of evolution in microbial populations[21,27,45] are first steps towards a more comprehensive understanding of rapid adaptive evolution and we have obtained partial answers to questions like (i) what are the relevant parameters, (ii) how gradual is evolution, (iii) and in what sense and to what degree can evolution be reproducible or predictable.[122-124] Whether these insights extrapolate to other systems and to different temporal or spatial scale is unclear and doubtful.

In addition to evolutionary dynamics, we have made substantial progress in understanding how proteins or molecular circuits evolve to acquire new functions and what the constraints on diversification are. Deep mutational scanning experiments[74] allow high-throughput characterization of protein function landscapes, directed evolution experiments have shown how proteins can acquire new functions, and sequencing of immune receptor repertoires has shown how the vertebrate immune systems prepares for future challenges and optimize response against recurrent challenges. Beyond individual proteins, however, predicting or even characterizing the perturbations remains challenging beyond the crudest of perturbations (e.g. transposon sequencing know-out libraries[125]).

Genetic diversity is easy to quantify these days by high throughput sequencing. But interpreting this diversity is difficult since (i) it remains difficult to associate genetic diversity with function and phenotype, (ii) most data are static, and (iii) limited to a minority of the interacting entities.

Recent metagenomic studies quantifying diversity, composition, and dynamics of microbial ecosystems are just scratching the surface and we have little understanding of the rules governing such communities. We tried to give examples and show how diversity is observed on many scales and also evolution acts across scales, from the molecular, through the genomic and organismal to the ecological. Despite a lot of experimental data and different theoretical approaches that lead to fascinating observations we still lack a unified framework for understanding diversity. Clearly, as evidenced by microbial communities, the events on one scale influence the observed diversity at another scale. Do we need to develop a detailed understanding on all scales to describe a given phenomenon or can we understand one scale without considering the others? Is there a well defined separation of scales when describing diversity, is everything deeply interconnected or is there are cascade of scales where one scale feeds into the next? In addition to a lack of understanding of the relevant organizational scales, we know little about the relevant time scales. Is the diversity we observe stable or inherently transient? Is change driven by properties of the dynamical system, or are systems adiabatically coupled to changing environments?

In summary, we have developed a reasonable understanding of dynamics on short time scales in systems where the "objective" of the evolving population is clear and static, e.g. the global influenza virus population. But our ability to generalize and extrapolate to longer time scales, more complex environments or ecosystems is limited to absent.

Acknowledgment

We are grateful to Benjamin Good and Thierry Mora for critical reading of the manuscript and providing valuable feedback.

References

1. N. R. Garud, B. H. Good, O. Hallatschek and K. S. Pollard, Evolutionary dynamics of bacteria in the gut microbiome within and across hosts, *bioRxiv*, p. 210955 (2017).
2. M. L. Bendall, S. L. Stevens, L.-K. Chan, S. Malfatti, P. Schwientek, J. Tremblay, W. Schackwitz, J. Martin, A. Pati, B. Bushnell, J. Froula, D. Kang, S. G. Tringe, S. Bertilsson, M. A. Moran, A. Shade, R. J. Newton, K. D. McMahon and R. R. Malmstrom, Genome-wide selective sweeps and gene-specific sweeps in natural bacterial populations, *The ISME Journal* **10**, 1589 (2016).
3. S. Zhao, T. D. Lieberman, M. Poyet, M. Groussin, S. M. Gibbons, R. J. Xavier and E. J. Alm, Adaptive evolution within the gut microbiome of individual people, *bioRxiv*, p. 208009 (2017).
4. D. S. Falconer, *Introduction to Quantitative Genetics* (Textbook Publishers, 2003), Google-Books-ID: R052PwAACAAJ.
5. J. Kingman, On the genealogy of large populations, *Journal of Applied Probability* **19A**, 27 (1982).
6. M. Kimura, Stochastic processes and distribution of gene frequencies under natural selection, *Cold Spring Harb Symp Quant Biol* **20**, 33 (1955).
7. R. R. Hudson, Gene genealogies and the coalescent process., *Oxford Surveys in Evolutionary Biology* **7**, 1 (1990).
8. S. M. Krone and C. Neuhauser, Ancestral processes with selection, *Theoretical population biology* **51**, 210 (1997).

9. Y. Iwasa, Free fitness that always increases in evolution, *Journal of Theoretical Biology* **135**, 265 (1988).

10. J. Berg, S. Willmann and M. Lässig, Adaptive evolution of transcription factor binding sites, *BMC Evolutionary Biology* **4**, p. 42 (2004).

11. G. Sella and A. E. Hirsh, The application of statistical physics to evolutionary biology, *Proc. Natl. Acad. Sci. USA* **102**, 9541 (2005).

12. R. B. Corbett-Detig, D. L. Hartl and T. B. Sackton, Natural selection constrains neutral diversity across a wide range of species, *PLoS Biology* **13**, p. e1002112 (2015).

13. H. Li and R. Durbin, Inference of human population history from individual whole-genome sequences, *Nature* **475**, 493 (2011).

14. S. Schiffels and R. Durbin, Inferring human population size and separation history from multiple genome sequences, *Nature Genetics* **46**, 919 (2014).

15. D. B. Weissman and O. Hallatschek, Minimal-assumption inference from population-genomic data, *eLife* **6** (2017).

16. R. N. Gutenkunst, R. D. Hernandez, S. H. Williamson and C. D. Bustamante, Inferring the joint demographic history of multiple populations from multidimensional snp frequency data, *PLoS Genetics* **5**, p. e1000695 (2009).

17. A. O. Bergland, E. L. Behrman, K. R. O'Brien, P. S. Schmidt and D. A. Petrov, Genomic evidence of rapid and stable adaptive oscillations over seasonal time scales in drosophila, *PLoS Genetics* **10**, p. e1004775 (2014).

18. G. Sella, D. A. Petrov, M. Przeworski and P. Andolfatto, Pervasive natural selection in the drosophila genome?, *PLoS genetics* **5**, p. e1000495 (2009).

19. J. E. Barrick, D. S. Yu, S. H. Yoon, H. Jeong, T. K. Oh, D. Schneider, R. E. Lenski and J. F. Kim, Genome evolution and adaptation in a long-term experiment with Escherichia coli, *Nature* **461**, 1243 (October 2009).

20. S. F. Levy, J. R. Blundell, S. Venkataram, D. A. Petrov, D. S. Fisher and G. Sherlock, Quantitative evolutionary dynamics using high-resolution lineage tracking, *Nature* **519**, p. 181 (2015).

21. F. Zanini, J. Brodin, L. Thebo, C. Lanz, G. Bratt, J. Albert and R. A. Neher, Population genomics of intrapatient HIV-1 evolution, *eLife Sciences* **4**, p. e11282 (2016).

22. WHO | Global Influenza Surveillance and Response System (GISRS).

23. V. N. Petrova and C. A. Russell, The evolution of seasonal influenza viruses, *Nature Reviews Microbiology* **advance online publication** (2017).

24. V. V. Ganusov, N. Goonetilleke, M. K. P. Liu, G. Ferrari, G. M. Shaw, A. J. McMichael, P. Borrow, B. T. Korber and A. S. Perelson, Fitness costs and diversity of the cytotoxic T lymphocyte (CTL) response determine the rate of CTL escape during acute and chronic phases of HIV infection, *J. Virol.* **85**(20): 10518–10528 (2011).

25. D. N. Levy, G. M. Aldrovandi, O. Kutsch and G. M. Shaw, Dynamics of hiv-1 recombination in its natural target cells, *Proc. Natl. Acad. Sci.* **101**, 4204 (2004).

26. R. Shankarappa, J. B. Margolick, S. J. Gange, A. G. Rodrigo, D. Upchurch, H. Farzadegan, P. Gupta, C. R. Rinaldo, G. H. Learn, X. He, X.-L. Huang and J. I. Mullins, Consistent viral evolutionary changes associated with the progression of human immunodeficiency virus type 1 infection, *Journal of Virology* **73**, 10489 (1999).

27. B. H. Good, M. J. McDonald, J. E. Barrick, R. E. Lenski and M. M. Desai, The dynamics of molecular evolution over 60,000 generations, *Nature* **advance online publication** (2017).

28. L. Tsimring, H. Levine and D. Kessler, RNA virus evolution via a fitness-space model, *Phys. Rev. Lett.* **76**, 4440 (1996).
29. I. M. Rouzine, J. Wakeley and J. M. Coffin, The solitary wave of asexual evolution, *Proc. Natl. Acad. Sci.* **100**, 587 (2003).
30. M. M. Desai and D. S. Fisher, Beneficial mutation selection balance and the effect of linkage on positive selection, *Genetics* **176**, 1759 (2007).
31. B. H. Good, I. M. Rouzine, D. J. Balick, O. Hallatschek and M. M. Desai, Distribution of fixed beneficial mutations and the rate of adaptation in asexual populations, *Proc. Natl. Acad. Sci.* **109**, 4950 (2012).
32. R. Neher and B. Shraiman, Statistical genetics and evolution of quantitative traits, *Rev. Mod. Phys.* **83**, 1283 (2011).
33. R. A. Neher, B. I. Shraiman and D. S. Fisher, Rate of adaptation in large sexual populations, *Genetics* **184**, 467 (2010).
34. R. A. Neher, T. A. Kessinger and B. I. Shraiman, Coalescence and genetic diversity in sexual populations under selection, *Proc. Natl. Acad. Sci. USA* **110**, 15836 (2013).
35. D. B. Weissman and N. H. Barton, Limits to the Rate of adaptive substitution in sexual populations, *PLOS Genetics* **8**, p. e1002740 (2012).
36. E. Cohen, D. A. Kessler and H. Levine, Recombination dramatically speeds up evolution of finite populations, *Physical Review Letters* **94**, p. 098102 (2005).
37. E. Brunet, B. Derrida, A. H. Mueller and S. Munier, Phenomenological theory giving the full statistics of the position of fluctuating pulled fronts, *Physical Review E, Statistical, Nonlinear, and Soft Matter Physics* **73**, p. 056126 (2006).
38. J. Schweinsberg, Coalescent processes obtained from supercritical Galton-Watson processes, *Stochastic Processes and their Applications* **106**, 107 (2003).
39. M. M. Desai, A. M. Walczak and D. S. Fisher, Genetic diversity and the structure of genealogies in rapidly adapting populations, *Genetics* (2012).
40. R. A. Neher and O. Hallatschek, Genealogies of rapidly adapting populations, *Proc. Natl. Acad. Sci.* **110**, 437 (2013).
41. R. A. Neher, C. A. Russell and B. I. Shraiman, Predicting evolution from the shape of genealogical trees, *eLife* **3** (2014).
42. K. Kosheleva and M. M. Desai, The dynamics of genetic draft in rapidly adapting populations, *Genetics* **195**, 1007 (2013).
43. F. Horns, C. Vollmers, C. L. Dekker and S. R. Quake, Signatures of selection in the human antibody repertoire: Selective sweeps, competing subclones, and neutral drift, *bioRxiv*, p. 145052 (2017).
44. E. Toprak, A. Veres, J.-B. Michel, R. Chait, D. L. Hartl and R. Kishony, Evolutionary paths to antibiotic resistance under dynamically sustained drug selection, *Nat. Genet.* **44**, 101 (2012).
45. O. Tenaillon, A. Rodríguez-Verdugo, R. L. Gaut, P. McDonald, A. F. Bennett, A. D. Long and B. S. Gaut, The molecular diversity of adaptive convergence, *Science* **335**, 457 (2012).
46. D. H. Morris, K. M. Gostic, S. Pompei, T. Bedford, M. Łuksza, R. A. Neher, B. T. Grenfell, M. Lässig and J. W. McCauley, Predictive modeling of influenza shows the promise of applied evolutionary biology, *Trends in Microbiology* (2017).
47. M. Łuksza and M. Lässig, A predictive fitness model for influenza, *Nature* **507**, p. 57 (2014).
48. M. Lescat, A. Launay, M. Ghalayini, M. Magnan, J. Glodt, C. Pintard, S. Dion, E. Denamur and O. Tenaillon, Using long-term experimental evolution to uncover the patterns and determinants of molecular evolution of an Escherichia coli natural isolate in the streptomycin-treated mouse gut, *Molecular Ecology* **26**, 1802 (2017).

49. J. Barroso-Batista, A. Sousa, M. Lourenço, M. L. Bergman, D. Sobral, J. Demengeot, K. B. Xavier and I. Gordo, The first steps of adaptation of escherichia coli to the gut are dominated by soft sweeps, *PLoS Genetics* **10** (2014).

50. J. M. Barroso Batista, Adaptation of E. Coli to the mouse gut, PhD thesis, 2016.

51. A. Giraud, Costs and benefits of high mutation rates: adaptive evolution of bacteria in the mouse gut, *Science* **291**, 2606 (2001).

52. J. Barroso-Batista, J. Demengeot and I. Gordo, Adaptive immunity increases the pace and predictability of evolutionary change in commensal gut bacteria, *Nature Communications* **6**, p. 8945 (2015).

53. V. Mustonen and M. Lässig, From fitness landscapes to seascapes: non-equilibrium dynamics of selection and adaptation, *Trends in Genetics* **25**, 111 (2009).

54. I. Cvijović, B. H. Good, E. R. Jerison and M. M. Desai, Fate of a mutation in a fluctuating environment, *Proc. Natl. Acad. Sci.* **112**, E5021 (2015).

55. A. Melbinger and M. Vergassola, The impact of environmental fluctuations on evolutionary fitness functions, *Scientific Reports* **5**, p. 15211 (2015).

56. T. Mora and A. Walczak, Quantifying lymphocyte receptor diversity, ArXiv:1604.00487 (2016).

57. T. Mora, A. M. Walczak, W. Bialek and C. G. Callan, Maximum entropy models for antibody diversity, *Proc. Natl. Acad. Sci.* **107**, 5405 (2010).

58. J. Desponds, T. Mora and A. M. Walczak, Fluctuating fitness shapes the clone-size distribution of immune repertoires, *Proc. Natl. Acad. Sci.* **113**, 274 (2016).

59. E. Kussell and S. Leibler, Phenotypic diversity, population growth, and information in fluctuating environments, *Science (New York, N.Y.)* **309**, 2075 (2005).

60. O. Rivoire, Informations in models of evolutionary dynamics, *Journal of Statistical Physics* **162**, 1324 (2016).

61. A. Skanata and E. Kussell, Evolutionary phase transitions in random environments, *Physical Review Letters* **117**, 1 (2016).

62. O. Rivoire and S. Leibler, The value of information for populations in varying environments, *Journal of Statistical Physics* **142**, 1124 (2011).

63. A. Mayer, V. Balasubramanian, T. Mora and A. M. Walczak, How a well-adapted immune system is organized, *Proc. Natl. Acad. Sci.* **112**, 5950 (2015).

64. V. Mustonen and M. Lässig, Fitness flux and ubiquity of adaptive evolution, *Proc. Natl. Acad. Sci. USA* **107**, 4248 (2010).

65. T. J. Kobayashi and Y. Sughiyama, Steady state thermodynamics in population dynamics, *Physical Review Letters* **115**, p. 238102 (2015).

66. Y. Sughiyama, T. J. Kobayashi, K. Tsumura and K. Aihara, Pathwise thermodynamic structure in population dynamics, *Physical Review E - Statistical, Nonlinear, and Soft Matter Physics* **91**, 3 (2015).

67. T. J. Kobayashi and Y. Sughiyama, Fluctuation relations of fitness and information in population dynamics, *Physical Review Letters* **115**, 4 (2015).

68. C. Jarzynski, A nonequilibrium equality for free energy differences, *Physical Review Letters* **78**, 0 (1996).

69. G. E. Crooks, Nonequilibrium measurements of free energy differences for microscopically reversible Markovian systems, **90**, 1481 (1998).

70. G. E. Crooks, The entropy production fluctuation theorem and the nonequilibrium work relation for free energy differences, *Physical Review E – Statistical, Nonlinear, and Soft Matter Physics* **60**, 2721 (1999).

71. G. Lambert and E. Kussell, Quantifying selective pressures driving bacterial evolution using lineage analysis, *Physical Review X* **5**, 1 (2015).

72. V. Mustonen and M. Lässig, Adaptations to fluctuating selection in Drosophila, *Proc. Natl. Acad. Sci.* **104**, 2277 (2007).
73. K. Crona, A. Gavryushkin, D. Greene and N. Beerenwinkel, Inferring genetic interactions from comparative fitness data, *eLife* **6**, p. e28629 (2017).
74. D. M. Fowler, C. L. Araya, S. J. Fleishman, E. H. Kellogg, J. J. Stephany, D. Baker and S. Fields, High-resolution mapping of protein sequence-function relationships, *Nature Methods* **7**, 741 (2010).
75. E. Neher, How frequent are correlated changes in families of protein sequences? *I2Ni*, **91**, 98 (1994).
76. M. Socolich, S. W. Lockless, W. P. Russ, H. Lee, K. H. Gardner and R. Ranganathan, Evolutionary information for specifying a protein fold, *Nature* **437**, 512 (2005).
77. W. P. Russ, D. M. Lowery, P. Mishra, M. B. Yaffe and R. Ranganathan, Natural-like function in artificial WW domains, *Nature* **437**, 579 (2005).
78. W. Bialek and R. Ranganathan, Rediscovering the power of pairwise interactions, arXiv preprint arXiv:0712.4397 (2007).
79. M. Weigt, R. A. White, H. Szurmant, J. A. Hoch and T. Hwa, Identification of direct residue contacts in protein-protein interaction by message passing, *Proc. Natl. Acad. Sci. USA* **106**, 67 (2009).
80. M. Figliuzzi, H. Jacquier, A. Schug, O. Tenaillon and M. Weigt, Coevolutionary landscape inference and the context-dependence of mutations in beta-lactamase tem-1, *Molecular Biology and Evolution* **33**, 268 (2016).
81. A. Birgy, Paysage adaptatif des bêta-lactamases TEM-1 et CTX-M-15, PhD thesis, 2017.
82. J. D. Bloom, J. J. Silberg, C. O. Wilke, D. A. Drummond, C. Adami and F. H. Arnold, Thermodynamic prediction of protein neutrality, *Proc. Natl. Acad. Sci.* **102**, 606 (2005).
83. M. A. DePristo, D. M. Weinreich and D. L. Hartl, Missense meanderings in sequence space: a biophysical view of protein evolution, *Nature Reviews Genetics* **6**, 678 (2005).
84. C. S. Wylie and E. I. Shakhnovich, A biophysical protein folding model accounts for most mutational fitness effects in viruses, *Proc. Natl. Acad. Sci.* **108**, 9916 (2011).
85. H. Jacquier, a. Birgy, H. Le Nagard, Y. Mechulam, E. Schmitt, J. Glodt, B. Bercot, E. Petit, J. Poulain, G. Barnaud, P.-a. Gros and O. Tenaillon, Capturing the mutational landscape of the beta-lactamase TEM-1, *Proc. Natl. Acad. Sci.* **110**, 13067 (2013).
86. N. Halabi, O. Rivoire, S. Leibler and R. Ranganathan, Protein sectors: evolutionary units of three-dimensional structure, *Cell* **138**, 774 (2009).
87. O. Rivoire, Elements of coevolution in biological sequences, *Physical Review Letters* **110**, 1 (2013).
88. S. Cocco, R. Monasson and M. Weigt, From principal component to direct coupling analysis of coevolution in proteins: low-eigenvalue modes are needed for structure prediction, *PLoS Computational Biology* **9** (2013).
89. D. S. Marks, L. J. Colwell, R. Sheridan, T. A. Hopf, A. Pagnani, R. Zecchina and C. Sander, Protein 3D structure computed from evolutionary sequence variation, *PLOS ONE* **6**, p. e28766 (2011).
90. R. M. Adams, J. B. Kinney, T. Mora and A. M. Walczak, Measuring the sequence-affinity landscape of antibodies with massively parallel titration curves, *eLife*, p. 1601.02160 (2016).
91. D. M. Weinreich, N. F. Delaney, M. A. Depristo and D. L. Hartl, Darwinian evolution can follow only very few mutational paths to fitter proteins, *Science (New York, N.Y.)* **312**, 111 (2006).

92. K. Murphy, P. Travers and M. Walport, *Janeway's Immunology*, 7 edition edn. (Garland Science, 2007).

93. M. Kuraoka, A. G. Schmidt, T. Nojima, F. Feng, A. Watanabe, D. Kitamura, S. C. Harrison, T. B. Kepler and G. Kelsoe, Complex antigens drive permissive clonal selection in germinal centers, *Immunity* **44**, 542 (2016).

94. J. D. Bloom, P. A. Romero, Z. Lu and F. H. Arnold, Neutral genetic drift can alter promiscuous protein functions, potentially aiding functional evolution, *Biology Direct* **2**, p. 17 (2007).

95. F. J. Poelwijk, D. J. Kiviet, D. M. Weinreich and S. J. Tans, Empirical fitness landscapes reveal accessible evolutionary paths, *Nature* **445**, 383 (2007).

96. A. N. McKeown, J. T. Bridgham, D. W. Anderson, M. N. Murphy, E. A. Ortlund and J. W. Thornton, Evolution of DNA specificity in a transcription factor family produced a new gene regulatory module, *Cell* **159**, 58 (2014).

97. D. W. Anderson, A. N. McKeown and J. W. Thornton, Intermolecular epistasis shaped the function and evolution of an ancient transcription factor and its DNA binding sites, *eLife* **4**, 1 (2015).

98. H.-X. Liao, R. Lynch, T. Zhou, F. Gao, S. M. Alam, S. D. Boyd, A. Z. Fire, K. M. Roskin, C. A. Schramm, Z. Zhang, J. Zhu, L. Shapiro, N. C. S. Program, J. Becker, B. Benjamin, R. Blakesley, G. Bouffard, S. Brooks, H. Coleman, M. Dekhtyar, M. Gregory, X. Guan, J. Gupta, J. Han, A. Hargrove, S.-l. Ho, T. Johnson, R. Legaspi, S. Lovett, Q. Maduro, C. Masiello, B. Maskeri, J. McDowell, C. Montemayor, J. Mullikin, M. Park, N. Riebow, K. Schandler, B. Schmidt, C. Sison, M. Stantripop, J. Thomas, P. Thomas, M. Vemulapalli, A. Young, J. C. Mullikin, S. Gnanakaran, P. Hraber, K. Wiehe, G. Kelsoe, G. Yang, S.-M. Xia, D. C. Montefiori, R. Parks, K. E. Lloyd, R. M. Scearce, K. A. Soderberg, M. Cohen, G. Kamanga, M. K. Louder, L. M. Tran, Y. Chen, F. Cai, S. Chen, S. Moquin, X. Du, M. G. Joyce, S. Srivatsan, B. Zhang, A. Zheng, G. M. Shaw, B. H. Hahn, T. B. Kepler, B. T. M. Korber, P. D. Kwong, J. R. Mascola and B. F. Haynes, Co-evolution of a broadly neutralizing HIV-1 antibody and founder virus, *Nature* **496**, p. 469 (2013).

99. J. A. Weinstein, N. Jiang, R. A. White, D. S. Fisher and S. R. Quake, High-throughput sequencing of the zebrafish antibody repertoire, *Science (80-.)* **324**, 807 (2009).

100. G. Georgiou, G. C. Ippolito, J. Beausang, C. E. Busse, H. Wardemann and S. R. Quake, The promise and challenge of high-throughput sequencing of the antibody repertoire, *Nat. Biotechnol.* **32**, 158 (2014).

101. H. S. Robins, P. V. Campregher, S. K. Srivastava, A. Wacher, C. J. Turtle, O. Kahsai, S. R. Riddell, E. H. Warren and C. S. Carlson, Comprehensive assessment of T-cell receptor beta-chain diversity in alphabeta T cells, *Blood* **114**, 4099 (2009).

102. E. H. Warren, F. a. Matsen and J. Chou, High-throughput sequencing of B- and T-lymphocyte antigen receptors in hematology, *Blood* **122**, 19 (2013).

103. A. Six, M. E. Mariotti-Ferrandiz, W. Chaara, S. Magadan, H.-P. P. Pham, M.-P. P. Lefranc, T. Mora, V. Thomas-Vaslin, A. M. Walczak, P. Boudinot, E. Mariotti-Ferrandiz, W. Chaara, S. Magadan, H.-P. P. Pham, M.-P. P. Lefranc, T. Mora, V. Thomas-Vaslin, A. M. Walczak and P. Boudinot, The past, present and future of immune repertoire biology – the rise of next-generation repertoire analysis, *Front. Immunol.* **4**, p. 413 (2013).

104. M. Shugay, O. V. Britanova, E. M. Merzlyak, M. a. Turchaninova, I. Z. Mamedov, T. R. Tuganbaev, D. a. Bolotin, D. B. Staroverov, E. V. Putintseva, K. Plevova, C. Linnemann, D. Shagin, S. Pospisilova, S. Lukyanov, T. N. Schumacher and D. M.

Chudakov, Towards error-free profiling of immune repertoires, *Nat. Methods* **11**, 653 (2014).

105. E. S. Egorov, E. M. Merzlyak, A. A. Shelenkov, O. V. Britanova, G. V. Sharonov, D. B. Staroverov, D. A. Bolotin, A. N. Davydov, E. Barsova, Y. B. Lebedev, M. Shugay and D. M. Chudakov, Quantitative profiling of immune repertoires for minor lymphocyte counts using unique molecular identifiers, *J. Immunol.* **194**, 6155 (2015).

106. Y. Elhanati, A. Murugan, C. G. Callan, T. Mora and A. M. Walczak, Quantifying selection in immune receptor repertoires, *Proc. Natl. Acad. Sci.* **111**, 9875 (2014).

107. M. V. Pogorelyy, Y. Elhanati, Q. Marcou, A. L. Sycheva, E. A. Komech, V. I. Nazarov, O. V. Britanova, D. M. Chudakov, I. Z. Mamedov, Y. B. Lebedev, T. Mora and A. M. Walczak, Persisting fetal clonotypes influence the structure and overlap of adult human T cell receptor repertoires, arXiv:1602.03063 (2016).

108. M. V. Pogorelyy, A. A. Minervina, D. M. Chudakov, I. Z. Mamedov, Y. B. Lebedev, T. Mora and A. M. Walczak, Method for identification of condition-associated public antigen receptor sequences, *bioRxiv*, 1 (2017).

109. J. Jovel, J. Patterson, W. Wang, N. Hotte, S. O'Keefe, T. Mitchel, T. Perry, D. Kao, A. L. Mason, K. L. Madsen and G. K. Wong, Characterization of the gut microbiome using 16S or shotgun metagenomics, *Frontiers in Microbiology* **7**, 1 (2016).

110. B. E. Wolfe, J. E. Button, M. Santarelli and R. J. Dutton, Cheese rind communities provide tractable systems for in situ and in vitro studies of microbial diversity, *Cell* **158**, 422 (2014).

111. R. Mac Arthur, Species packing, and what competition minimizes, *Proc. Natl. Acad. Sci.* **64**, 1369 (1969).

112. S. Azaele, S. Suweis, J. Grilli, I. Volkov, J. R. Banavar and A. Maritan, Statistical mechanics of ecological systems: Neutral theory and beyond, *Reviews of Modern Physics* **88**, p. 035003 (2016).

113. A. Posfai, T. Taillefumier and N. S. Wingreen, Metabolic trade-offs promote diversity in a model ecosystem, *Phys Rev Lett* **118**, p. 028103 (2017).

114. J. O. Haerter, N. Mitarai and K. Sneppen, Theory of invasion extinction dynamics in minimal food webs, *Phys. Rev. E* **97**, p. 022404 (2018).

115. M. Tikhonov, R. W. Leach and N. S. Wingreen, Interpreting 16s metagenomic data without clustering to achieve sub-otu resolution, *The ISME Journal* **9**, 68 (2015).

116. C. K. Fisher, T. Mora and A. M. Walczak, Variable habitat conditions drive species covariation in the human microbiota, *PLoS Computational Biology* **13**, 1 (2017).

117. S. J. Biller, P. M. Berube, D. Lindell and S. W. Chisholm, Prochlorococcus: the structure and function of collective diversity, *Nature Reviews Microbiology* **13**, 13 (2014).

118. R. Braakman, M. J. Follows and S. W. Chisholm, Metabolic evolution and the self-organization of ecosystems, *Proc. Natl. Acad. Sci.* **114**, E3091 (2017).

119. G. Biroli, G. Bunin and C. Cammarota, Marginally Stable Equilibria in Critical Ecosystems, arxiv (2017).

120. D. A. Kessler and N. M. Shnerb, Generalized model of island biodiversity, *Physical Review E* **91**, p. 042705 (2015).

121. M. J. Rosen, M. Davison, D. Bhaya and D. S. Fisher, Fine-scale diversity and extensive recombination in a quasisexual bacterial population occupying a broad niche, *Science* **348**, 1019 (2015).

122. S. Gould, *Wonderful Life: The Burgess Shale and the Nature of History* (1989).

123. V. Orgogozo, Replaying the tape of life in the twenty-first century, *Interface Focus* **5**, p. 20150057 (2015).

124. M. Lässig, V. Mustonen and A. M. Walczak, Predicting evolution, *Nature Ecology & Evolution* **1**, p. 0077 (2017).
125. T. van Opijnen and A. Camilli, Transposon insertion sequencing: a new tool for systems-level analysis of microorganisms, *Nature Reviews Microbiology* **11**, 435 (2013).

Discussion

D. Fisher Just a comment on where we are getting to in the condensed matter physics analogy. The theory that Richard talked about and some of the experiments are sort of getting up to the ideal and periodic solids. And our understanding of those might become good enough where the deviations from the theory start pointing towards much more interesting biology including the ecology as in some of the things Richard showed. I guess another success is that you, Richard, are now invited to the WHO meetings that decide on flu vaccines! Are there other comments or questions on these quantitative issues in evolutionary experiments and related things?

S. Eaton So I wondered if it is the case really if there is no ecology in those different tubes of *E. coli*. I just wondered if one could see such a signature: maybe you could have two separate populations developing that as long as the other is present, they outcompete everybody else like if they compartmentalized the metabolism in some way. Would you see these signatures in the sequencing that you are doing? Are you sequencing individual clones or the whole population?

M. Desai We sequenced whole population samples and one of the surprising conclusions was that you indeed see striking signatures of the spontaneous emergence of co-existing clones. There appear to be mutations that shift the equilibrium between them, but nevertheless at least two, and in some cases more than two co-existing clones arise. Presumably due to some kind of ecological interaction. In one case, it has been previously realized that there were co-existing clones due to an acetate cross-feeding mechanism. We found nine other examples out of the twelve lines. So, it is just all over the place and we have no idea what the mechanism is. Yet we know in at least a number of cases that it is not the same mechanisms, but it is an open question what it is.

A. Murray Just one comment about something Richard said about evolution in the real world. I think it is a reasonable conjecture that evolution under conditions where there is rapid adaptive radiation and where there are novel niches formed, might actually look surprisingly like what happens in the laboratory where mutations that inactivate genes lead to beneficial phenotypic changes.

E. Siggia To broaden the discussion slightly, I like to recall another thread of evolutionary theory which is more phenomenological than what has been discussed so far, and which is to look directly at phenotypes and phenotypical evolution. A good example is provided by this old paper of Nilsson and Pelger who showed that in morpho-space so to speak there is a path that would take a photo-sensitive surface and turn it into a vertebrate eye via a path that would continuously and monotonically improve the visual acuity. There is also work from Paul Francois, from his thesis in Paris and continuing in his lab at McGill, which looked at phenotypes and showed there is a monotonic path with respect to some fitness that would lead to interesting phenotypic processes, but always described at the phenotype level. There is also work from Leslie Valiant at Harvard who rephrased some algorithms in machine learning as evolutionary algorithms and then showed that certain classes of functions can be learned quickly and other classes of functions are impossible to learn that way. The whole subject was sort of encapsulated by Chris Adami, who calls it "evolution of the fleetest", bringing in the notion that if you could achieve a reasonable solution rapidly, you might see this type of evolution.

D. Fisher I think some of those things will come back in the general discussion too.

Prepared comments

M. Desai: Characterizing the landscape on which evolution takes place

To make any predictions about how evolution will proceed, we need to know something about how the specific details of a particular biological system determine the set of possible evolutionary trajectories. This space of evolutionary possibilities is often visualized as a "fitness landscape", which represents the map between genotype and fitness. Of course we cannot hope to exhaustively characterize the fitness landscape in even a single system. However, since all biological systems were created through the evolutionary process, there is some hope that there may be some general properties of fitness landscapes that are common across a wide range of systems.

With this hope in mind, much effort has been devoted to characterizing fitness landscapes in a number of specific systems. Since it is impossible to measure the fitness of every possible genotype, many of these studies have instead attempted to measure "combinatorially complete" landscapes involving a small number of specific mutations (typically less than 10). For example, Dan Weinreich measured the fitness of each of the 25 possible combinations of 5 mutations involved in resistance to a bet-lactamase antibiotic. He found that the landscape was not particularly "smooth". Instead, there is extensive epistasis in this system (i.e. the effect of a given mutation depends strongly on which genotype it arises in), which results in local peaks and "dead ends". Similar studies in other systems have come to varied conclusions, but epistasis and context dependence do appear to be fairly widespread.

However, it is not clear whether these small fitness landscapes involving just a few loci are really representative of the space of evolutionary possibilities more generally. Some other empirical work has expanded the space under consideration by measuring partially complete landscapes involving somewhat larger numbers of mutations (e.g. all pairwise combinations of several thousand yeast deletion mutations). However, even this is only a very limited view of a complete fitness landscape, which is actually an extremely high-dimensional object, where there are often billions of mutational "directions" that could be taken from any particular genotype. It may often be the case that many of these directions are not really independent, so the effective dimensionality of the landscape may be somewhat smaller. Still, the dimensionality of any complete landscape is likely to be so high that the conclusions from existing limited studies may not be very generalizable.

A key challenge is therefore to find better ways to characterize the high-dimensional genotype-fitness map. One possibility is to take the same perspective that evolution does: a local one. At any given moment, evolution cannot "see" the full landscape. Instead, we can think of evolutionary dynamics as wandering in the dark with a dim lantern. At any given moment, a population consists of a cloud of relatively closely related genotypes. This cloud can explore nearby genotypes by acquiring additional mutations, but it can only explore in this very local sense. The relevant quantities are thus the nature of the cloud of related genotypes (which determines how far in genotype space the population can explore) and the distribution of fitnesses of the nearby accessible genotypes (along with the mutation rates at which these genotypes are created).

We need new ways of characterizing the fitness landscape in terms of these local statistics: mutation rates and distributions of fitness effects. We also need to understand better how these local statistics of the landscape change as the population evolves: to what extent can an individual mutation change the nearby landscape? Several examples of such effects have been observed (e.g. mutations can dramatically change future mutation rates) but it is not very clear how common they are. We also need a better understanding of how key features of evolutionary dynamics and population genetics depend on particular aspects of these maps. For example, what are the key local statistics that determine observable quantities such as the expected substitution rates or site frequency spectra?

Finally, it is important to remember that there are many effects such as time-varying environments or ecological interactions that can complicate any fitness landscape picture. While these effects can be included in generalized versions of the fitness landscape, relatively little is known about how these generalized landscapes typically look or how these complicating factors affect evolutionary dynamics. It will be important to try to find ways to quantify these factors, by identifying and measuring some of the key local statistical properties of these more general landscapes.

Note from the editors: At this point came the rapporteur talk by Aleksandra Walczak. See the joint rapporteur report by Richard Neher and Aleksandra Walczak.

Discussion

D. Fisher Thank you Alex. I actually thought it was a very optimistic presentation since you went a good way taking some vague questions and making good questions out of them. We can take a few comments or questions now and then go on to the prepared remarks.

G. Süel Since you touched on the work by Rama Ranganathan on proteins and statistical coupling between amino acid positions, I thought it might be good to add that in fact it is not a large percentage of residues that are coupled statistically. It is typically around 14–15%. I think that is actually important as it shows that there is a core module that appears to be critical for function and the rest of the residues are less important for the function. As was shown long time ago, you can mutate most positions of any protein and the function will not be changed. It is also important to point that stability does not equal function in proteins. There are a lot of disordered proteins known, and in fact Rama Ranganathan and Stanislas Leibler had a paper on protein sectors (*Cell* **138**(4): 774–786, 2009) where they showed that the melting point of a protein and its function can be orthogonal. This means that you can have slightly melted proteins that can still function. It is potentially important to think about these things because what we see in proteins, may not be just restricted to proteins. You might indeed think about gene regulatory networks in a similar way with "chro-modules" that are more preserved next to other things that can be mutated more easily.

E. Koonin I would like to bring up a very classical issue, coming back to Daniel's favorite point about time scales. In the first rapporteur talk, we heard primarily about mechanisms of micro-evolution: what is going on in a very short time scale even within a single human body. In Alexandra's talk, we heard, an appropriately much more vague account of evolution on longer time scales. I would like to probe the feeling around the table on the very notorious question whether the same mechanisms that dominate micro-evolution are also the most important ones in macro-evolution. Is there such a thing as macro-evolution that is qualitatively different from micro-evolution? In the traditional evolutionary biology community one tends to say no: the change in allele frequencies (as in micro-evolution) will account for everything. True or false?

B. Shraiman Along these lines, on the micro-evolutionary scale we are often worrying about growth, fast growth, as it seems fitness equals fast growth. On the macro-evolutionary time scale, however, fitness equals survival. Ultimately, when we are looking at the species and try to understand the diversity of species, one thing that they have in common is not that they all grow very fast, but is that they survived, whereas most branches of the tree have gone extinct. The question becomes: under what circumstances is faster growth a good strategy for long-term survival?

E. Koonin We will come back to that.

J. Lippincott-Schwartz I would just like to make a comment. Animal and plants evolution have chloroplasts and mitochondria which themselves have DNA that is being propagated through the system. This needs to be considered when you are looking at microbial versus animal/plant evolution.

C. Brangwynne We saw some beautiful talks on bacterial biofilms. One of the things that was at least tangentially alluded to is this idea of bacterial persisters. These are temporarily antibiotic resistant dormant states. I was thinking of this in the context of evolution and evolutionary strategies. In that case, the advantageous strategy is to basically "play dead".

D. Fisher There are several comments here that are associated with fitness and evolution on different time scales and on more complex aspects that are going on such as the co-evolution that Jennifer was commenting on. One thing that really holds back the subject of evolution — aside from the tendency to quote dead white males — is that fitness is regarded as a singular term, which is tremendously misleading. I hate using "omes" for things, but I would much prefer people to call it "fitnome" because it is really function of all kinds of things depending on time scales and so on...

Prepared comments

I. Gordo: Mammalian gut, a microbial jungle that (i) was misunderstood, (ii) appears to be important for health, (iii) could/should be controlled

During the past years, laboratory petri dish experiments have helped us observe evolution in real time in microbes and test theoretical predictions on microbial adaptation, predictions to which many physicists have contributed.

In recent years and with the advances in NGS (next generation sequencing), it is becoming clear that those predictions can occur in natural systems, e.g. with the patterns predicted by the theory of clonal interference being recurrently observed. Many of the observations made have been in the context of infectious diseases.

Much less is known about microbial evolution in the context of health. A particular system that has attracted much attention recently is the mammalian gut. It is now becoming clear that in the gut hundreds of bacterial species can coexist within a host and that they collectively or sometimes individually can contribute to a healthy physiological status of their hosts. This leads to the current need and will to understand this system with sufficient depth to be able to control it, such that one may restore health when diseases associated with gut microbiota dysbiosis occur.

For example, answers to questions such as:

- How much evolutionary change occurs during the lifetime of a healthy mammal?
- How much of that change is due to de novo mutation, recombination, genetic drift and/or natural selection?
- If natural selection is important: What is the typical effect of a new beneficial mutation in the intestinal tract?
- Does microbial evolutionary change differ in health vs disease?

They are fundamental questions that need to be answered for the proper understanding of this natural ecosystem.

Experiments using laboratory mice and specific strains of bacteria are starting to show that clonal interference can be pervasive in the guts of healthy mice, and that new mutations with quite large selection coefficients can readily spread to high frequencies within a population. Here on the left figure is an example of recent work showing how quickly new mutations with strong fitness effects can spread within a population of a commensal strain of Escherichia coli, tagged with fluorescent neutral markers, when it colonizes the gut microbiota of a mouse.

The extent to which a similar clonal interference pattern occurs in a more natural system is yet to be understood.

Remarkably very few studies of population genetics, variation within a species, of bacteria living inside each of us have been done. This figure on the right shows an example of what can be observed when typing, with particular molecular markers, E. coli clones from weekly fecal samples taken from a healthy human. Each color represents a different lineage and the bricks represent lineages that carry antibiotic resistance determinants (the human host is not undergoing any antibiotic treatment). As you can see a remarkable amount of strain variation is observed. In addition, rapid changes in frequency of the strains can be detected and the causes of such pattern are yet to be determined. This kind of data also informs us on how much we may miss in current microbiology studies where typically a single clone is sampled from a host at a single time point.

In sum, I think that we are in the beginning of a new era in the study of microbial evolution where in vivo studies and healthy hosts are to receive as much attention as in vitro studies and disease conditions.

A. Perelson: From blackboard to bedside: How physics can lead to life-saving drugs and disease cures

Physics teaches us that simple models can frequently capture the essence of complex phenomenon. One case in point is the interaction between an infecting virus and the host immune system. Immune systems are extraordinarily complicated containing more cells than there are neurons in the brain, but being distributed throughout the body, having no central controller and communicating by both cell-to-contact and via long-range secretion

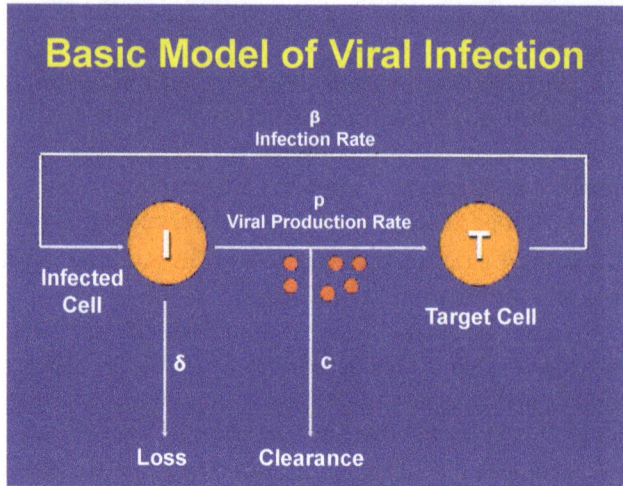

Fig. 1.   The basic viral dynamic model. Virions (red circles) infect target cells, T, and with infection rate constant β generate infected cells, I. Infected cells produce new virus particles, virions, at rate p per cell, and are lost by either death or in the case of HCV possibly by getting cured at rate δ per cell. Virions are cleared from the circulation at rate c per virion.

of molecules. Nonetheless, simple models have provided enormous insight into how HIV survives within a host and causes AIDS and how HIV and hepatitis C virus should be treated with combination drug therapy. There are no broad-spectrum antibiotic like drugs for treating viral infections. In fact, there is only one chronic viral infection that can be cured by antiviral therapy and that is a rather recent development.

Viruses cannot replicate on their own. They must enter cells and exploit host cell resources to replicate. A simple model of virus infection is shown in Figure 1, where virus V enters a cell susceptible to infection, i.e. a target cell, T, leading with rate constant β to the generation of an infected cell, I. Infected cells then produce virus at rate p per cell, which can be cleared from the body at rate c per virus. Infected cells are also killed by the virus or by the host immune system at rate δ per cell. This simple cartoon of a virus infection, when turned into mathematical equations has been called the standard model of viral infection. The equations defining it are as follows:

$$\frac{dT}{dt} = s - dT - \beta VT,$$

$$\frac{dI}{dt} = \beta VT - \delta I,$$

$$\frac{dV}{dt} = pI - cV. \tag{1}$$

These equations can describe the kinetics of acute infection during which time virus grows exponentially to a peak value then declines and finally stabilizes at a steady state value.[a] The parameter values determine whether the virus is cleared, i.e. whether at steady state virus is cleared, i.e. $V = 0$, or whether persistent infection occurs, i.e. $V > 0$. In the case of HIV, there are no known individuals who have spontaneously cleared the infection.

After a month or more of infection the virus has reached its set-point, i.e. its non-zero steady state value. Patients are usually treated with antiretroviral drugs after the set-point has been reached. The basic model has also been successfully used to analyze the effects of therapy with antiretroviral drugs. For example, reverse transcriptase (RT) inhibitors block the ability of HIV to productively infect a cell. HIV protease inhibitors (PI) cause infected cells to produce immature non-infectious viral particles, V_{NI}. Thus, in the presence of these drugs, the model equations become,

$$\frac{dT}{dt} = s - dT - (1 - \varepsilon_{RT})\beta VT,$$

$$\frac{dI}{dt} = (1 - \varepsilon_{RT})\beta VT - \delta I,$$

$$\frac{dV_I}{dt} = (1 - \varepsilon_{PI})pI - cV_I,$$

$$\frac{dV_{NI}}{dt} = \varepsilon_{PI}\, pI - cV_{NI}. \tag{2}$$

where ε_{RT} and ε_{PI} take on values between 0 and 1 and represent the efficacies of RT and PI inhibitors ($\varepsilon = 1$ being a 100% effective drug). Further, V_I and V_{NI} are the concentrations of "infectious" and noninfectious virus, respectively, and $V = V_I + V_{NI}$ is the total virus concentration.

From Figure 1 one can immediately see that if a drug causes the production of non-infectious virus, then the infection process will be perturbed, but neither virus production from infected cells nor viral clearance will change. Therefore, to a first approximation, the viral level should not change unless the existing infected cells die rapidly and are not replaced because newly produced virus is non-infectious. Thus, Figure 1 suggests that the lifetime of productively infected cells as well as the rate of clearance of viral particles should be deducible from a careful analysis of the viral decline caused by administration of a protease inhibitor. Solve Eqs. (2) under the simplifying assumption that the number of target cells remains constant at its pre-therapy steady state value, $T_0 = c\delta/\beta p$, and that therapy is with a 100%

---

[a]M. A. Stafford, L. Corey, Y. Cao, E. S. Daar, D. D. Ho, A. S. Perelson. Modeling plasma virus concentration during primary HIV infection. *J. Theor. Biol.* **203**(3): 285–301 (2000).

effective protease inhibitor ($\varepsilon_{PI} = 1$, $\varepsilon_{RT} = 0$). The solution is,

$$V(t) = V_0\, e^{-ct} + \frac{cV_0}{c-\delta}\left(\frac{c}{c-\delta}\,(e^{-\delta t} - e^{-ct}) - \delta t\, e^{-ct}\right),\qquad(3)$$

where V_0 is the set-point viral load before initiation of therapy.[b] This solution only depends on three parameters, V_0, c and δ, where c is the viral clearance rate and δ is the death rate of productively infected cells. Allowing the target cell concentration, T, to vary necessitates using numerical methods to predict $V(t)$ but does not substantially alter the outcome of the analysis. Fitting either the analytical solution above or the numerical solution of Eq. (2) to the viral decline data obtained from patients on PI therapy allowed me to get minimal estimates of the viral clearance rate and the death rate of productively infected cells. If drug therapy were not 100% effective then new infections would still be occurring and these rates would need to be higher to explain the observed data. Nonetheless, these initial estimates transformed our view of HIV from a slow virus infection taking 10 years or so leading to the death of a patient to a rapid infection with virus having a half-life of six hours or less and infected cells producing virus living a day or two. Thus giving us the first view of a very dynamic battle between the virus and the host and one in which drug therapy could cause dramatic shifts in favor of the host. Further modeling of this type led to the first estimates of how fast HIV was evolving and becoming drug resistant and showed the need at that time for combination therapy with three or more drugs to control the infection. The subsequent introduction of combination therapy transformed HIV infection from being a death sentence to a treatable chronic disease.

The basic model was subsequently used to model hepatitis C virus (HCV) infection and treatment. The first drugs developed to treat HCV, such as type I interferon, blocked the ability of infected cells to produce virus, and thus changed the parameter p in Eq. (1) to $(1 - \varepsilon)p$ where ε was the drug efficacy. It turned out that HCV like HIV is cleared from the circulation very rapidly and thus blocking viral production leads to profound and rapid viral declines. As shown by Neumann *et al.*[c] using the basic model one could then estimate the efficacy of a new HCV drug candidate with data from clinical trials that only lasted a few days. This ability to rapidly screen compounds for in vivo efficacy help speed the development of new drugs that ultimately led to our ability to cure HCV. The basic model and variants have also been used by many workers to study many other viral

[b] A. S. Perelson, A. U. Neumann, M. Markowitz, J. M. Leonard, D. D. Ho. HIV-1 dynamics in vivo: virion clearance rate, infected cell life-span, and viral generation time. *Science* **271**(5255): 1582–6 (1996).

[c] A. U. Neumann, N. P. Lam, H. Dahari, *et al.* Hepatitis C viral dynamics in vivo and the antiviral efficacy of interferon-alpha therapy. *Science* **282**(5386): 103–7 (1998).

infections including influenza, West Nile virus, Zika virus, hepatitis B virus, measles and cytomegalovirus to name a few. An entire field of study has grown around these activities and it has been called both viral dynamics and virophysics.

S. Quake: DNA sequencing in biophysics

Over the past decade there have been fantastic advances in DNA sequencing technology which have been driven by biophysics and which have also created new measurement tools for biophysical questions. DNA sequencing uses a few important ingredients: the ability of DNA polymerase to copy DNA molecules, clever synthetic chemistry to create nucleotide analogs with useful properties (such as radioactive or fluorescent labels and the ability to terminate DNA polymerase extension), and a method to read out the states of molecules. Historically, this read-out method was based on analytical chemistry and used gel electrophoresis to separate molecules based on size. The new generation of DNA sequencing technologies continue to use the basic ingredients of DNA polymerase and clever nucleotide analogs — although in this case the chemical modifications include removable labels and reversible chemical termination. Most importantly, the read out method has shifted to a biophysical approach: repeated imaging of large fields of view by optical microscopy as the DNA templates are gradually extended. This has created a revolution by increasing throughput dramatically, and DNA sequencing speeds have increased by orders of magnitude while the costs have decreased by orders of magnitude. As a particularly dramatic example, the cost to sequence a human genome went from approximately $1 billion with gel electrophoresis to approximately $1 thousand by optical biophysics: six orders of magnitude in cost decrease and corresponding improvements in speed as the time to sequence has been reduced from years to hours.

There have been many important conceptual consequences to this improvement in measurement technology. First, the genomes for many organisms, including humans, have been sequenced to a level of precision that enables identification of all genes. This truly puts an important bound on genetics — the number of genes is not only finite and countable, but they have been enumerated. This greatly restricts the ability to create models which postulate arbitrary genes to accomplish any given biological phenomenon. Furthermore, any genome can be sequenced, so one is no longer restricted to working with model organisms and it is possible to choose organisms with unusual biological properties and dive deeply into their genetics. Finally, these DNA sequencers can be used for more than just measuring genomes or genotypes — they can also be used as molecular counters to enumerate RNA molecule distributions (RNA can be converted into DNA through the use of the reverse transcriptase enzyme). This so-called

RNAseq approach provides a powerful method to measure the phenotype of an organism or cell at a given point in time.

After the feat of creating reference genomes for a variety of organisms, these sequencing technologies began to be applied to other problems in genetics. One area of interest was to understand the genetic diversity and variation between individuals of the same species — whether they are human or microbes! This very quickly led to the ability to test quantitative models of evolution and selection; these models are intimately connected to many deep problems in statistical physics, and it has been quite useful to bring those tools to bear by viewing biological populations in the same way that ensembles are studied in statistical physics. At the simplest level, neutral theories of evolution map directly onto random walk statistics. Significantly, the application of these models has expanded into areas beyond conventional population genetics. Scientists have realized that important aspects of human biology — such as viral infection, the relentless growth of cancer and the diversity of antibodies created in the immune system — represent important examples of evolution within our bodies which can be quantitatively analyzed with sequencing technologies in order to understand their behavior in the context of evolutionary models. There is a long theoretical history of trying to make this connection, but there had been very little if any connection with experiment. This has now changed thanks to the advances in DNA sequencing technologies and we expect it to continue to be a fruitful area of study.

The ability to use DNA sequencing technologies to measure organism phenotype is also enabling the deep understanding of cell state and cell type. This is most dramatically seen in the emerging frontier of single cell transcriptomics, in which virtually every messenger RNA molecule in a given single cell can be enumerated by sequencing. Other technologies besides sequencing play an important role in such measurements, notably the application of the physics of microfluidic devices. Such data enable one to embed each cell in a roughly $\sim 20,000$ dimensional space (corresponding to the number of genes in the human genome) and to ask questions about the fundamental nature of cell type and identity. These data will provide the basis for new theories of development, physiology, and homeostasis based on deep biophysical understanding of the nature of the cell.

Discussion

D. Fisher One thing with sequencing is that we seem to be more limited about ideas than about what one can do with the currently available technology or in a number of years. We still have time for some general discussion before the break. After that, we will come back to some of the bigger picture aspects of evolution.

E. Koonin I want to come back for a second to the issue of sequencing and continue on what Steve Quake said and bring up the subject of metagenomics. It seems to me that it is critical, not only the increase in the capacity of the sequencing method, which is huge of course. But also the ability to characterize the diversity in the entire biosphere in pretty much an unbiased way, independent of culturing, growing, etcetera. I think this creates, combined with methods for analysis of those sequences, opportunities that were unthinkable even a decade ago or even less.

U. Alon When you look at these presentations, you see that genotype space is characterized by a vast diversity of mutations. An emerging theme in this whole conference is that in contrast when you look at phenotypes, you see low dimensionality. One example from Alan Perelson is this rate of HIV clearance which is captured by a four-parameter model. So if we have a mapping from very high diversity in sequence to a low dimensional phenotypic space, we can do much better in understanding the geometry of phenotypic space. One direction we can think of is sexual reproduction and the differences between individuals in a species (with all our differences and all our traits) that forms a cloud in phenotype space. We can then ask the question: is this cloud in line with the low dimensionality of the phenotype space between different species? Since the 80's there are different examples for evolution along the lines of "least genetic resistance". So if you exaggerate the difference between two individuals, you will get the characteristics of a different species. That suggests something about the polymorphisms that are common in genetic variation that lead to diversity in phenotypes that are pushing you like vectors along this low dimensional manifold rather than perpendicular to it.

D. Fisher That's a point where I think Boris will come back to.

S. Chu There, some things struck me about the questions that are being posed. Maybe we have some technology, maybe it has already been done but if you want to know how evolution works in a big community, the gut is a great example. You can sequence stuff and you can put on fluorescent tags. You may not be able to cultivate individually all the stuff in these thousands of interacting bugs in your gut. But you can then, by putting in place fluorescent tags, identify the carcasses of the bacteria that come out. Then you can do something to change it (I started eating Chinese vegetables for breakfast instead of eggs and toast, and I can guarantee you my gut changed). Then you can see how the multiply interacting microbes in our gut get affected. We have all the pieces now and you don't have to be able to cultivate all the individual bacteria. It is just a suggestion, maybe people are already doing this.

A. Walczak People have done this, yes. There are clear differences between vegetarian and high animal food consumers.

I. Gordo Because of the process of evolution being highly repeatable in these experiments, you actually can tune the diet to increase the frequency of a strain. We have been able to do that in animals to increase the abundance, but not with vegetables, using a specific sugar.

S. Chu I was actually suggested something quite different as the question was not the survival of the fittest individual organism, but the survival of the fittest interacting organisms. And then, if you start tracking hundreds of these interacting organisms, might you be able to say something different?

A. Chakraborty There is a recent paper where they've done what you say, Steve, in two populations. One in Finland and one on the Russian side of the border where they are genetically very similar. One population has huge amounts of auto-immune diseases. The other has not. They've been able to identify the community of interacting microbes that are the key difference. So, these sort of studies are happening now in humans.

D. Fisher We should take a break. I would just like to make a mini-summary. Some of the things that Alex talked about was evolution at the level of single proteins and things that one can do to manipulate small numbers of mutations, and one comment was that a lot of evolution is already happening when it is not even trying. Also that is certainly an important part of evolution, but here we may already have some level of understanding. And then there is what has just been discussed now, which is that enormous amount of things can be done to track what is going on. But it is not quite clear what the questions are, never mind how one is going to get the understanding. And having been involved together with Stephen in studying the immune system, I would say at this point that what we have learned from being able to sequence the immune repertoire is very limited because we cannot really connect at this point the genotypes with the phenotypes. The fact that we cannot do that is associated with the huge redundancy in many different ways of being able to bind to the same thing (see also the work of Alan Perelson). After the break we will have a bit more on evolution, and then turn to more general discussion questions.

D. Fisher We are going to start up again. So this session is going to start off by going away from the short term evolution, getting to the bigger picture evolution, and then it is going to evolve into general questions. So the first remark is from Nicole King, who I guess is going to be the one to touch with reality on this part of the session.

Prepared comments

N. King: Transitions to multicellularity

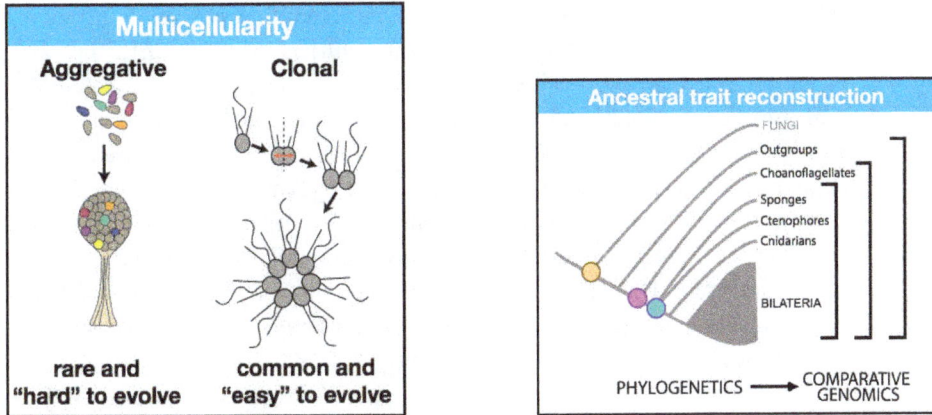

Of the major transitions in evolutionary history, three have captured the most attention — the origin of life, the origin of the eukaryotic cell, and the origin of multicellularity. Unlike the origins of life and the eukaryotic cell, transitions to multicellularity have occurred over 30 times in evolutionary history, leading to well-known lineages like the animals, fungi, plants and slime molds, but also organisms about which we know far less, like the brown and red algae, the filastereans, and the ichthysporeans. Because most of these transitions occurred between 50 million to 1 billion years ago and were not preserved in the fossil record, they cannot be understood through direct observation and experimentation, but instead must be inferred through comparative approaches constrained by phylogeny.

All animals stem from a single transition to multicellularity, but what do we know about animal origins, and in what ways do they reveal general themes of transitions to multicellularity?

I'd like to make three points about animal origins that might be of interest to physicists. First concerns the long time scale and potential role of oxygen in animal origins. Although life first evolved over 3.5 billion years ago (bya), animal fossils did not appear until a little over 600 million years ago, not long after the appearance of other multicellular fossils and coincident with an increase in atmospheric and subsurface marine oxygen levels (as revealed through the work of Andy Knoll and others). One hypothesis has been that low oxygen levels prevented the evolution of large multicellular life forms until around 1 bya. More generally, this observation emphasizes how physical and chemical aspects of the environment constrain the niches into which organisms can evolve.

Next, when thinking about evolutionary transitions to multicellularity, it is important to consider whether any given transition was through an aggregative or clonal mechanism. Animals evolved through a transition to clonal development rather than aggregative multicellularity, with a single cell dividing repeatedly and the sister cells remaining physically attached. This means that each of the cells in the multicellular form has the same genotype as all the other cells. The few examples of complex multicellularity (e.g. animals, plants, and fungi) all stem from clonal multicellularity in which cells are not experiencing genetic conflict. In contrast, organisms with aggregative multicellularity have evolved relatively rarely and do not display comparable levels of morphological complexity, perhaps because "cheaters" can arise in a heterogeneous genetic environment. In this way, I tried to make the point that our ability to meet to discuss the physics of living material was made possible because our ancestors achieved multicellularity through a clonal process of development.

Finally, despite the fact that DNA and cellular features are not preserved in the fossil record of the first animals, we can reconstruct the cellular and genomic landscape of animal origins by comparing the cell biology and genome of living organisms within a phylogenetic framework. Genes and cellular features found only in animals and shared among most animal lineages, were likely present in their last common ancestor (blue circle). Other features may be more ancient and shared either with their closest relatives (choanoflagellates; magenta circle) or with more evolutionarily distant lineages (such as Fungi, orange circle). Through these types of analyses, we have learned that most animal genes evolved long before the origin of animals, while a small subset is restricted to animals. Interestingly, these animal-specific genes are essential for regulating key processes in development.

Moving forward, what are the big questions in animal origins that might be of particular interest to the physicist?

(1) How did the cell biology and genome of the first animal constrain subsequent stages in animal evolution?
(2) What were the special conditions that allowed animals to evolve? Why didn't animals evolve earlier? Why haven't animals evolved more than once? Could they evolve again?
(3) Can we determine the minimal molecular machinery required to make an animal?

P. Rainey: The evolution of individuality

My comment concerns the evolution of individuality and I include mention of plausible ecological conditions facilitating the transformation of matter into life.

I use the term individuality in a Darwinian sense: A Darwinian individuality participates in the process of evolution by natural selection. It does so by virtue of being a member of a population of Darwinian individuals where entities vary one to another, reproduce and where offspring resemble parental types.

A starting point and motivation for inquiry is recognition of life's hierarchical structure. This can be visualised as a set of Matryoshka dolls. The largest doll represents a multicellular organism. The multicellular organism is comprised of cells that are defined by the first nested doll. Inside cells are organelles (the second nested doll); delve into the nucleus and one finds chromosomes (the third nested doll), these are further divisible into genes (the smallest doll). One can conceive of additional nested entities ending at the point where life arises from matter. Above the level of the largest doll exists eusocial organisms.

The reason for giving prominence to these stages is that each defines a level where there is some capacity for replication. Wherever such capacity exists there is possibility of a Darwinian (adaptive) process. Selection therefore acts at each of these levels, although most effectively at the highest available level.

The nesting of levels also suggests that through evolutionary time, lower level entities have been subsumed within higher level self-replicating structures. John Maynard Smith and Eörs Szathmáry termed these events major evolutionary transitions[a] — transitions in which natural selection shifts focal level. Of course natural selection cannot simply shift focal level: its capacity to shift is dependent upon the emergence of those properties that define Darwinian individuality (variation, reproduction and heredity).

So how do these properties emerge? It is commonly assumed that variation, reproduction and heredity are given – properties that are so fundamental to life that they do not require explanation. This is marginally true of variation, but it is not true of either reproduction or heredity. These are derived traits and require evolutionary explanation.

It is tempting to invoke natural selection as cause, but to do so is to commit a logical fallacy. It is not possible to invoke that requiring explanation (e.g., reproduction) as the cause of its own evolution. Recognition of this fact raises a dilemma. The challenge is to explain how Darwinian properties arise from non-Darwinian entities, by non-Darwinian means. This problem haunts not just the emergence of life from non-life, but each of the major evolutionary transitions.

One solution is to argue that this perspective is too stringent and that there are likely to be opportunities for co-option of existing traits that ren-

---

[a] J. Maynard Smith, E. Szathmáry: *The Major Transitions in Evolution.* (Oxford: Freeman; 1995).

der this less of a dilemma than I pose. While I am sympathetic to this view, there is merit in a 'take nothing for granted / assume nothing' stance because it tests the bounds of possibilities.

A solution that takes nothing for granted is to recognise the crucial role of the environment — the fact that organisms are not as separate from their environment as we tend to think. Certain ecological conditions can scaffold Darwinian properties on otherwise 'unwitting' particles, in the same way as reproduction of viruses is scaffolded by the host. A simple example of ecological scaffolding requires nothing more than particles, patchily distributed resources (that discretise particles into collectives) and a means of particle dispersal (which approximates a means of patch-level reproduction). Dispersal between patches is crucial for long term persistence because resources within patches will eventually be depleted.

Such environmental conditions give rise to a population structure that causes selection to act on patches (patches are collectives of particles). By virtue of the environment, patches give rise to patch-level offspring. If patches are founded by single particles, the offspring patches bear close resemblance to parent patches. This sets the scene for the evolution of traits adaptive at the patch (collective) level. In time it is likely that exogenously scaffolded Darwinian properties become endogenous features of the evolving collectives.

Implicit in the scaffolding process are birth / death events that take place at the level of lineages of collectives. This is vital because it establishes a second (longer) timescale over which selection works (the shorter timescale being the doubling time of individual particles). The longer time scale tends to select against rapid growth, allowing exploration of phenotypic space and causing, in the long-term, the reproductive fate of particles to align with the collective. This leads to a "decoupling" of fitness and the emergence of a new kind of biological entity.[b]

Finally a comment on the emergence of life from matter and a push to argue that ecological scaffolding holds the key to understanding how geochemistry becomes replicating biochemistry. A most compelling case comes from the study of alkaline vent systems on the ocean floor.[c] At these vents environmental conditions are far from equilibrium. Gradients of redox, pH and temperature are generated by the flow of heated sea water through serpentine rock. In the context of porous iron-sulphur-encrusted carbonate cells that form the fabric of the vents, conditions exist that are energetically favourable for the reduction of carbon dioxide and concentration of chemical

---

[b]K. Hammerschmidt, C. Rose, B. Kerr, P. B. Rainey: Life cycles, fitness decoupling and the evolution of multicellularity. *Nature* **515**:75–79 (2014).

[c]W. Martin, M. J. Russell: On the origin of biochemistry at an alkaline hydrothermal vent. *Phil. Trans. R. Soc. Lond. B* **362**:1887–1925 (2007).

reactions, products and substrates. Beyond being energetically plausible, the porous structure of the vents allows the possibility of a Darwinian process to happen at the level of competing chemical reactions. The existence of such a process marks the beginning of life.

Discussion

D. Fisher Thank you Paul. I want to take the chairman's prerogative and ask Nicole sort of a direct question. So I guess one thing which has been found in the lab recently by Mike Travisano and Will Ratcliff is how easy it is to evolve a very primitive form of multi-cellularity in yeast where it evolves into a snowflake which reproduces by fragmentation and one can then select upon the size of the fragments and so on. So you have obviously something much more sophisticated as far as the kind of thing going on in mind as to what one might be able to at least conceive of doing in the lab, to try to get something which is sort of more animal like. Could you say just a little bit about what you would envision at least in an imaginary sense of what one might be able to do?

N. King Right. I think the success of those experiments speaks of how easy it is to get simple undifferentiated multi-cellularity even though they are actually seeing some baseline differentiation. But to be an animal requires preprogramming, all animals start with an egg and a sperm, they form a blastula, they undergo gastrulation. So those I see as a minimum. And the ideas, we know what are the genes that are required in animals to orchestrate those types of cell differentiation processes and morphogenetic processes. Could we envision, moving that machinery into an organism that cannot do that and would that be sufficient? Actually listening to Eric's talk made me a little bit pessimistic because there are these other roles such as the regulation of tension and elasticity in the cytoplasm and we do not know what encodes that.

A. Murray I mean one thing that might be worth saying about that you can evolve yeast under different conditions to be macroscopically visible aggregates. Under those conditions, the chemical environments of the cells in the middle and on the outside have to be different, right? So if you propagate things like that you will select to do the best in the environments they are in. So there will be the succession of evolution of metabolism and the response to metabolism and eventually this can give rise to something that looks like differentiation. You can imagine there is then a stage in evolution where you switch from metabolically driven differentiation to using second messengers or things like that. I think that is actually a very interesting avenue to explore.

N. King One quick comment. Which is remember that most microbial eukaryotes differentiate in time in response to different environments and so it is not hard to imagine temporal cell differentiation becoming spatial and that I think is an important part of animal origins.

J. Lippincott-Schwartz So I have a question related to Pau's talk. So, how big does the pond have to be in order for all of these parts to start self-organising a system that is big enough to trigger the evolutionary dynamics that you describe? Is there a lower limit where it just won't happen?

P. Rainey How big the pond as opposed to how big the number of individual entities?

J. Lippincott-Schwartz Both.

P. Rainey Right, the critical number of entities... In the experiments we have done, we had as few as 120 entities and you would say that is far too few. Rich Lenski wouldn't think of doing selection on a 120-bacteria. But within relatively short periods of time, we really do see phenomenal things happening including developmental cycles, even the reproductive division of labor. How is that possible with so little variation? Well this process of lineage selection, which is exemplified by what I showed, when there is a death there is the possibility of a birth. Selection is simply rewarding those types that persist. There is a lot of evolution going on within each lineage. Some math and some models show what is happening, despite the relatively small number of individual entities, because selection is not favouring fast growth, I touched upon this, there is the possibility of exploring a great amount of phenotypic space. So, you know, phenotypes that would be unachievable if fast growth was the only thing in town, now become possible. So this wandering around, sort of following pathways of least resistance is how we think about it as well, seems to generate stuff that we don't fully understand. So I think, in terms of the possibilities for life, one could imagine many other kinds of ecological structures that would be more effective than the pond scenario, for example.

C. Brangwynne This is a totally naive question, but I think it is being addressed in bits and pieces here and there but it goes back to the biofilms. Basically it is unclear to me what is multi-cellular and what is uni-cellular. The idea of independent evolution of multi-cellularity strikes me as interesting but if you have individual cells that are communicating and they are genotypically very similar and I don't necessarily see that as critical for distinction. This is just a comment.

N. King I'll take a stab and then maybe you could. There is a formalised definition of individuality and major transitions are thought to be shifts in the definition of individuality. So individuality typically describes the unit of selection, I don't know if we want to go there, but the definition of a truly multi-cellular organism is one in which the cells have their function dedicated to the individual and they won't survive away from the multi-cellular organism and that definition becomes slippery when you start talking about colonies that evolved under selection very recently or the types of things we talk about that can go in and out of multi-cellularity.

D. Fisher I think there is this question of time scales here also with the bacteria. If I remember correctly, and some of you probably know, that Vaughn Cooper was doing some evolution experiments on biofilms, then they became planktonic and went back to biofilms and the evolution occurs so fast that by the time the biofilm was grown they were already genetically differentiated between the bottom layer and the top layer. So that is a big population of bacteria doing that.

E. Wieschaus What struck me, Nicole in your presentation, was the emphasis on clonality. There are obvious reasons for why you can imagine that as a unit of selection. But what I was trying to figure out is whether clonality, even though it implies the same genotype, implies the same state of the cells. Within an organism, I would like to have multiple functional states of the cells. Can one imagine evolving that from a clonal mass of cells that are genetically identical but have different states due to random fluctuations or due to positions? Within something that is analogous to a biofilm, could you build on that to make more elaborate structures? Is that how it works?

H. Levine For Dictyostelium, a kind of slime mold, that is exactly what happens: you start out with cells that are identical and then differentiate to stable phenotypes even though they are in some sense genetically identical. The difference between clonal and non-clonal is very much what people think in terms of where that endpoint was. I am just wondering if there is some experimental program or synthetic biology tools that can really address the question of what goes wrong if we try to extend the complexity of something that is inaccurately defined. I just wonder if we can test it.

N. King Absolutely! In fact, we have fewer experiments with aggregative multi-cellularity, there are fewer examples and we cannot really differentiate between those two right now. I think that is plausible, yes.

G. Süel So, I just wanted to add to that. Yes indeed, in biofilms for example, you get distinct cell types. There are cells that for example can form spores, there are cells that can become dormant but are expressing a hundred and fifty genes to take up extra-cellular DNA, that is what Herbie was referring to in terms of components. So there is actually two distinct cell types that emerge within a clonal single-species bacterial community and I think the point that I wanted to make is that, with respect to even the pond and so forth in evolution I think one of the critical stages in evolution to me is when cells figured out that they can actually control their environment or manipulate the environment in some way because I agree that you cannot dissociate the organism from its context, which is the environment. But the critical point I think, or at least one of the critical points, is when cells started to realise that by secreting things or by their very act of replicating or metabolic activity, that other cells were now affected by their behaviour. That is definitely something that is playing a critical role inside the biofilm where you get actually different regions with different gene expression patterns observed, not because necessarily the environment is in any particular way structured, but because of the activity of the cells.

D. Fisher I think it is probably a good point to mark and to start bringing in the couplings to the environment and so on, or Nicole, do you have some sort of one line...

N. King I was just going to re-emphasise what Andrew said which is that there is some cell differentiation that comes as a result of physical properties and maybe that links to this discussion.

D. Fisher Eugene Koonin, who is the last prepared remarker of the meeting, I guess.

Prepared comments

E. Koonin: Frustration and competing interactions as the basis of biological evolution

Competing interactions between all kinds of biological entities and at all biologically relevant levels of organization permeate all of biology. Arguably, the lowest level of biological complexity is the folding of nucleic acid and protein molecules into unique, biologically functional three-dimensional structures. From the evolutionary perspective, the beginning of life can be most plausibly associated with the appearance of the first RNA molecules (ribozymes) endowed with RNA polymerase activity. Not unexpectedly, laboratory experiments that attempt to select for RNA molecules with this activity shows that it can be achieved only by structurally elaborate RNAs. The competition between short-range and long-range interactions is plainly apparent in protein and RNA molecules and is the defining factor of folding that underlies all molecular functions, without exception.

Moving up a level, in macromolecular complexes that perform most if not all, biochemical functions in cells and viruses, the competition between interactions within individual macromolecules and those between subunits that lead to complex formation is equally obvious. Examples abound, suffice it to point out the conformational changes in ribosomal proteins upon the ribosome subunit formation, in transcription factors upon DNA binding, and in virion proteins during morphogenesis of virus particles.

On another plane of biological organization that is unique to living matter, competing interactions can be conceptualized as selection pressures that act in opposite directions. Such conflicting selective processes are a key, inherent component of host-parasite coevolution. The frustrated state of a host-parasite system is caused by a complex interplay of conflicts between the parasite replication, the host replication and the interactions that stabilize the host-parasite system as a whole.

Deep analogies seem to exist between these biological conflicts and frustrated states in physics, such as striped glasses. Such systems are characterized by frustration whereby interactions leading to local minimum free energy states compete with a different set of interactions that underlie the global minimum, resulting in emergent phenomena. In both biological and physical systems, frustration effects appear to drive the evolution of complex patterns. These analogies open the road to using the methodology developed for the study of frustrated states in condensed matter physics to analyze biological evolution.

The concept of competing interactions and frustration naturally apply to evolutionary transitions and, more generally, major evolutionary innovations. The concept of major transitions in evolution (MTE) developed by Maynard Smith and Szathmary defines a distinct class of innovations that involve evolutionary transitions in individuality (ETI). A classic example of a MTE is the origin of multicellular organisms from unicellular life forms but MTE, although not numerous, punctuate the entire history of life. The key tenet of the MTE theory is that, within its framework, the transitions are construed not simply as innovations but rather meet strict criteria that make them akin to phase transitions in physics. The signature feature of MTE is ETI, which involves a change in the level of selection, e.g. from a single cell to an ensemble of cell (a multicellular organism). The second signature of MTE, sensu Szathmary, is that each transition is associated with the emergence a new type of information storage, use and transmission (e.g. multicellularity is linked to the rise of epigenetic informational systems). Competing interactions and/or levels of selections are immediately apparent in each MTE.

Starting from the most obvious, evolution of multicellularity involves the intrinsic conflict between selection forces that act at the level of individual cells and those that are manifest at the level of cellular ensembles or tissues. Obviously, to maintain the functionality of a multicellular organism, the proliferation of individual cells has to be tightly controlled. These competing interactions are essential for the development of a complex multicellular organism but cancer and aging also result from the frustration caused by the competition. Moving back in time, we know little about the origin of the first cells. It is nevertheless difficult to imagine an evolutionary scenario in which the emergence of cells was not preceded by an evolutionary stage at which all genetic information was encoded in small elements that resembled modern mobile genetic elements. Subsequent evolution involved accretion of such elements to form large genomes similar to those of modern prokaryotes. Under this scenario, the emergence of cells involved competition between the "interests" of individual genetic elements and those of ensembles of such elements that formed cellular genomes. Once again, replication of individual elements has to be subdued for the cell to function. The next MTE, the origin of eukaryotes, is associated with endosymbiosis that gave rise to the mitochondria and hence involved the inevitable conflict between the endosymbiont and the evolving eukaryotic cell that required coordinated reproduction of the host and the symbiont. The same conflict is inherent in the evolution of photosynthesizing algae via the cyanobacterial endosymbiosis that gave rise to the chloroplast. The frustration caused by host-symbiont conflicts in these MTE was resolved by the formation of the stable symbiotic associations, but the conflicts linger, e.g. in the form of mitochondrial diseases and frequent lysis of mitochondria that in some organisms result in insertion of mitochondrial DNA into the host genome.

The later MTE that led to the emergence of eusocial animals and societies also clearly involve competition between individuals and collective, or between collectives at different levels of organization. Generally, it appears that for ETI, which is the defining feature of MTEs, competing interactions between entities at different organizational levels are the intrinsic driving factor.

Taken together, all these observations seem to indicate that competing interactions could be the universal factor underlying all evolution. Accordingly, application of the apparatus used in condensed matter physics to analyze striped glasses and similar states to develop a formal theory of conflict-driven evolution could make a substantial contribution to evolutionary biology.

Discussion

A. Walczak So do you think there is a link between time scales and length scales, that there is a time scale that sets the scale for which we observe the result of frustration?

E. Koonin I think that these types of processes are definitely relevant both when we are talking about the conflicts between different length scales, spatial scales, and time scales. I suppose there is a link between the two although perhaps that link is not entirely hard.

R. Neher I want to step back a little bit and contrast two different ways of looking at these major transitions in biology. At some level, Paul or Nicole, you study specific systems and try to find some plausible scenario of how these things could have happened. Eugene tries to find overarching principles. And then there is the question on how does all of that helps us to organise and understand the diversity that we see now. What are the similarities, are these completely distinct problems, or is there a common denominator, what are the rules, the parentheses that tie all of this together?

T. Gregor Just coming back to Eugene's talk. I really like the concept of this frustrated system. Can you just briefly comment on, towards the bottom of your table, what that actually entails? It wasn't clear.

E. Koonin I think we can go through all the levels within the same framework. Multi-cellularity is very clear in terms of the conflict between, so to speak, interest of individual cells and cells collectively, we all know what happens when the interest of individual cells win. The very same pertains to the evolution of any population as well as community: Local minima for individuals or lowest level selection units and higher level selection units. I believe that the tragedy of the commons is actually a very closely related phenomenon, where immediate interest of individuals come in conflict with the global stability of the system over longer time scales.

D. Fisher It's like communism.

E. Koonin Yes perhaps, I think like the evolution of any society even though we better go through that aspect at the bar. So I think it is quite easy to follow these conflicting interactions at different levels of selection throughout that spectrum.

J. Lippincott-Schwartz Perhaps one way to think about evolution on earth through these different stages, archaea and bacteria.

N. King One origin of life.

J. Lippincott-Schwartz One origin of life but a proto-bacteria in an archaeic system. When you go to multi-cellularity one can speculate that it involves a clonal cell system encapsulating a bacterial/archaea population, a gut. So it is a symbiosis at all levels of evolution between really fundamentally different strategies for how life has originated as bacteria or archaea.

N. King You know that I'm sympathetic to that idea because all of these multi-cellular transitions occurred on a planet already dominated by bacteria. The first animals probably didn't have a gut so they probably had stable associations with bacteria. I agree that interactions with bacteria are central and I didn't talk about all our stuff.

A. Murray So yes I want to appeal to sort of rigour again and the distinction between the TARDIS and astronomy and the Large Hadron Collider and things like that. So that when we talk about evolution, the further back we go the more we conjecture and we get to a stage where the conjectures are large and the amount of evidence that supports the conjectures is modest and I think it is super important to recognise that and the distinction between some of the things that Richard and Alexandra were talking about, where we can do experiments and test explanations, and things where we make inferences about the past, I think that it is crucial that we admit that we do not have the TARDIS.

D. Fisher This is actually a really good point for transitioning on to the last bits of the meeting. So the same could be said about astrophysics and the universe rather recently, probably at the time when everyone here already had their PhD perhaps. But Cosmology was a field that mainly seemed almost philosophical and one could tell stories about things, it was very difficult to know what to do and then it became a real scientific field with predictions and falsifiable things, and further predictions and what to look at and how that coupled into a lot of other things that were seen in astrophysics and so on. So I think, you know, one of things that we could hope is that one is starting to move towards the direction in trying to understand evolution. So with that, I'm going to turn things over for a bit to Boris. The ultimate goal for this meeting, not the ultimate goal of all of science, is on fundamental questions, and Cosmology will not be one of these, but we will at least get started soon after the beginning of the earth, we will not go back much further. And then we will open to a more general discussion of what the important questions are and comment on the question, trying to turn them into better questions.

Closing Session

Wrap up

B. Shraiman So let's proceed with our overambitious and slightly pompous goal of articulating fundamental questions for the future. My role here is to kick off the discussion of the questions that we promised to articulate. I want to thank many people who contributed suggestions before midnight last night. Some of them contributed many questions! Among those questions there were a number of recurring themes and our co-chairs and I brainstormed past midnight and until Hotel Metropole's bar closed, trying to distill the questions alongside with many thoughts that have been brought up during our meeting. As you will see, it is going to be a rather raw spirit, which needs some more refinement.

Before introducing the list let me state the criteria for the admission to our club of Fundamental Questions. The question has to be BIG with the definition of "big" here being "general" and capable of opening new avenues for description and understanding. But the question also has to be well-posed in the sense of generating informative and falsifiable theory. We want new hypotheses that can stimulate new rounds of experiments and move science forward, which will also require going back, modifying the hypothesis and so on until eventually hopefully getting to the truth. Finally, because of the nature of this particular gathering, we wanted to focus on those questions where Physics ideas and methods can actually contribute novel approaches as opposed to many other questions and things that our Biology friends in this room and outside can do without Physics' help.

Some Fundamental Questions about Living Matter:
- How Complex is Living Matter?
- Evolvability of Phenotypes?
- Reproducibility of Evolution?
- Dimensional Reduction/Center Manifold?
- From Genes to Geometry?

- Size determination?
- Physical instabilities under biological control?

Let me start with the first question that I want to elaborate a little bit as an example of the process of trying to formulate these questions. So, how complex is living matter? We keep saying that living matter is complex. Eugene [Koonin], I think in our very first discussion period, brought up this matter of complexity. It is not a exact quote but it went something like this: "you people are all talking about pathways, but why pathways, there are just lots of genes, lots of enzymes, it is just a big entangled mess and it's very complex". Daniel Fisher immediately said, "no, no, no, there is evolution and evolution is working to make it simple". So can we try to distill this into a workable hypothesis?

For example, let us think of a quantitative phenotype as a function of a genotype and ask how complex that function is. How many genotypes are there? Well, four possible bases at any position on a chromosome and let us say there are n-positions with n very large, so four to n'th power and it is a gigantic space! So in principle, if you want to define any one phenotype as a function of a genotype you will need 4^n numbers to specify it, which is a lot of numbers. That would perhaps be what we call "complex". But on the other hand, if, perhaps, all of these little bases were all contributing additively, this function of ours would be defined with only n parameters and we could call that "simple". And of course there is a lot of room between 4^n and n. There is everything in between, there is polynomial complexity, e.g. n^2 etc, etc. Perhaps complex interactions and hence complex dependence on genotype are limited to small subsets of genes coding for interacting proteins that contribute to the same "functional module". Andrew Murray here is a great champion of modularity. With modular genetic architecture, overall complexity is limited by the size of the module (so n does not get too large) and modules by assumption would be weakly interacting, so complexity is kept under control by genetic compartmentalization. The question of complexity is a big question for sure, but is it well posed? Can hypotheses be falsified? I would argue here, that the answer to that is "Yes".

Perhaps we cannot map a phenotype as a function of a genotype for n equal to 10^9, or even 10^3, but it has already been done for n equal to five! We can imagine that n can be pushed to 10 and perhaps a bit beyond. That would be enough for us to actually be able to measure the complexity of a given quantitative phenotype (for example, defining complexity by the power spectrum of the discrete Fourier transform of the phenotype on the hypercube corresponding to the n-locus genotype). One could then proceed to compare different phenotypes: Is there a typical behavior? Is the answer the same for a single cell organism and a multi-cellular organism? Are some phenotypes more complex than others? And the complexity question is not just a topic for after dinner conversation. Depending on what the answer is, it could actually affect how one interprets Genome Wide Association Studies for example, so it could actually inform us on the genetics of disease. (A technical remark here is that in contrast to GWAS where the space of genotypes is a sample of naturally occurring variation, the study proposed above has envisioned artificially constructed genetic variants.) One can discuss this all at much greater length, but given our time constraints, let us instead at least briefly touch upon the other questions on the list...

A related question that come up often in our discussions, was the question of phenotypic evolvability: Can one give "evolvability" a quantitative definition? What makes a given phenotypic aspect more or less evolvable? More generally, one must remember that phenotype is not just one number, but a collection of many relevant properties, e.g. for microbes, growth rates in many different conditions. One could begin by asking: How far can life get in the multi-dimensional phenotypic space with some fixed number of mutations? What is close (or far) in the phenotype space, may or may not be close (or far) in genetic space in the sense again of the number of mutations, transpositions, duplications, the kind of stuff that genomes do very easily. It is a big question, but still, there are plausible paths forward. We talked today about morpho-space and there were clearly some actionable items to follow up on, yet it is by no means a unique context where evolvability question can be pursued. We also talked today about reproducibility of evolution. What is reproducible? On what time scale? What is predictable? Can we distinguish evolved from accidental? I guess, that discussion is still very fresh in our minds, so I will move on.

Skipping to the next question on the list, it may have been yesterday that we talked about the question of how many relevant variables there are in the cell. So of course again, things are very complex, you know, DNA, RNA, proteins, many proteins. There are a lot of variables, but the question is, what are the relevant ones? There seems to be some indication that there are lots of things but not all of them matter. Is there a sense in which we can actually identify the relevant variables, get an idea of the dimensionality of that relevant space? These are words, but one can try to translate them into actionable statements with suitable math ideas. There is the dynamical systems notion of a "center manifold", or perhaps more properly, the "slow manifold". It corresponds to a hierarchy of time scales with fast dynamics establishing certain relations between variables and confining slow dynamics to the low(er) dimensional "slow manifold". Slow dynamics is then defined by the "relevant variables" that define position on that reduced dimensional manifold. There are ways in which one can try to interpret data to determine the dimensionality of the manifold and that could be used as a tool for identifying these relevant variables, and for knowing when we have identified enough of them to get at least a useful approximation. In his report, Terry Hwa gave a very interesting example of a small number of relevant variables describing physiological state of a microbe. I think we also talked about cells, particularly in the context of stem cell fate determination. People like to think of different states of stem cells and transitions between these states. We heard that the response of a cell to particular stimulus depends on the past history of the signaling that that cell has experienced. We can study things conditional on history or we can say, well, that history is reflected again in some internal variable and if we have identified the internal variable, we are cooking with gas. Perhaps sometimes the reduced dimensionality space can be further clustered into states and descriptions can be reduced to transitions between states, whether deterministic or stochastic. Many current approaches to cell differentiation assume underlying discreteness of physiological "states": are these the stable attractors — catchment basins of the Waddingtonian landscapes — on the reduced dimensional manifold? Siggia in his report, and many others too, touched on these ideas and outstanding questions.

This morning we were talking about morphogenesis and developmental dynamics encoding shape. Genes and maternal factors in the embryo set up initial conditions for the dynamical process that structures the embryo and ultimately generates the physical shape. So how does this dynamics encode shape? How do genes encode the dynamics? Shape, I think, is a great example of a "physics phenotype". Nipam [Patel] just told us about very interesting things happening in the development of this little crustacean, which is basically an animal Swiss army knife, with all its many different appendages, and there is a way of genetic switching, transforming one "tool" into another. It was a description of binary phenotypes, so there were walking legs and swimming legs. But if you were to look a little closer, there might perhaps be everything in between. What is the space of appendage morphologies that can be reached by genetic tweaking? To describe these shapes we need to graduate to the next level of description, describing the geometry and once we learn to do that, we will learn more about development. In his report, Maha defined "shape" by separating it from the matter of "size" — same shape objects can be scaled to each other — but size determination is another very general question that we had a lot of discussion about. These discussions were getting more into specific mechanisms, but the question is general and perhaps the number of mechanisms is finite and again applicable broadly in different contexts.

Finally, a common theme that we have seen is that biology often does its business by controlling physical instabilities. This is true for liquid-liquid phase separation forming "membraneless organelles" that Cliff [Brangwynne] told us about in his report, and of the coiled morphology of chick's gut in Maha's report. A physical instability of course can be an ally or an enemy. So the question is, what are the properties that Biology keeps far from the transition, keeps far from instability? Versus, what properties are kept close to criticality to gain control?

I think this is going to be it. I will leave a list of placeholders for these different questions and open it up for discussion. I'm sure many of you have other burning suggestions for other questions, so we can go deeper and we can go broader.

D. Fisher I think these questions do have two different characters. Some of them are very clear questions and some of them are still struggling, like the first one. You know, to make good questions one might already address out of a bigger picture, questions. There are other ones like that that are too vague, like how does the brain work, but which one might try to build down from. There may be other ones on this list that one would like in some way to build up from and make it a somewhat bigger question. So I'm going to give this thing here to Andrew Murray, to chair the discussion, but since I'm still sitting here, and officially I'm chair, I'm going to feel free to interrupt by pushing my button regularly. I could follow someone who unfortunately one has to hear a lot about in the US and tweet instead but I will refrain from that.

B. Shraiman I actually asked these two gentleman to co-chair this ... so a biologist and a physicist and they're friends, and they agree most of the time.

D. Fisher Which is disturbing, believe me ...

A. Murray I guess maybe I could start with having one response to the first question and I think one image that was particularly helpful was the image that Michael Desai had at the lab. So I think one of the things that is complicated about living matter is how fragile it is. We've talked a lot about robustness but the guess is ... I mean what we know is that you can't change more than an order of thirty base pairs without having a fifty-fifty chance of killing a yeast cell, which is a relatively robust organism. One guess is that if one could just turn fluxes and pathways up and down by factors of ten, many of those, if you did five of them at once, would also be lethal. So I think this doesn't answer the question on how complex is living matter, I think it suggests that evolution, diverse and wonderful though it is, has explored a remarkably small fraction of what is available, because you can only change a small number of things at once.

A. Walczak I have a question. So in your attempt to summarise did you figure out what are the key things that still need to be measured and explored by measurement and what we have measured enough and we should just now go home and think?

A. Murray I guess I can say what I said to Daniel and Boris and others last night about this. So ... many of you are physicists and there is an interesting question about whether we can rely on you to understand life. I would argue we've measured quite a lot of things already, there are many more to measure. I would sort of encourage you to go off and sit quietly somewhere and come back and tell us that you understand life and you now have a series of experiments and measurements you would like us to do, to test whether you are right. Because one of the things I argued with Boris about last night is whether the attempt to explain life, on the assumption that you are not currently capable of doing it, is going to generate interesting new sorts of Physics. I do not mean new modes of atomic interaction or new forces or anything, but in the same way that condensed matter introduced new collective concepts. I think it is an interesting question, of whether the efforts of physicists and biologists together trying to explain Biology will require us to think about Physics and possibly even Mathematics in novel ways. So I don't think we've measured everything, but I don't think we have even thought about what we need to measure.

A. Walczak I wouldn't wait in a corner, you may not live long enough.

J. Howard I think there is a concept here that overlaps many of these questions and it is the concept of emergence or collective phenomena. It is sort of strange not seeing that concept, you know, articulated here. Even though it is a little bit, maybe, overused concept but it is, it really is a fundamental concept and it is behind many of the things that we have been trying to understand and which is how the small things make the large things. You know, I think it also pertains all the way up to the very top ... you know ... evolution, because after all the phenotype ... you know, selection is operating at the emergent level, whereas variation is happening at the small level. I don't see how we can understand evolution or any of these processes unless we can understand how emergent phenomena arise. It seems that is what physicists are very good at doing, you know, especially condensed matter physicists.

W. Bialek So, in those discussions last night I was sometimes at the center, so rather than complaining that my favourite thing didn't make it onto the list, what I'll do is to say that I think some of these questions are maybe even more general than you gave them credit for, and we should remember that. So for example, in the case of going from genes to geometry, you had the very specific example of morphogenesis in mind but of course there is a community of people who worry about how one goes from genes to behaviour. Behaviour is the geometry of trajectories, right, in some space of postures of the animal or the sounds of your voice, whatever right. And so the question of how do genes encode stereotype movements of animals or how do patterns of connections among neurons encode trajectories of movement. This issue of going from something that is ... sort of ... discrete and on the relevant time scale static to the encoding of something which is continuous, geometrical and in some cases dynamical, because you pointed out that the shape emerges over time. I think that is actually even broader, it is a much broader problem than the morphogenesis problem.

In thinking about physical instabilities, we tend to be attracted by sort of exotic and recent examples but can you ask a very simple question about the stability of folded proteins. We often talk about why they are stable, but remember that the natural units for measuring their stability is not per protein molecule but per amino acid residue, and then they are barely stable at all. It is much less, it is vastly less than kT per amino acid residue. So I think that the notion that biological systems are somehow near transitions has something that has been implicit in a lot of things that people just haven't drawn out. The last thing I would add is in response to Andrew's comment about, you know, what we're all supposed to go do. I think that one thing that ... if I can be allowed one more philosophical remark ... one thing that characterises, let us say the community of people sitting here and what has happened over the last decade that is different than the decade before is that I think that most of the theoretical physicists that have been spending their time thinking about biological problems have been very closely engaged with experiments at different levels. This back and forth leads to a sharpening of the theoretical questions, and that sharpening of the question leads to a suggestion for a new experiment. I think actually that's not a bad shape.

D. Fisher Just a very brief comment. I also worry that things have gone into the direction where stepping back quite far from the experiments is not sort of done, or is maybe frowned on. One really has to have a whole spectrum of that, and I think that sociology maybe discourages that stepping far back as well.

A. Murray I agree in spades that it is about the emergence of things. That relates to what Bill just said about it is less than kT per amino acide but that is because proteins are collections of amino acids, polymerised as a string. I think one of the really crucial questions for the community of biologists interested in talking to physicists and vice versa is ... there are two possibilities: One is that there will be general principles that describe emergence but the mechanism, the molecular mechanisms will be wildly different and so we will deal with general principles which biologists are familiar with under sort of exploration with selection, which is the principle of all sorts of events inside cells and also Darwinian evolution. But they occur in mechanistically different ways. How much we will be able to unify beyond sort of enjoying each other's company will be modest. There is another possibility that there are genuinely, not only principles that are deep but sort of mechanisms that are deep as well and we will approach at least the character of a physical law. I think, my prejudice will be that it will look less that way but I don't think we know the answer.

A. Perelson I just like to make a comment about something that I think is missing from here. Many of us are starting to think of Biology as sort of an information science: That there is information that is coded, that has to be passed on from generation to generation. People doing molecular biology now come to sit at their computers and type in information and get proteins synthesised, DNA synthesised. You don't necessarily have to go into the wet-lab, you can have other companies make all of this stuff and do things. But it is the information content that is important. And I don't see it as part of any of these questions or characterisations. I think somehow what distinguishes living matter from other types of matter is that there is an information content that is passed on and changed by evolution but we need somehow to articulate that within this set of questions and lists.

B. Shraiman Well I would defend by saying that actually information appears in many of those questions, right? Starting from phenotype as a function of genotype, coding of geometry. It enters just about everywhere. But if you have another way of formulating the question that specifically focuses on this and makes it an actionable item, I think that would be a wonderful entry.

E. Koonin So, addressing what Andrew said a few minutes ago and also Paul Rainey's talk, in particular regarding the apparent god-given nature of Darwinian properties. A fundamental question that I want to ask both is what is life, in a certain sense? But not philosophically off course. What are the defining features of that matter that we consider living? In particular, is a very simple ... seemingly simple property namely replication fidelity above a certain threshold that allows preservation of individuality the necessary and sufficient condition for subsequent evolution and development of what we call life. Are there additional essential requirements such as a particular type of genotype/phenotype mapping or particular type of ... how do you call it ... fitnome, c-scape or something like that. And coming back to what Andrew said, this is quite explorable, definitely in computer simulations but also in a test tube in particular in the context of ribozymes. I think that such very general questions might be somewhat within reach.

M. Elowitz I just want to pick up on Alan's comment and Boris' response. I think one of the big things that has changed is our ability to synthesise DNA and that means we have the ability to construct things. So, I think a theme that kind of cuts across maybe some of these questions but maybe is its own question is how many of the interesting processes that go on within cells or during development can we rationally program and our challenge is basically to ask: Can we take each of the most interesting processes that we see in cells and program them to varying extent from scratch, I think that would be kind of a key challenge.

H. Goodson Also to follow up on what Alan was saying, one way to phrase the question he was asking more concisely would be: How can information theory explain some of the otherwise inexplicable aspects of biological systems? In signal transduction I think there are some very clear places where biologists are just like "why? what?" The information theorists have actually some pretty specific explanations to why it is the way it is.

C. Brangwynne I also worry that some of the ways we are thinking about this is so centred on our, you know, the examples that we have. But, you know, think about what is happening potentially in exoplanets and then I just thought, five or ten years from now if artificial intelligence does take off as Elon Musk warns us and these robots start to build themselves and kill us all off and then these questions will seem very silly.

U. Alon This is a comment. These questions are wonderful actually and I photographed them, to think about them. I want to make a comment on a different level. When Physics meets Biology it is a meeting of two cultures and that is an opportunity for us to reflect not only on the results but on the process of the way we do science. This Solvay conference has very many women, has very many young people, it is different. Biology is learning from Physics, for example, the publication: how to put things on to ArXiv, solving a lot of problems that we have with publications being co-opted by commercial interests in the journals. Can we turn science unto itself and study our own process? How can we as a group of people searching for the unknown … teach ourselves how we can communicate effectively, collaborate effectively, use language effectively, use differences effectively like here we have people who believe in optimality and people who believe in non-optimality. In many fields you can just self-report: I'm a selectionist, I'm not a selectionist. So to avoid these conflicts which occur at the bottom. Can we teach ourselves how to accept an idea that challenges us at the level of identity. So we get a fast system saying "aah" and then maybe a more reflective system saying "wait, what process can we have to give this more of a chance?". We have so many problems now. There are systems in science that don't align with our values of trust for example, how we fund science. If I give you a hundred thousand dollars I trust you do the best science you can. Are we basing our funding on trust? Truth-telling, we are not seeing the truth. We know how to tell the truth, that is a huge virtue. Are our systems based on that? Can we turn science on itself? Based on these meetings of cultures is to find out how we can redesign science and think about our processes to better align with our values. I think it is a historic opportunity

C. Marchetti I just wanted to go back to the point raised by Joe actually because it seems to me that this question of dimensional reduction and emergent behaviour are really closely tied together, they are also one and the same.

Because I think in Physics I think essentially what here is called dimensional reduction really only works when we describe phenomena that actually exhibit emergent behaviour and this idea of trying to identify, you know, relevant variables, I think it is also important to keep in mind ... I mean its ... the other physicists laugh obviously but it is important to keep in mind that identification must also depend I think on the phenomenon we are trying to describe. It is not one that we can make just a priori for any given phenomenon or for any given system. I really think it depends on what questions we are asking.

D. Fisher A comment on the relevant variables. So the sort of framework and the whole way... conceptual way of understanding a lot of general things in Physics as the renormalisation group of Ken Wilson. But my father used to make the comment, but I didn't understand it at the time but now I really understand, that by far the most important contribution of Ken was the notion of irrelevant variables, that there are a lot of things that don't matter. I think that is one of the things that gets forgotten often, that is, which things don't matter, or they only matter a little bit or they could be different ... Epigenetic, some epigenetic aspects of regulation, ... some gut part of me feels: ok Biology could work fine without that, it happens to have that but maybe it is not so crucial. It could have done things in some other way. So I think that in some way ties to a lot of these ... also it helps evolution work. The fact that everything doesn't matter. Lots of ways of doing it, most of the details of how evolution ends up finding doesn't necessarily matter.

W. Bialek A quick follow up to both of these remarks. Again, in Neuroscience there has been a lot of interest in the use of dimensionality reduction ideas, either in looking directly at behaviours or looking at the activity of networks of neurons and thinking about the selectivity of neurons for the features of sensory input, whether you should think about dimensionality reduction there. Without going into details, what I would say is that at least in my experience the cases that have been the most successful are the ones where you have a reasonably strong theoretical framework that tells you why the dimensionality reduction might happen, rather than just trying to sift through the data looking for it. And so I guess this sort of reinforces the points that you are making that, you know, there are cases in which you expect this to work and there are other cases in which you don't. And as we wander into more complex problems we need those guides because otherwise, you know, you'll discover as Ruelle has taught us, that if you just go looking you'll discover that everything is six-dimensional because six is the logarithm of a million ... you just try to sample the points.

A. Murray Are there other big questions that we have missed?

B. Halperin I'm not trying to address the bigger question but just to go back to dimensional reduction for a moment. I'm not sure in a highly nonlinear space what it means by this dimensional reduction. So we can just think about something like a glass ... or a spin glass let us say where we know that there ... we believe that there are in certain regimes many different ground states or states which have almost exactly the same free energy and are very far apart from each other and do not look like each other in any way. Some of them might be bigger dimensional states and some are smaller. So in other words if you are looking at achieving some kind of phenotype or some kind of ... let us say ... there may be lots of different ways of doing it some of which have twenty dimensions or some have five hundred dimensions, I'm not even sure that is a well-defined ... if you have discrete minima, I'm not even sure what it means, how you characterise the dimensions.

B. Shraiman I think that is probably the question that I have to try to answer. Well of course, not every system is reducible and a spin glass may be a good example of that. However, let us think dynamical systems: So there is a well-defined way in which you can reduce a multi-dimensional system to an attractor, if there is an attractor. Then the motion on the attractor is going to be effectively predictive, the motion on the central manifold will approximate the dynamics of the system. Now whether the cell actually behaves that way is a question. But there is a way to ask this question and get an answer and maybe it is yes or maybe it is no, maybe the answer is three, maybe five or twenty seven ... Yet Terry for example, in his talk presented some evidence for the existence of this dimensional reduction, at least evidence suggesting that of something like this happening. So Terry, do you want to continue?

T. Hwa I want to inject a slightly different view. So normally we think of a dimensional reduction as a way to kind of simplify the description. We as human beings when we try to analyse some data, some phenomenon. But I would like to suggest that maybe this is something that Biology is doing. Evolution is not infinitely powerful, given finite time it cannot just change everything into everything, right? So maybe Biology has through evolution learned some tricks to help itself to reduce the dimensionality that we don't know of and that we see bits and pieces of it when we analyse data. So to me, understanding Biology, a big part is to understand this kind of special things that collapse the complexity for the reason of, you know, growth and survival, evolvability and all that. That Biology has discovered these things and to me personally the joy of coming to study Biology is that this is the kind of thing you couldn't possibly search if you just get an infinite number of parameters. But here actually life examples can help us to guide and get to this new type of science, maybe expansion of Physics or joining of Physics and Biology.

B. Shraiman In fact, continuing this line of thought. Terry just brought us back to how complex the living matter actually is. So the real thought is that it is perhaps not quite as complex as it could have been.

T. Hwa I think, when I first came to Biology and for many of us, that we are attributing a lot of glorious things to Biology; everything is evolvable, everything is optimal. Maybe it is simpler than that. If we do not give it a chance we will never find this simplicity. So maybe we have to find some tricks that actually guide us to think in different ways and then there will be new science.

B. Shraiman So maybe I want to close this gathering with a toast to Emergent Simplicity!

Address by the Director of the International Solvay Institutes M. Henneaux

This conference has been a great meeting. The program was wonderful. As an outsider, I could clearly see that the discussions went superbly, in the Solvay tradition. Our scientific committee was right to choose biophysics, The Physics of Living Matter: Space, Time and Information in Biology as the subject of the 27th Solvay conference.

Since this is the first time that biophysics is chosen as a theme for a Solvay Conference on Physics, we shall contact you next week to get your opinion. Your feedback will be very much appreciated.

I would like to express the gratitude of the Solvay Institutes to Boris, the conference chair, for all the careful work that went into the scientific organization of the conference. The preparation started almost two years ago. This tells a lot! We are also grateful to the session chairs and the rapporteurs, who were actively involved in this careful preparation. I would also like to thank you, all the participants, who made the scientific discussions proceed vividly in the spirit of the Solvay conferences. Without your effective participation, the conference would not have been a success. As you know, we will publish proceedings. Since discussions are central, they will be reproduced in the proceedings. Transcribing heated discussions into a good written text requires a lot of editorial work. We are grateful to all the auditors who accepted to help us in this important enterprise, and to Alexander Sevrin, the secretary of the scientific committee, who is organizing the editorial work. You will hear from him in the coming days! The success of the 27th conference makes us look forward with confidence to the next Solvay Conference on Physics! I can tell you that the scientific committee already started to discuss its theme, but it is too early to make any revelation.

With these optimistic closing words about the future of the conferences, I wish you a very nice trip back home.

www.ingramcontent.com/pod-product-compliance
Lightning Source LLC
Chambersburg PA
CBHW081509190326
41458CB00015B/5326